UNDERSTANDING AND PREDICTION

SYNTHESE LIBRARY

MONOGRAPHS ON EPISTEMOLOGY,

LOGIC, METHODOLOGY, PHILOSOPHY OF SCIENCE,

SOCIOLOGY OF SCIENCE AND OF KNOWLEDGE,

AND ON THE MATHEMATICAL METHODS OF

SOCIAL AND BEHAVIORAL SCIENCES

Managing Editor:

JAAKKO HINTIKKA, *Academy of Finland and Stanford University*

Editors:

ROBERT S. COHEN, *Boston University*

DONALD DAVIDSON, *Rockefeller University and Princeton University*

GABRIËL NUCHELMANS, *University of Leyden*

WESLEY C. SALMON, *University of Arizona*

VOLUME 94

STEFAN NOWAK

UNDERSTANDING AND PREDICTION

Essays in the Methodology of Social and Behavioral Theories

D. REIDEL PUBLISHING COMPANY

DORDRECHT-HOLLAND/BOSTON-U.S.A.

Library of Congress Cataloging in Publications Data

Nowak, Stefan.
    Understanding and prediction.

    (Synthese library; v. 94)
    Includes bibliographies and index.
    1. Social sciences – Methodology.
    2. Sociology – Methodology.
    3. Social sciences – Statistical methods.
    I. Title.
H61.N63   300'.1'8   75–44179
ISBN 90–277–0558–5

Published by D. Reidel Publishing Company,
P.O. Box 17, Dordrecht, Holland

Sold and distributed in the U.S.A., Canada, and Mexico
by D. Reidel Publishing Company, Inc.
Lincoln Building, 160 Old Derby Street, Hingham,
Mass. 02043 U.S.A.

Printed in The Netherlands

To the Memory of

STANISŁAW OSSOWSKI

Teacher and Friend

SUMMARY TABLE OF CONTENTS

TABLE OF CONTENTS

PREFACE

One of the more characteristic features of contemporary sociology is an increasing interest in *theories*. More and more theories are being developed in various areas of social investigation; we observe also an increasing number of verificational studies aimed primarily toward the verification of various theories. The essays presented in this volume deal with theories too, but they approach this problem from a *methodological perspective*. Therefore it seems worthwhile in the preface to this volume to make a kind of general declaration about the author's aims and his approach to the subject of his interest, and about his view of the role of methodological reflection in the development of sciences.

First let me say what methodology cannot do. It cannot be a substitute for the formulation of substantive theories, nor can it substitute for the empirical studies which confirm or reject such theories. Therefore its impact upon the development of any science, including the social sciences, is only indirect, by its undertaking the analysis of research tools and rules of scientific procedures. It can also propose certain standards for scientific procedures, but the application of these standards is the domain of substantive researchers, and it is the substantive researchers who ultimately develop any science. Nevertheless the potential impact- of methodological reflection, even if only indirect, should not be underestimated. Its role is especially important in those disciplines where there is considerable disagreement with respect both to the question of what a theory is or should be and to the question of what constitutes the conditions of acceptability or rejection of theories in a given science. Needless to say, sociology is one such discipline.

One of the basic questions faced by any methodologist who would like to propose certain rules and standards for a particular scientific discipline is the degree of generality, the degree of abstractness of the analysis he intends to undertake. Related to this is the problem of the degree to which he may use as the source of his methodological standards the results of general reflection on the methodology and philosophy of sciences, especially if this reflection is based on the experience of other sciences.

In my opinion we may study the methodological features of theories of a given kind, and any methods of empirical investigation in any discipline where they seem to be most developed, but the solution of the problem of 'applicability' of such a method to another area of investigation should be preceded by our answering two other questions:

(a) What is the *exact nature of the problem* to which the given way of theorizing or the given method constitutes or delivers the answers, and for what other areas of studies and in what other disciplines do the problems so defined seem to be valid?

(b) What *empirical assumptions* do we have to make in order to 'legitimize' such a method of investigation or theory construction?

When we find in two or more disciplines problems having a similar logical structure, and a commonly acceptable set of empirically valid assumptions underlying the method used for solving these problems, then we can say that the sciences in question are methodologically similar in the given respect at the given level of generality of our comparison. If either the problems are different, or the empirical assumptions of a certain method are valid in one discipline and invalid in the other, we have to conclude that with respect to this particular comparison these sciences are methodologically different.

It seems that certain questions when formulated in a most general abstract way appear in most or at least many sciences, and that the most general assumptions of the validity of certain methods seem to be fulfilled in various domains of empirical reality. Thus we have to say that many (if not all) empirical sciences have *some* methodological features in common. But still *not all* the questions will be valid for all sciences, and not all the assumptions which turn out to be true in some can be accepted in others. Therefore (besides the rules of formal disciplines, like logic or mathematics, which seem to be obligatory for all the sciences) we may only say that we find certain general patterns of problem formulation which are *fairly general*, and certain empirical assumptions of similar character which seem to constitute a *fairly general framework* for our scientific thinking. But when formulated at this general level such rules, even if valid in many sciences, are *obviously insufficient* in any of them, and this applies to the same degree to social and natural sciences. Many sciences may be similar both with respect to their problems and with respect to methods, if we look at these similarities in sufficiently general terms; but such a general aspect cannot constitute a sufficient basis for the development of any science.

It seem that one of the basic rules for methodological reflection if such reflection is to be of practical use in a concrete science is to try to see both the common features of different (though by no means all) disciplines and at the same time to see all of their methodological 'peculiarities'.

The essays presented in this volume are an attempt to look at the problem of construction and verification of social and behavioral theories from this double perspective: they try to specify certain more general features of scientific thinking with respect to which the social sciences are by no means unique, and at the same time they try to look at their special features inherent in both our particular human perspective at the social phenomena and in important and unique features of the reality studied by them.

The twelve essays in this volume can be classified into three groups. The first three deal with the problem of formulation and operationalization of *sociological concepts*, the main emphasis being that 'understanding' of both the social behavior of man and the 'meanings' of the social environment in which people live is a necessary premise both for the formulation of theoretical concepts and for the proper choice of indicators for our verificational studies. At the same time I try to demonstrate that this 'understanding' is compatible with the more general rules of concept formation in science. The next four essays deal with different aspects of the *verificational process and inductive analysis* in the social sciences. Here again the basic assumption is that we do not have to give up the principle of causality in our thinking about social processes, but at the same time we have to be 'realistic' and admit that most (if not all) of the empirical findings we have in our discipline are of a statistical character, and additionally that they are limited in their validity to one particular society or to a certain cultural area. These essays propose a solution based on the acceptance of the postulate of causality on the one side, and by the historical or cultural variations of social reality and the probabilistic character of most of the observed sequences on the other. The remainder of the essays deal primarily with the *problems of theory formation*, the main stress being on the analysis of the logical structure of certain approaches to theory construction.

Even if the range of problems discussed in this volume is fairly broad, it is not my aim to present a general overview of the problems of theory formation and verification in the social sciences. Many important problems of theory construction have been left aside, such as the application of the functional approach to the study of whole institutions or societies. Another group of problems not discussed are those dealt with by the 'structuralist' approach to theory construction in social science. Finally, the reader will not find here the problems posed by the theories of social dynamics and change. The essays in this volume constitute probes into those problems of the methodology of social theories in which their author was interested, nothing more.

Many of the essays constitute the continuation of ideas I have discussed in papers previously published either in Polish or in English, in books and periodicals. I have drawn especially from my book *Studia z metodologii nauk spolecznych* (Studies in the Methodology of the Social Sciences, Warsaw, 1965), in which earlier versions of some of these essays were published. Nevertheless, after ten years, I found it impossible to reprint or translate any of the essays without serious changes, and the same applies to almost all the other studies based on my former publications. Most have been rather basically rewritten, and in several cases this involved not only new argumentation but a basic change in my views. Several of the essays are being published for the first time.

The new essays were written and the entire volume was rewritten and prepared for publication during my stay at the Center for Advanced Study in the Behavioral Sciences at Stanford, California, where I spent the academic year 1974 − 75. I would like to express my gratitude to the entire staff of the Center not only for all the help and research and secretarial facilities I received there, but also, or perhaps even primarily, for the wonderful social climate they create for the fellows. There were too many members of the Center's staff whose help was important to me during my stay at the Center to mention their names here. The same applies to the many cofellows at the Center who discussed with me the various essays as they were being written and who made important comments on them.

Nevertheless I would like to mention one person from the Center's staff without whom preparation of the English version of this book would scarcely have been possible for someone who is not a native speaker in this language: namely, Miriam Gallaher, who 'polished' from an editorial point of view most (unfortunately not all) of the essays in this volume. But I would like to stress that her contribution goes beyond strictly linguistic help. As the first − intelligent and critical − reader of these essays she pointed out a number of unclear passages and in many cases proposed their reformulation, with an obvious improvement of the text independently of its linguistic side.

But even though these essays were written (or rewritten) during my stay at the Center for Advanced Study, many of them as I noted above are a continuation of my earlier writings. And here I would like to express my gratitude to all my Polish colleagues – sociologists and philosophers – who participated in discussions of some of them or expressed their views about the author's proposals in private conversations with him. Here again there were too many to list each of their names individually.

There is nevertheless one person who must be mentioned in the preface to this volume, because he played a decisive role in shaping both the interests in and the approach to the methodology of social sciences presented here. This is Stanisław Ossowski, the late professor of sociology at Warsaw University. He used to teach his students that the more we are involved in social problems and the more we hope to use our knowledge to improve and transform society, the more we should be interested in the quality of the tools we use for implementing our goals, i.e. in the quality of social theory and the methods of investigating social phenomena. This book is dedicated to his memory.

STEFAN NOWAK

CHAPTER I

CONCEPTS AND INDICATORS IN
HUMANISTIC SOCIOLOGY*

1. THE PROBLEM OF 'VERSTEHEN'

A sociologist in the late thirties or early forties looking for a philosophical creed which could deliver him a set of guiding rules in his investigations would have faced an uneasy choice between two obviously incompatible orientations. One of these orientations stressed the *basic difference between social and natural sciences.* According to it social sciences deal with basically different subject matter than the natural ones — their subject matter consists of *meaningful* human behaviors and culturally meaningful objects. Therefore their goal (and at the same time their method) is *to understand these meanings* properly, and to come to certain generalized conclusions about them. The term 'understanding' (Verstehen) referring to the special character of the approach to social phenomena was introduced by W. Dilthey in the 19th century, but — as it was stressed by Alfred Schutz[1] and many others — the 'practice of Verstehen' is as old as society itself, if not older. Whenever in the history of mankind one man has observed the behavior of others, he would also try to guess the *purpose* of this behavior, the *motives* of the behaving persons, etc. He would also try to guess how others would perceive and react to his own behavior. When Dilthey told the humanists of the 19th century that in their studies of social and cultural phenomena they were using a special method of 'Verstehen', they found themselves in a situation similar to that of the hero of a play by Molière: he was told one day that he had been speaking prose all his life.

The idea of the distinctive nature of social sciences was accepted by most sociologists. Some, like Max Weber, made from it the methodological foundations of their theories. Weber wrote at the first page of his monumental work *Wirtschaft und Gesellschaft:*

Sociology (in the sense in which this highly ambiguous word is used here) is a science concerning itself with the interpretative understanding of social action and thereby with a causal explanation of its course and consequences.[2]

Following this concept of sociology Robert MacIver wrote:

> There is an essential difference from the standpoint of social causation between a paper
> flying before the wind and a man flying from a pursuing crowd. The paper knows no
> fear and the wind no hate, but without fear and hate the man would not fly nor the
> crowd pursue.[3]

But at the time MacIver wrote these words there were already many scientists of the 'empiricist' or 'operationalist' schools ready to challenge them. The source of this orientation can be found both in developments in the philosophy of science and in modern trends in other sciences, especially in psychology where we observe the rise of Watsonian behaviorism and it continuation by theoreticians of S-R learning theories. All these trends had the same goal: to pursue the idea of the theoretical and methodological unity of sciences. The theoretical unity had the form of some naive dreams about a pyramid of all the sciences unified into a consistent reductive structure with physics as the basic foundation of all human knowledge. The postulate of methodological unity stressed the notion of the objectivity of scientific procedures in its many aspects. Thus the philosophers of the Vienna Circle stated that scientific propositions are meaningless unless we know how to test them. More important for the social sciences was the attempt to extend this objective approach to the construction of scientific concepts in the form of the 'operationist' theory of concept formation proposed by P. Bridgman.[4] According to Bridgman, a concept is useless unless we know how to measure or identify its referent. But, what was even more important, *the results of measurement were held to completely exhaust the meaning of scientific concepts.* What is 'unoperational' is meaningless.

The doctrine of operationism was a kind of natural theory of concept formation for early behaviorism. It was also soon accepted by empirically oriented sociologists, such as Charles Stewart Dodd and G. Lundberg. And here the basic consequence was the rejection of the dualism of sciences postulated by Dilthey. The early empiricists had stressed that the traditional interest of the humanities in the subjective aspects of social phenomena was responsible for the underdevelopment of social sciences. In a direct reply to MacIver's statement quoted above G. Lundberg wrote:

> The doctrine that man is the one unique object in the universe whose behavior cannot
> be explained within the framework found adequate for all others is a very ancient and
> respectable one. We merely make the contrary assumption here. From the latter point
> of view a paper flying before the wind is interpreted as behavior of an object of

specified characteristics reacting to the stimulus of specified characteristics within a special field of force. Within this framework we describe the man and the crowd, the paper and the wind. The characteristics of these elements would never be the same in any two cases of wind and paper or man and crowd. But it is the faith of science that sufficiently general principles can be found to cover all these situations, and that through these principles reliable predictions can be made of the probability of specific events.[5]

And then he questioned the usefulness of any interest in the psychological aspects of social reality:

Today however, a considerable number of students of social phenomena are still firmly convinced that the phenomena with which they have to deal cannot be explained without, for example, a category called 'mind' which carries the whole vocabulary of subsidiary terms; (thoughts, experience, feelings, judgements, choice, will, value, emotion, etc.)... Any attempt to deal in other words with the behavior which these words are used to represent meets with the most determined resistance on the ground that 'something has been left out'. And what has been left out? Why, 'will', 'feelings', 'ends', 'motives', 'values', etc. Arguments or demonstrations that the behavior represented by these words is accorded full recognition within the framework of 'physical' sciences are to some apparently as futile as were the arguments against phlogiston to Priestley. He just knew that any system which left out phlogiston was eo ipso fallacious. I have no doubt that a considerable part of the present content of the social sciences will turn out to be pure phlogiston.[6]

In general one can say that our hypothetical social scientist, looking for philosophical grounding would have had to make a choice between two sociologies. One of these would give him a good 'understanding' of what social phenomena are without promising to substantiate this understanding by any controllable research procedures and without promising to be useful in predictions. The other proposed complete scientific rigorism at the cost of abandoning all interpretative understanding of social reality. We can agree that it was not an easy choice to make.

Fortunately enough the dilemma — observation or 'understanding' — turned out to be a spurious one. Philosophers of science soon understood that the operationist doctrine of concept formation doesn't work even in the field of natural sciences, because it would lead to elimination of theoretical thinking in them. It turned out that even in the most empirical natural sciences operational definitions cannot exhaust the *whole meaning of theoretical concepts* — most scientific theories have also to postulate the existence of some 'latent', hypothetical properties to be used in the explanations of phenomena at the observable level. Therefore the philosophers came to the conclusion that even in the language of natural sciences we need two kinds of concepts. Those of one category are defined in purely 'observational language', and refer to the 'observational level' of the area of reality studied by the given science, to phenomena which can be assessed by 'pure observation'. Concepts of the other category refer to unobservable 'latent' or 'hypothetical' properties, the existence of which is postulated in order to obtain a better explanation and more fruitful prediction of phenomena on the observable level. While the concepts of the first category belong — to use the terminology of Rudolf Carnap — to the 'language of observation', concepts denoting hypothetical latent properties are used for the construction of explanatory hypotheses and theories. Some writers (like Carnap)[7] called the latter 'theoretical concepts', others like McCorquodale and Meehl[8] called them 'hypothetical constructs'.

The concepts denoting 'latent properties' must of course be related to phenomena on the observable level, but this relation is usually much more complicated than was presented in the operationist doctrine — according to which the scientific concepts *had to be defined completely* on the basis of a definite set of measurement operations. Propositions connecting the language of theory and the language of observation Carnap calls 'correspondence rules'. These correspondence rules may sometimes be partial definitions of a theoretical concept, but they may also be elements of theory which connect phenomena on observable and unobservable levels by *indirectly verifiable empirical relations*. Whatever their nature, the observable properties seldom exhaust the whole meaning of the given concept. Theoretical concepts usually have some '*surplus meaning*' so that besides the observable properties which have been used for their 'operationalization' or for testing hypotheses formulated in the language of 'latent mechanism' some other empirical derivations can be made. Each such new derivation of observable events is a new test of the correctness of our assumptions about the nature of the hypothetical structures or mechanisms.

The evolution of empirical investigations in the behavioral sciences took a similar direction. We can observe a slow departure from the postulates of strict behaviorism and operationism in those orientations which at the beginning strongly accepted the methodological assumptions of operationism. This can be nicely observed in the development of S-R learning theory. At its beginning the behaviorist assumption eliminated from the field of interest of psychology any reference to subjective, or more generally to the unobservable, processes or dispositions. Operationalism was a natural doctrine of concept formation for this orientation, and the idea that all concepts of psychological theory are to be completely definable by certain research conditions and observable aspects of human (or rather: animal) behavior seemed quite obvious for early behaviorists. But we can detect even in the relatively early stages of this theory attempts to 'smuggle' in some notions of traditional psychology, the meanings of which were obscured by the 'strictly operational' way of defining them. Certain properties and processes, even when defined 'operationally', were supposed to exist in the heads of experimental rats as 'intervening variables' of the learning process. Hull's theory of learning is full of such intervening variables, and some of the variables, even when completely operationally defined, had names which unavoidable suggested traits we know from our own inner experience. Then — step by step — even the meanings of these concepts became more and more independent from their observable indicators. Very often they were only *theoretical terms without any definitions*. These terms were simply taken from traditional psychology or from everyday language and given the same explanatory function to connect stimuli to subsequent behavior as in the 'understanding' approach to the study of individual behavior. Thus Tolman postulated the existence of a 'cognitive map' of the maze in the head of a rat which find an exit from the maze. Mowrer introduced such intervening variables as 'fear' or 'expectation' in order to explain how habits are formed in the learning process. Other theoreticians had to assume that, at least for human beings, the increase or decrease of subject's 'self-esteem' and other 'internal rewards and punishments' may play the same role that food and sexual rewards had played in early learning experiments.

When contemporary learning theoreticians discovered that observing the behavior of a 'model' may have behavioral consequences similar to those of subject's own rewards and punishments, the process of 'rehumanization' of the discipline was almost completed.

The development of sociology was somewhat different because here the radical postulates of the early empiricists could not be applied to practical empirical research and theorizing. The complexity of the traditionally acknowledged objects of sociological investigation made practically impossible the translation of concepts referring to 'meaningful' social units into strictly operational definitions free of any reference to social meaning. Nevertheless, the pressures of empiricism were strong enough to cause sociologists to look for 'operational definitions' for their units and variables of observation. In practice, it appears that the sociologist took a *term* referring to a behavior or situation, the 'social meaning' of which was more or less obvious, then tried to discover the proper operationalization for this concept that is, to specify its *observable indicators*, so as to use these indicators in his study. This procedure has been clearly presented by one of the founding fathers of contemporary empirical sociology, P.F. Lazarsfeld.

Let us see how social scientists establish devices by which to characterize the objects of empirical investigation. There appears to be a typical process which recurs regularly when we establish 'variables' for measuring complex social objects. This process by which concepts are translated into empirical indices has four steps: an initial imagery of the concept, the specification of dimension, the selection of observable indicators, and the combination of indicators into indices.

He then specifies these steps in the following way:

The flow of thought and analysis and work which ends up with a measuring instrument usually begins with something which might be called imagery. Out of *analyst's immersion* in all the details of theoretical problems he creates a rather vaque image of construct... The next step is to take this original imagery and divide it into components. The concept is specified by an *elaborate discussion* of the phenomena out of which it emerged... After we have decided on these dimensions there comes the third step: finding indicators for the dimensions.[9]

From the point of view of Lundberg's philosophy each of these steps is loaded with the 'phlogiston of social meanings', *Nevertheless it would be premature to say that empiricist philosophy had no inhibitive impact upon contemporary empirical sociology* with regard to the *theoretical use* of these meanings. When we look both at Lazarsfeld's description and at the practice of most contemporary sociological researchers we can see that the 'understanding imagery' constitutes only a necessary step toward formulation of the operational definition of the concepts itself. Once we have our set of indicators, *the concept itself may be more or less forgotten*. Actually one could say that there have rarely been any concepts at all if *by concept we mean a clear and unambiguous specification of the meaning of the given scientific term*. Rather, what we usually find in contemporary empirical sociology are certain *terms* which refer to some more or less vaguely meaningful aspects of social reality. Once they have been 'operationalized' in terms of our indicators, indices and scales, these operational definitions *seem to play the role of theoretical definitions* both in our research and in our theories. It is as if a physicist were to think of his theory as relating only to the results of his measurements understood in terms of readings of research instruments, nothing more.

Of course to make the procedures described by Lazarsfeld theoretically complete, we would have to travel the same path but in the opposite direction: from the results of our measurements and the relationships between them, to the concepts our measures operationalized, and then to our final conclusions about the phenomena to which these theoretical concepts apply. But for that purpose the *concepts must be defined in theoretical terms with sufficient precision*; the 'initial imagery' is not enough. We have to specify also as clearly as possible both the logical and the empirical relations between these concepts and the indicators operationalizing them, so that we will know to what degree the classification of our objects in terms of indicators corresponds to their classification with respect to the concepts of our theoretical language. It is needless to say that this is rather seldom the case in empirical sociology.

Moreover, the character of the concepts used should 'reflect' our thinking about the nature of social phenomena. If we accept (as most social scientists seem to accept) that we are dealing in the social sciences with objects, the 'unobservable', hypothetical side of which may be equally as important and interesting to us as the observable one, and that a large area of this hypothetical side of social reality is made up of the social meanings of the observable behaviors, then we should stop being 'ashamed' to reflect all these social meanings in the conceptual language of the social sciences. The goal of this paper is to discuss certain aspects of concepts formation in sociology and in the behavioral sciences with the main stress on the role of concepts defined with reference to 'subjective meanings' in the language of sociological theories. I will also try to discuss this problem in a way that will permit us to use the concepts so defined in empirically testable social theories.

2. OBSERVABLE AND HYPOTHETICAL PROPERTIES IN SOCIOLOGICAL CONCEPTS

Let me begin by saying that in the social as well as the natural sciences we need in our conceptual language terms referring to observable properties and events and terms referring to unobservable properties and events. The unobservable hypothetical properties refer primarily to psychological states, of persons and especially the way people perceive, think about and feel about the social world around them.

In order to be accurate and precise, we should distinguish *three categories of concepts* in the language of social enquiry.

The first category comprises those concepts which refer to *purely observable* characteristics of human behavior without any definitional reference to its unobservable, psychological aspects. When we say that someone 'is running' or that within the last year he has 'changed his place of his stay' we have in mind only strictly physical aspects of his behavior or existence. Some other variables of social enquiry, like 'sex' or 'age', may also belong to this category of concepts if defined in strictly observational terms (e.g. they are not understood as 'social age' or 'sex roles'). Such strictly observational concepts exist, of course, in our sociological language, but they are relatively rare and in general I do not think we can make any interesting social study if the problem is formulated only with the use of purely observational concepts.

The second category of concepts comprises those which are defined purely in terms of some *latent properties,* with the behavioral, observable aspects of the meanings of these concepts being completely 'open'. If we define a certain act an 'act of revenge' we have in mind only the psychological, motivational aspect of this action without any explicit definitional reference to its observable aspects. Both the act of murder and an unfavorable review of our colleague's last book may be treated as a case of the 'act of revenge', if we infer the existence of certain motivations behind them. When Max Weber defined 'fight' as a social interaction wherein each of the actors tries to realize his will against the will of the other actor in this relation, the definition was completely 'open' with respect to its behavioral aspects and the fight was defined *only* in terms of a certain configuration of motives of the participants. In general we can say that the observable 'indicators' in this category of concepts do not belong to their meanings; they are definitionally completely external to the indicated phenomena, and are related to them by a non-definitional empirical relation.

Finally, the third category of concepts comprises those which refer by their meaning to both the observable and the unobservable aspects of social behavior. If we say that someone is 'escaping', we have in mind both his overt behavior and his motivations. If we say that someone is 'anti-semitic', we have in mind both his overt behavior and his attitude toward Jews.

This last category of concepts deserves special attention, because it includes what one might call *syndromatic concepts.* While in some concepts in the empirical sciences the *simple enumeration* of a set of properties exhausts the whole meaning of the given concept (whether it is defined in observational or in non-observational language), this is not the case with syndromatic concepts. The meaning of the syndromatic concept involves reference not only to the properties denoted by the given concept, but also to certain interrelation between these properties. Here I would like to stress especially certain relations between phenomena from the observable and the hypothetical levels, though of course, a syndromatic concept may postulate certain relations among a cluster of phenomena on one specific level only. Thus in the case of the concept 'anti-semitism' we can assume that the motivations of 'anti-semitics' are causally related to their behavior in such a way that we can use the behavior as an indicator of the unobservable dispositions, or conversely, we can explain the behavior by these dispositions. In general, the syndromatic concepts seem to have the nature of very simple *theoretical models of social phenomena.*

Needless to say, all these types of concepts apply as well when we pass from the level of individual behavior to the level of groups and social aggregates. All the properties, events and processes occurring on these higher levels of social aggregations and social systems are – at least partly – definitionally reducible to properties and behaviors of the individuals composing these groups,[10] and among them to the psychological dispositions and psychological processes characteristic of human beings. Therefore, all three kinds of variables noted above can serve as definitional references for processes occurring on the collective level necessarily belong to one of these categories. The concept of 'average age' as a group characteristic is obviously of the first category, being reducible to external observational properties of individuals only. The concept of 'group integration' – when 'group integration' means a system of 'mutual bonds and sympathies' – is of the second category, since it is defined in terms of attitudes of group members. But the concept of 'cooperation', referring both to patterns of behavior and to certain accepted rules of behavior, is of the third kind. It involves by definition observable and hypothetical properties of group members. And, of course, if we treat the syndromatic concepts from the collective level as theoretical models of social reality, their structure will be much more complex than the structure of syndromes denoting individual human behavior.

To say that concepts referring to (or including) the psychological aspects of social reality in general and to 'social meanings' in particular constitute 'theoretical constructs' of the social sciences is to stress the *methodological similarities* between the social and natural sciences. But we should not forget about the *differences* too.

The first difference is that in the natural sciences the hypothetical character of theoretical constructs is usually temporary. The development of new observational tools and techniques of assessment often transforms these constructs into more and more observable entities. For example, the invention of the Wilson's chamber made elementary particles 'more obserable'; genes, which had for a long time been 'very hypothetical' entities, have become more and more 'observable' because of the electronic microscope. Many examples of this kind of development could be given. In the social sciences, however, the situation is quite different. Everything that goes on in other people's minds is intrinsically hypothetical to us, and all the developments in neurophysiology, even when they point quite clearly to localizational aspects of certain psychological processes and to what kinds of neurological connections are involved, bring us no closer to a direct observation of the inner experiences of other persons. As it was stated by many writers the bridges between subjective feelings and their objective counterparts are still purely correlational, and we cannot imagine, at least for the time being, that they will ever be crossed in such a way that we will observe other people's feelings in the same way we observe our own.

If we look at this problem from a different perspective, the situation is quite the reverse. The physicist has only a hypothetical notion of an electron, as described by the statistical formulas of quantum mechanics, and he may only dream that perhaps in the future the objects of this interest may become more visible – if he does not believe too strongly in the Heisenberg principle of indeterminancy. On the other hand, the sociologist studying cultural processes in a certain population has direct access to them inasmuch as he can treat himself as a sample from this population. *To the degree that he assumes a similarity of feeling and thinking between himself and the studied population*, a social scientist may say that he has direct access to the phenomena denoted, at least by some hypothetical constructs of his scientific language.

Other differences are even more striking. In the natural sciences – at least at the beginning of the development of a given field of study – the strictly observable characteristics of the objects and phenomena, and the observable similarities and differences between them, usually delimit the field of study and deliver definitional references for the concepts which will be used to build a theory. If we later introduce some hypothetical, latent properties into the conceptual apparatus of the given discipline, this is usually because we are not able, using observational language alone, to build a sufficiently efficient explanatory theory for the observable phenomena.

This is by no means true in the social sciences. Here we usually start at the very beginning of any study with notions which have an explicit or at least implicit reference to subjective, i.e. hypothetical, aspects of the studied phenomena. When we decide to study, for example, the 'family', 'bureaucratic organization', the 'price system', 'religion', 'social stratifiction' or the 'life aspirations of the members of a society', we have to understand the meanings of these terms in the way they are understood by the members of the studied communities — because we more or less 'see', or aim to 'see', this reality in the way they do. When we want to assess the existence or occurrence of social phenomena, we are usually interested both in their subjective side and their observable course. In a rather peculiar way, due to our common sense Verstehen of the nature of social reality, we are not limited, even at the level of the most non-scientific observation of social reality, to its merely extraspectional aspects; rather, along with these aspects we 'perceive' at least some latent entities which structure them, whereas in the natural sciences we usually come to the discovery of latent entities and forces at a rather late period in the development of the science. If we look at the problem from this particular angle it turns out that our everyday language is more 'theoretical' with reference to social phenomena than to natural phenomena. Due to the possibility of our understanding human behavior, we start our analysis of social facts with concepts whose meaning also includes some *explanatory* notions about observable aspects of this behavior.

3. CONCEPTS DEFINED WITH HUMANISTIC COEFFICIENT IN THE LANGUAGE OF SOCIOLOGY

I stressed above that, due to the 'dual nature' of social reality, the concepts of the social sciences may include in their meanings reference both to observable phenomena and to their hypothetical, psychological aspects.

When the representatives of 'humanistic sociology' stressed the importance of Verstehen, they had in mind not just any, but a special kind of psychological counterparts of observable behavioral sequences, namely those the existence of which the acting persons are more or less aware of because they can observe them in their own *introspection*. These conscious psychological states may constitute the individual and social meanings of their behavior.

The special importance of the properties denoted by concepts defined in introspectional terms, and of theories formulated in 'meaningful' terms, is based upon the asumption that man is a conscious being and therefore that human values, people's knowledge about the conditions they live in, their perceptions and 'definitions of social situations' constitute – at least partly – those situations. These meanings are also the *components* of social phenomena at the level of human aggregates, composing ideologies, cultures, and other areas of 'social reality'. At this level these phenomena are also constitutive forces shaping human interactions, determining social structures and being involved in their dynamics and change.

On the basis of the social meanings attributed to certain observable sequences we often classify into the same conceptual categories events which are, from a strictly observational point of view, quite dissimilar. Thus we say, for example, that baseball and chess belong to the same general class, 'games', but in order to define any of these games as well as the more general class we have to make some reference to their 'meaningful side' and to certain similarities of their social meaning. On the other hand, we may classify certain behavioral sequences which externally seem quite similar into different conceptual categories, because different meaningful contexts are attributed to them by the members of the studied populations. Thus, we distinguish between marital and extramarital sexual relations even though from a strictly behavioral point of view they may be quite similar, and treat a boxing match and a fight between two hooligans as different kinds of events, etc. Our everyday language, which is the starting point for any more scientifically refined sociological language, reflects the importance of such distinctions with quite good accuracy, acknowledging these hypothetical entities as *definitional components of elements of social reality* and revealing the structure of these meaningful-behavioral syndromes in our definitions.

For these reasons, as stated by Florian Znaniecki, the facts to which these concepts refer are the primary data of social enquiry. He wrote:

The primary empirical evidence about any human action is the experience of the agent himself supplemented by the experience of those who react to his action, reproduce it or participate in it. The action of speaking a sentence, writing a poem, making a horseshoe, depositing money, proposing to a girl, electing an official, performing a religious rite as empirical datum is what it is in the experience of the speaker and his listeners, the poet and his readers, the blacksmith and the owner of the horse to be shod, the depositor and the banker, the proposing suitor and the courted girl, the voters and the official they elect, the religious believers who participated in the ritual. The scientist who wants to study these actions inductively must take them as they are in the human experience of those agents and reagents; they are his empirical data inasmuch as and because they are theirs. I have presented this elsewhere by saying that such data possess for the student a humanistic coefficient.[11]

Let us call concepts which define the social phenomena processes and situations in as nearly as possible the same way as they are perceived by the people participating in them *'understanding constructs'* or *concepts defined with a humanistic coefficient*. When a historian of legal systems tries to understand the meaning of the norm of 'partrimonial power' in Ancient Rome, he will reconstruct as accurately as possible all the rights and obligations which were clustered in the role of 'pater familias' according to the opinions of the citizens of Rome. When Max Weber, looking for the 'ideal type' of economic ethics in early capitalism, reconstructed the value system of Benjamin Franklin, he defined this concept according to rules which Znaniecki later called the 'approach with a humanistic coefficient', i.e. according to the inner experience of a person who was for him the exemplification — or better, 'the ideal type' — of the notion he introduced. Some definitions of 'nation' exemplify concepts defined 'with a humanistic coefficient'. Such a concept is therefore defined in a way that corresponds to the rules of construction of *analytic definitions*, i.e. definitions which report or reconstruct the meaning of a term among its actual users and in which, moreover, the users' subjective experience is one of the elements of the definition.

It should be noted here that such concepts or constructs not only refer to phenomena which are accessible to the direct observation of the persons involved, but are also usually defined in a way that permits the scientist to *relate the subjective aspects of their meanings to his own introspection* or, what is equally important, to relate them to some current notions which are sufficiently intersubjective to make these concepts a means of intellectual communication. The scientist's introspection (or the assumed similarity of subjective meanings of the given term among the social scientists using it) is in the case of 'understanding concepts' the ultimate reference for the proper understanding of the 'meaning' of these concepts. In some cases he has to define a phenomenon from his own subjective experience in all its complexity. In other cases he is able to construct a concept the meaning of which, although ultimately referred somehow to his introspection, denotes a phenomenon he has never experienced psychologically in all its complexity. Max Weber was able to define the notion of 'charismatic power' although we doubt that he himself was involved in a power structure of this type. The anthropologist may perfectly understand and define a system of magic beliefs although he does not share those beliefs.

Let me now specify certain basic functions of 'understanding-constructs' defined according to the rules of 'humanistic coefficient' in social enquiries.

Their first function is that they permit us as sociologists both to understand for ourselves and to present to others *what a society is and how it works.* By this I mean our understanding of the system of basic values and perceptions that govern the behavior of peopple in a given society, the specification of these values and perceptions into norms and roles, their patterns of interrelation in the structure of the society, how its members use them in guiding their behaviour, their beliefs, philosophy, religion, etc. When a sociologist thinks about his own society, he is quite likely to forget about this function because these things are as taken for granted by him as by any other member of his society. But when an anthropologist comes upon an 'undiscovered' tribe in New Guinea, the transformation of *meaningless into meaningful* is his basic and primary task; all his other theoretical tasks can only be based on this elementary knowledge of social reality. As long as he does not understand the society in terms of its basic rules of social and cultural functioning, he is unable to formulate any other, more specific questions for his study.

But suppose we are dealing with our own society or with societies so close to our own that this 'elementary understanding' can be more or less validly assumed. What, then, is the utility of understanding constructs in our research and theorizing?

First, they apply when it comes to the *specification of units* of our analysis, i.e. the 'objects' to which our propositions will refer. These units may be human individuals, in which case their 'physical distinctness' from the rest of the world is visible. But they may be 'higher level units' like groups, organizations, social systems, etc, and here in order to specify the boundaries of such units we have to take the 'humanistic coefficient' into account when looking at the social reality. If we want to study the family or local party organization we have to understand the boundaries of the unit in the way specified by the 'subjective-objective' interrelations within it, and the 'anatomy' of the given society as a whole.

These units have to be characterized by certain *variables* because the formulation of the research problem assumes that we would like to find the relationships between some variables characterizing these units. Here again, needless to say, the concepts specifying the variables will have to take into account the 'humanistic coefficient' of social reality.

This does not mean, of course, that a social scientist will have to use only the same concepts for the classification of social reality as his subjects do. The postulate of theoretical fruitfulness of the concept may lead the scientist to *redefine* some part of the social reality. The specification of research units and of the research problem is by no means identical with the specification of the *explanatory theory* that will be used, and the concepts this theory will apply. We can understand 'social role' in a way that refers to the subjective perceptions of the 'role obligations' of the 'role sender' and 'role receiver'; but in order to explain how the role behaviors are 'reinforced' we can use the conceptual apparatus of e.g. behavioralistically oriented S-R learning theory. After having defined the notion of 'authoritarian ideology' in terms of certain beliefs held by certain people, we can explain its origin in the language of psychoanalysis, the concepts of which do not have direct correspondence to the conscious subjective experience of the people we are studying. *Neither the nature of the object nor the objective of the study has to be specified by the same concepts that are used in the explanatory theory.* The scientist is entitled to use any concepts he needs forhis theoretical analysis. But here again it turns out that the understanding constructs play a great role as indirect *definitional references* for most of the new concepts the scientist would like to use. Let us consider the most typical categories of such concepts.

Znaniecki stressed the great importance of seeing social reality 'through the eyes of its participants', W.I. Thomas spoke of the 'definition of a situation' by the people involved in it. But a scientist may come to the conclusion that the 'definition of a situation' held by someone is obviously *false:* there is a great difference between *thinking* that one has a million dollars and really *having* it. There is a great difference between thinking that

one is popular in a certain group and actually being liked by others. Having
a million dollars and being liked by others are both 'meaningful phenomena'
but the meanings are of a different order from those in a self-definition of
one's own social situation. They are 'objective' in the sense that they are
external to the person involved. They may influence his behavior in a way
unexpected by him. If one has incorrect notions about his bank account he
may still sign a check, but it will overdraw his account. The check will not
be cashed by the bank, with all the consequences of this act falling upon the
signer. If one overestimates his popularity in a group, he may try to initiate
certain actions within the group, but it is rather doubtful that he will
succeed.

In like manner, our natural surrounding determines our actions both by
the way we perceive and define it, and *by its own physical characteristics*.
Suppose there are two men in the desert dying from thirst. One thinks that
water lies to the north, the other is equally sure it lies to the south. Each
will go in the direction he thinks the water is, but which of them will survive
(or neither) depends on the correctness of the definition of the situation.

To the degree that the course and effect of human action depends not
only upon the way people perceive their social and natural surroundings but
also upon the objective characteristics of these surroundings themselves,
these characteristics have to be reflected in the conceptual language of social
theories. The best way of conceptualizing them may be sometimes to
present them in the form of 'objectively valid' definitions of situations – to
the degree, of course, that the scientist himself believes he is able to perceive
the situation in a more objective way than the actors themselves. These
definitions may thus be different from the subjective definitions of the same
situations by the actors.

But even if the scientist's concepts denote only the meaningful aspects of
the behavior of the social actors and not the characteristics of their social or
natural surroundings, this does not imply that they have to be defined with
a 'humanistic coefficient', The theoretical concepts may take into account
some *abstract properties* of the social structure and cultures which, *in the
way the scientist defines them, are not perceived and experienced by the
members of the given society*, even if the attitudes and values of the
members constitute the ultimate definitional reference for these concepts.
'Authoritarianism' is a property which in some of its meanings is attributed
to a system of beliefs; therefore the focus of its ultimate reference is

somehow 'inside the human mind' — but most people who are classified as 'authoritarian' as a result of their answers to the F-scale questionnaire will be completely unaware that they are 'authoritarian'. The concept is meaningless to them. The members of a group may know that they rather like each other, or that most of them accept the goals the group — but they are usually unaware to what degree their group is 'integrated' or 'cohesive'. Such notions as 'authoritarianism' and 'cohesiveness' refer to some abstract properties of the phenomena. The definitions of these concepts are not based directly on the subjective experiences of those to whom they are applied, although ultimately they refer to these experiences.

Bronislaw Malinowski was quite aware of the fact that scientific concepts may be more or less distant from the subjective definitions of certain situations by their participants, when he wrote in his diary:

New theoretical point: (1) Definitions of a given ceremony, spontaneously formulated by the Negroes. (2) Definitions arrived at, after they have been 'pumped' by leading questions. (3) Definitions arrived at by interpretation of concrete data.[12]

The abstract character of such concepts and their 'disagreement' with the feelings and perceptions of the members of the community to which they refer may be of a *temporal character* only. A synthetic definition of a certain concept proposed by a scientist may later on, when it is well known to the members of the society, come to have its referents fully in the consciousness of the members of the society, because they have learned and accepted the concept as their own 'definition of the situation'. When Marx defined a social class as a group of persons who are bound by a common position in the system of ownership of the means of production, this was a purely synthetic definition based upon theoretical analysis of class systems; it had no clear cut correspondent in the consciousness of members of most class societies. At that time, according to the general understanding, classes were distinguished by many other criteria, but not by the relation to the means of production. But today, to those representatives of the working class who accept the Marxist ideology, the Marxian definition of proletariat is a definition with a 'humanistic coefficient'. Another example is the impact of the conceptual scheme of psychoanalysis upon the behavior of those who try — more or less successfully — to perceive and define their motivational mechanisms in psychoanalytic terms. These are only two of many manifestations of the well-known fact that social theory — and therefore also social concepts — may become an active element of social reality. This internalization of abstract constructs has certain important consequences for the problems of theory formation in the social sciences, among others fot the well-known problem of 'self-fulfilling' and 'self-destroying' prophecies which I shall not discuss here.[13]

Finally, we can introduce into the language of the social sciences concepts which are defined in a way that is free from indirect reference to 'subjectives meanings'. First, as is obvious, we will need in our theories concepts which refer to certain objective characteristics of both human beings and their natural surroundings. But we may also need certain psychological constructs which are free even from indirect reference to the subjective experience of our 'subjects', Many strictly *dispositional concepts* of psychological theories like 'intelligence' could be cited here. Or, we may include certain concepts which refer to the *unconscious level of psychological processes*. The methodological character of the latter concepts is quite interesting because many are defined so that their meaning has a certain reference to processes we know from our own introspection (e.g. 'unconscious envy'). But at the same time the concept postulates that the phenomena denoted by it (viz. 'unconscious' processes) cannot be, at least in normal situations, due to the existence of 'censorship' experienced in the introspection.

4. UNDERSTANDING CONSTRUCTS IN DEFINING THE DEPENDENT AND INDEPENDENT VARIABLES OF SOCIAL THEORIES

Let us consider now some problems of concept formation in the social sciences with special attention to the definitions of *concepts for explanatory and predictive theories*. For the sake of simplicity, I shall not discuss here whole systems of theoretical propositions, but will concentrate only upon particular generalizations which according to a simple scheme relate certain independent variables to the dependent ones. The dependent variable refers to the phenomena we would like to explain in our theory. The independent one may constitute their *causes*, if we seek to have a causal explanatory theory.

The conceptual scheme of any theory with respect to its dependent variables is the result of interaction between the *scientist's goal* and the *objective regularities* he finds in the studied reality. The goal defines the initial phenomena he would like to explain. The objective regularities reveal whether these phenomena are uniform with respect to the mechanisms governing them, and therefore whether they 'deserve' from a strictly methodological standpoint to be denoted by a common concept and if so how this concept should be defined.

The 'dual nature' of social reality poses certain interesting problems with respect to the way we define the social 'units' and their characteristics we would like to explain. One can say that at the beginning we usually have to understand them with a 'humanistic coefficient', because we have to start from somewhere. Therefore we treat them as meaningful-behavioral syndromes, giving to these concepts the meanings attributed to them by the members of the studied population and defining them accordingly. But we soon discover that 'common sense social theory' reflected in syndromatic concepts of everyday language is far from being perfect. First, these syndromes do not include many additional variables which should be specified in an empirically adequate theory, and therefore we start to 'elaborate' them by introducing the new variables. What is more important, it often turns out that these observational-inferential syndromes denote classes of phenomena such that no good theory can be formulated which could explain them in their totality. These concepts are simply not good enough to be dependent variables of any theory. Then the theoretician faces the problem of *redefining* the objective of his study, defining a set of variables which are for both *important* and *explainable* by a theory.

Let us concentrate first on 'what is important' because this usually involves directly or indirectly the *value assumptions of a theory*. Suppose the scientist has found that the relationships between the behavioral and the meaningful sides of social reality are of such a kind that in his theory he has to concentrate primarily on one or the other, but cannot explain these two levels at the same time. What is more important here then? How should he define the object of his study?

In order to make this alternative clear, let us assume that we are dealing with *applied* social studies, i.e. those in which the value assumptions of the given study are fairly clear: the study is undertaken in order to find out certain facts or test certain theories, which are important from the standpoint of a certain 'goal' which the scientist or his 'sponsor' would like to pursue. Now with respect to this, let us specify two extreme – 'ideal' in the

Weberian sense — types of value orientations which may determine the exact specifications of the object of our interest, and the two kinds of studies corresponding to them.

Studies of the first kind may be characterized as based upon *self-centered value orientations*. In such a study the research problem is based upon narrowly understood values and goals of the scientist or — as is more often the case — on the goals and values of his 'sponsor', who is interested only in the consequences the studies reality will or may have for him directly. From such a problem perspective, the *only thing that really counts is the overt behavior* of the members of the studied populations, and especially behavior which may be consequential for the scientist or his sponsor. The sponsor does not have to be an individual, either. It may be a group which would like to study the behavior of members of *another* group in order to predict (and eventually to 'control') the behavior which may be of practical importance for the 'collective sponsor' directly.

In such a study the scientist may of course be aware that in order to predict efficiently the behavior of the members of the studied population in the respect important to him, he will also have to use concepts and theories that will take into account the 'mental states' and psychological processes of his subjects, because he knows that explanatory theories which do not use hypothetical constructs are usually of little practical value. He may even decide that some of the concepts of his study will have to be defined with the use of 'humanistic coefficient', because the way the people see their social situation may be useful for the prediction of their behavior. But his interest in the psychological aspects of their behavior is here of only *instrumental character*. He would be equally happy to have a theory which would not include such constructs. From the standpoint of his own value system he classifies the studied reality primarily with respect to its strictly behavioral side, distinguishing which behavior is 'desirable' and which is 'undesirable' because of its external consequences. The basic *dependent variables of such studies may be defined in a strictly behavioristic way*. He treats the studied reality as 'external environment', the laws of functioning of which have to be discovered for the better adaptation of this environment to the goals and values determining the problem of his study. The psychological constructs in such a study are defined according to the same rules by which electrons or genes are defined in natural sciences; the main goal is to build a theory which will maximize the predictions of some strictly observably defined behavioral sequences. If a scientist were to define and use the notion of 'satisfaction with work' in a study the only goal of which is to discover the factors increasing the workers' productivity, this would be a typical example of problem formulation in which the concept would have instrumental, auxiliary functions only.

As a result of theoretical analysis in such a study the scientist may have to redefine the concepts he had at the beginning. The classes of 'desirable' and 'undesirable' behavioral sequences of the studied persons may turn out to be not uniform with respect to some explanatory mechanisms and thus to need a different classification. In any case, *to the degree that the scientist's value orientations have an impact upon the specifications of the goal of the study*, the only things which are important in such a study are strictly observable events and behaviors, and their observable consequences for those who determine the value assumptions of the studies. *And the behaviors seen from such a problem perspective constitute the basic dependent variables of such theories.*

It is difficult to find a clear case of such theorizing in the social sciences although many studies in applied sociology (e.g. in market research) come fairly close to it. But there is one striking example in psychology, namely the Skinnerian approach[14] to learning processes, which concentrates first upon the kind of behavior one would like to produce, and later on the rules of 'operant conditioning' which are able to bring about the desired results, of a strictly behavioral character.

External aspects of human behavior may be of primary interest in some theoretical situations, which do not assume clearly such self-centered value orientations. This is the case when in studying human behavior we try to discover its mechanisms only as a means toward a more extended theory in which this behavior will play the role of independent, i.e. explaining, variables. Suppose we are interested basically in the functioning of an organization or institution and would like it to function according to a certain pattern we accept. We know of course that in order to explain (or to improve) the functioning of our institution we have to take into account the behaviors of its members. Then, as a next step in our analysis, we try to discover what factors determine this behavior, and to develop a corresponding theory which explains and predicts them. But here again we are primarily interested in the manifest behaviors, and only secondarily – and in a strictly instrumental way – in their psychological correlates. For the airlines it is important to have *smiling* stewardesses, while the question as to whether they are actually *happy* is only instrumental with respect to the smiling. For an alienated dictatorial regime it is important to have obedient citizens. The question of why they obey is only an instrumental one.

In all such problem formulations we find a kind of basic dichotomy *as the underlying normative assumption:* 'we' who are studying the external social world in order to adjust it to our needs, or − if this cannot practically be done − at least to adjust our own actions to its 'iron laws' in order to maximize our utilities, and 'they', − the subjects of our study, whose manifest behaviors have to be predicted and evaluated from the standpoint of our adjustment needs. Surprisingly enough, we find such a 'utilitarian dichotomy' in Lundberg's book. He wrote:

All data or experience with which man can become concerned consist of responses of the organism − in environment. This includes the postulate of an external world and variations both in it and the responses to it... The broadest and most general classification of aspects of the universe which any species will make consists of those aspects which involve the adjustment needs of all or nearly all individuals...

...We delimit the total universe therefore into aspects, categories and classifications on the basis of the differential response with which we are compelled to adjust. This adjustment consists of course of observably different behaviors...

...The test of adequacy ('truth') of any system at any given time will in any event be determined by certain empirical tests notably whether the system affords a rationale for adjustments that have to be made and whether it aids in planning these adjustments...[15]

One could say that the dichotomy proposed by Lundberg looks different from that stated above, namely that 'we' may also refer to all human beings who have to interact in their adjustment with the external world, and that the needs of *all human beings* should constitute the reference for the definition of 'adjustment problems'. But once we say this, the impossibility of using a behaviorist philosophy for such problem formulation is in the social sciences becomes obvious, because then the *frames of reference for the adjustment needs of the scientist and those of his subjects became of equal importance in defining the problems of social studies.* This means we have to look at the social world from the subjective perspective of its participants, just as we look at it from our own perspective in the 'utilitarian' notion of natural studies. Let us call this second kind of study the *others-centered value approach to social phenomena.*

In the others-centered approach we take into account the thinking and feeling of others, their needs, wishes, desires and frustrations, as the starting point for the structuralization of social reality, *just because these pheno-mena are being experienced by our subjects*. Assuming that others feel more or less as he does, and assuming additionally that their feelings and thoughts are as important as his own and constitute a 'legitime' normative premise of social studies, the scientist may define certain psychological constructs in a way which corresponds to the feelings and thinking of his 'subjects' because they denote the *basic categories of his scientific interest*, the *basic depend-ent variables* of his study. He may find sometimes that the concepts he defines for this purpose happen to be the same as concepts defined for the purpose of prediction of strictly behavioral aspects of social reality, but again they may be quite different. Their cognitive function is relativized to the correct denotation of those phenomena which are primarily important for him: *the thinking and feeling of others*. The notion of 'satisfaction with work' defined from this problem perspective may be quite different from the notion defined within the framework of productivity-oriented industrial sociology. For in the other-centered study the scientist will take into account all the aspects of the perceived work situation which make people *feel more or less happy* with their job, independently of whether the phenomena so defined will have an impact upon their productivity or not, or whether they will be related to any other kinds of overt behavior.

The formulation of such research problems, along with the corresponding conceptualizations, is in a sense an extension of our everyday attitude toward persons for whom we have a strong feeling of *identification*. Parents usually would like their child to *feel happy*, and *not just to behave as if he were happy*, and they understand the notion of happiness from their own subjective experience. The 'indicators of happiness' interest them only insofar as they permit them to assess these *feelings* correctly. For a good teacher, it may be more important to be able to influence the ways his students feel and think than to be able to shape their overt behavior.

The normative criteria used for the formulation of such research prob-lems and corresponding concepts do not necessarily have to be based on *positive* identification of the researcher with his subjects. It is not the case in our everyday social life that we care only about the positive feelings of others. When I hate someone, I may be interested in discovering the conditions which may lead to his suffering, and by this I mean again certain of his *feelings* and *emotions* and not any overt behavior, just as I care about the subjective experience of persons whose feelings constitute a 'positive' value for me. We could imagine a sociologist who — being a psychopath or sociopath — decided to study the principles of constructing a society in which the people were as unhappy as possible; he would also have an 'others-centered value orientation', just as do those who want to maximize human happiness. Only their values would be different. And both categories would be different from those in which the psychological constructs constitute only the auxiliary tools for predicting overt behavior.

I have already discussed the degrees of similarity and of difference in the rules of concept formation in the social and the natural sciences. Here it might be said that *to the degree that the conceptualizations and theories of social science are rooted in others-centered value orientations and problem formulating, there is a basic difference between the social and the natural sciences.* In the natural sciences all that ultimately 'counts' for us as human beings are the observable consequences of all 'latent proceses'. Whether we treat the latent variables as 'heuristic fictions' only or we believe that they reflect 'real entities', these entities may never constitute for us a value per se because we cannot identify with electrons in the way we can (positively or negatively) identify with other people. In terms of the value reference of our studies the orbits of electrons are of a strictly instrumental character for us, but the *attitudes of other people may be either instrumental or ultimate values of social studies depending upon the attitude of the scientist himself.* In the natural sciences, any latent property which is not useful, at least indirectly, for prediction of strictly observable phenomena has to be rejected and replaced by another which is more useful for that purpose. In the social sciences if a concept adequatly denotes certain elements of the subjective experience of the studied subjects, it *may* have its place in a theory which tries to concentrate on the assessment and prediction of these 'mental states' of others, even if the usefulness of such a concept for the prediction of overt behavior would be minimal.

These feelings again may be defined strictly in the way the subject experienced them. Or the scientist may come to the conclusion that they should be redefined in order to be useful dependent variables of a theory. This implies all the problems discussed in an earlier section of this paper. *But the focus of our interest in all such cases lies primarily in the subjective experience of our subjects.*

Are theories of this kind, which take the 'latent properties' of human

experience as the basic dependent variables, 'scientifically legitime'? Legitimization of any scientific study is always dependent on our answers to two
questions: Do we want to know something, and are we able to know it?
With respect to the first question, it is enough to say that at least some
scientists care about or are interested in the thinking and feeling of others in
order to answer it positively. The second question will be discussed in the
subsequent sections of this paper, when I shall analyze the 'techniques of
Verstehen'.

I have presented above two extreme types of situations: one in which the
scientist focuses only on that part of the external world around him which
has to be changed so as to maximally satisfy his own goals, and the other in
which the needs of the studied subjects, their feelings and attitudes, in
themselves constitute a 'legitime' normative premise for the formulation and
conceptualization of the research problem. In a rather unexpected way this
simple and rather traditional dichotomy of the attitudes of people toward
other people seems to constitute the foundation for two different strategies
of theory building in the social sciences, each of them presenting different
rules for concept formation and delivering different criteria of their practical fruitfulness. The majority of social studies fall somewhere between these
two extreme categories, defining the objects of their interest in a way that
includes both certain manifest behavioral sequences and their subjective
meanings. But in cases where we have either to choose one of these sides or
to make clear why we want to study them both, it is useful sometimes to
reconstruct the value assumption of our study in order to find at least a
partial answer to our question.

Suppose now that we are aware of what we really would like to explain,
which implies that we know what behaviors or what elements of inner
experience of our subjects are the dependent variables of our projected
theory. The question now is the degree to which the *independent variables*
of our theory may also include certain understanding-concepts. Needless to
say, both 'common sense sociology' and more sophisticated social theories
are often formulated so that the antecedent variables of the theory explaining human behaviors refer to some conscious states and processes of the
'actor'.

Hypotheses and theories whose antecedents contain concepts denoting
conscious states of mind and conscious psychological processes can be
roughly classified into three categories: (1) theories which relate phenomena

in a certain way because the sciencist knows this relation from his own *introspection*, (2) theories which concentrate on the relation between the *motive* of a certain behavior and the behavior itself, and (3) theories which postulate that we can explain human behavior under the assumption that man is *rational*.

This typology does not exhaust all possible theories in the behavioral sciences which can relate antecedent conscious psychological states with subsequent behavior. We can include into these theories any proposition which turns out to be true: If it turns out that religious persons have preferences for a certain kind of T.V. program, there is no reason why this (not very important) generalization should not be included in our theory. If we discover that people don't like to have a 'dissonant' picture of a certain object, person, or situation and that they try to eliminate this dissonance by various 'cognitive' operations, this is an important theoretical proposition of fairly high generality.

Nevertheless there is a special reason to concentrate our attention on the three approaches distinguished above, this being that all these theories have one thing in common: they are developed *through our own (i.e. subjective)* understanding of the nature of the psychological mechanisms governing human behavior. In a certain sense they are self-evident for the scientist who develops them. This 'self-evidence' may mean either of two things:

(1) First, it can mean that the scientist's understanding of the rules of functioning of the human mind (including his own) permits him to develop more fruitful hypotheses, which implies the *heuristic* function of Verstehen.

(2) Second, it can mean that due to their 'self-evidence' these theories are *a priori empirically valid*, useful for explanations and predictions without prior empirical test of their validity.

Let us look at these three approaches to the construction of 'understanding theories', concentrating primarily on the problem of their 'self-evidence'.

5. THE VALIDITY OF INTROSPECTIONIST 'SELF-EVIDENCE'

The first thing we do when we try to understand and to explain someone's behavior is to relate this behavior to something that we know from our own experience, and especially to find in our own *introspection* something which might correspond to what we actually observe. This kind of explanation was the object of analysis by Theodore Abel in his classic paper 'The Operation Called Verstehen'. He stresses that in such explanations we use special kinds of propositions which he calls 'behavior maxims'. He wrote:

Behavior maxims are not recorded in any textbook of human behavior. In fact, they can be constructed ad hoc and be acceptable to us as propositions even though they have not been established experimentally. The relation asserted appears to be self-evident.

This peculiarity of behavior maxims can be accounted for only by the assumption that they are generalizations of direct personal experience, derived from introspection and self-observation...

We find, then, that in all its essential features the operation of Verstehen is based upon the application of personal experience to observed behavior. We 'understand' an observed or assumed connection, if we are able to parallel either one with something we know through self-observation does happen.[16]

In the same article Abel quotes the characteristic of 'emotional syllogism' which was described by F. Alexander:

Our understanding of psychological connections is based on the tacit recognition of certain causal relationships which we know from everyday experience and the validity of which we accept as self-evident. We understand anger and aggressive behavior as a reaction to an attack; fear and guilt as result of aggressiveness; envy as an outgrowth of the feeling of weakness and inadequacy. Such self-evident connections as 'I hate him because he attacks me' I shall call emotional syllogisms. The feeling of self-evident validity of these emotional connections is derived from daily introspective experience as we witness the emotional sequences in ourselves.[17]

It seems that the primary importance of emotional syllogism lies in the fact that it may deliver to us the hypothesis for *how* the people whose behavior we observe will react 'internally' to certain external stimuli and situations, and why they will react in the given way. In order to hypothesize how they will react, we try either to recall our own reactions to similar situations, or to 'construct' their feelings and thinking from certain elements we have gained from our own experience. I understand how a boy feels when he is reprimanded by his father, because I remember similar experiences from my own life. I can try to understand how a man feels when on trial by magnifying certain of my own reactions to much weaker social stimuli of a 'somewhat similar' type.

In his evaluation of Verstehen, Abel stresses rightly that knowledge based on the scientist's introspection *cannot constitute the method of verification* of a behavioral maxim understood as a law of science. It only reveals that such a connection is a *possible* but not a necessary or probable one:

In any given case the test of the actual probability calls for the application of objective methods of observation, e.g. experiments, comparative studies, statistical operations of mass data, etc.

And then he states that certain connections may be generally true even if we do not 'understand' them.

In this instance the operation of Verstehen does no more than relieve us of a sence of apprehension, which would undoubtedly haunt us if we were unable to understand the connection... The satisfaction of curiosity produces a subjective increment but adds nothing to the objective validity of our proposition.[18]

As I tried to state above, the 'sense of curiosity' to which Abel refers may, under others-centered normative assumptions motivate the selection of goals of social studies, quite 'legitime' in determining what we shall study. From this point of view, we are justifiably interested in 'understanding' what other people think and feel when doing this or that, or more generally in understanding the structure of the conscious psychological processes which 'mediate' between certain stimuli they experience and their reactions. If we were to come to the conclusion, for example, that in a certain stimulus-behavior sequence we always find the same psychological mediating process, this would be an interesting scientific discovery in the kind of sociology which believes that such 'humanistic interest' is legitime.

Quite another problem is the degree to which we can use our own introspection for this purpose, or knowledge yielded by the behavior maxims of 'emotional syllogisms'. Here I agree completely with Abel that neither the scientist's introspection nor the behavior maxims can be used as proof that our conclusions are valid, even highly probable, because they cannot be treated as general or probabalistic laws of science unless properly tested. From the standpoint, the scientist's introspection or the introspection of any single human subject implies only that such psychological processes *may* occur. The task of inductive study is to find out the additional subjective and objective conditions under which the given processes are highly probable.

But suppose now we come to the conclusion that certain behavior maxims are, given certain additional conditions, valid with a fairly high probability: most people under such conditions actually 'feel compelled' to behave in the given way. What is the status of such a proposition in behavioral theory?

First, we may treat it literally in the sense stated above, and use it for assessing how a particular person feels or thinks in a given situation, or what he *perceives to be the reasons* for his behavior. Thus we may use such generalizations for diagnostic assessment of the 'subjective side of social reality' to the degree we are interested in it.

But we may try to treat these behavioral maxims as *theoretically valid explanations* of the behavior itself. This would be based upon the assumption that the individual is able to assess correctly not only the relationships between the objective phenomena, but also the relationships among objective and subjective ones, or in other words, upon the assumption of the validity of introspective experience in assessing the structure of psychological processes governing human behavior.

Unfortunately, the validity of introspection in this function is at best only partial, i.e., at best people perceive correctly only some of the totality of objective causes of their subjective reactions, or only some of the psychological antecedents of their behavior. So even if we accept such partial validity of introspective evidence, such theories usually require 'elaborations' by additional variables. But in this case we are looking for variables the importance of which our subjects are unaware of, but which together with those processes they are aware of *really produce* their subsequent behaviors.

In these two *quite different functions* — the behavior maxims play of course only a heuristic role, but I would be far from underestimating them as a starting point for the development of 'understanding theories' of human behavior.

6. MOTIVES OF GOAL-ORIENTED BEHAVIORS

In theories built from behavioral maxims, any psychological connection is possible if we know that in has been reported by someone on the basis of his introspection. This is in clear contrast to two other kinds of theories, which are characterized by much greater internal logical consistency. They seem somehow to be not only introspectively but analytically *self-evident*. The first of these uses the notion of *motive* as the basic explanatory construct. For example, if someone tells us that he went to the cinema because he wanted to see the film, it seems absurd to question the validity of his explanation. When we hear that somebody ate a heavy dinner because he happens to like spaghetti, the information about the motive for his behavior seems to deliver sufficient explanation of this behavior.

Sometimes instead of the term 'motive' we find the term 'goal' in such explanations, with stress upon the *purposive nature* of the action; but here the term 'goal' means nothing else than a motive which consists of wanting a certain state of affairs and being willing to realize it by suitable actions. Then of course we can say that such a goal (motive) constitutes the cause of the action itself, and therefore a proposition connecting the goal with the action can be used for the causal explanation of the behavior.

Explanation of actions by their motives constitutes the basic assumption of Weberian methodology, which can be seen from the definition of sociology quoted in Section 1 of this chapter. Weber was also aware that this kind of explanation can be applied in situations both where the goals of the action is related to personal needs of the actor (Zweckrationales Handeln) and where it consists in the realization of some broader social values not directly related to the 'interests' of the actor (Wertrationales Handeln).

The extension of the notion of 'motive' to all *internalized social values*, which might eventually determine human behavior, to all norms which prescribe certain ends of our actions, or prescribe 'proper' behavior, makes the use of this approach fairly universal. First, whenever we find that someone really *wanted* to do something, and then *did* it, we are inclined to use this assessment as the explanation of consequent behavior. What is more important: since motives, wishes, desires, or feelings of obligations may be of different strengths, when we state that someone wanted *more* to do *A* *than* to do *B*, we are not surprised that later on he really did *A* and not *B*. In general, we use *hierarchies of motives or hierarchies of values* which the individuals internalized from their cultures as antecedents of propositions explaining the behavior corresponding to it. And then the question arises; to what degree and *in what sense* are such explanations valid? Obviously they are not based on general laws and we have to transform the singular propositions in to law-like statements, finding out the additional-internal and external conditions of their validity.[19] Even if we find such laws, we can interpret them in two different ways. The situation here is similar to the case of the 'behavior maxim', and the goal of our theory may be to find the additional conditions under which people who *believe that they are motivated* to do something, to realize a certain goal, will really do so. Then our interest focuses on *perceived motives* of behaviors. But we may go one step further in our theorizing and say either that the phenomena the actors observed in their inner experience formed the *real motives* of their behavior, or that they were *unaware of their real motives*, which were such and such. These two kinds of theories are by no means identical of course.

In the situations of the first kind, we are interested primarily in assessment of strictly temporal sequences. We say for example, that people who *think* they really hate sex will react violently to the distribution of pornographic literature in their country and are likely to vote for severe laws against it. In this case we establish a *sequential relation* between certain values people believe in and their consequent behaviors. Needless to say, such a theory reveals certain interesting connections between psychological and behavioral aspects of social reality, and may also be used for the purpose of prediction.

But when the validity of given motivational explanations of individual behavior, or even the validity of generalizations that relate declared motives to subsequent behaviors, are open to challenge, then we have a quite different notion of the term 'motive': for here we mean a *real psychological factor* involved in *causing* a behavioral sequence and directing it toward a certain *end* (goal). This motive may be perceived by the actor more or less correctly, or it may not be; the only thing that counts is its role in causing subsequent behavior. To call something a *real motive* is to assume its role in a causal process.

It is obvious that the validity of such a motivational explanation depends upon the validity of the causal theory we are able to apply in support of it. And such a causal theory, as does any theory, needs independent empirical evidence. It is also obvious that the action of most motives in this sense is conditional with respect to certain conditions under which they may (or are likely to, or almost certain to) cause the given behavior. The 'motive' will then be relative to the conditions under which certain psychological states operate, but will be defined ex post on the basis of empirically tested theories.

Here again we have to admit that in spite of all that we know about the factor of self-deception in the perception of our motivational mechanisms, many such perceptions are at least partly valid. In many cases we really know why we did something, or intend to do something. In such case, the assessment of the motives of our behavior is valid, but we are unable to trace the whole explanatory scheme leading from the motive to the subsequent behavior. Therefore even if we can explain our behavior by our real motive, it is only a partial explanation. Complete explanation is possible only within a broader theory in which the real motive constitutes only a part.

As I said above, the motivational explanation *seems* to have the character of analytical self-evident theories. This is because the term 'motive' *by definition means the real psychological cause* of the behavior. Therefore its application presupposes that we assess correctly the causal connection between a particular psychological state and the behavior caused by it. The general class of *motives* may be used validly for (partial) explanations of corresponding behaviors, provided that the particular *psychological causes* of the behaviors are validly included in the class.

In a theory of motivation, particular motives should be defined in a way that is independent of the meaning of the behavior caused by them. The same applies not only to the 'motive' itself, but to what might be termed the 'motivational potential' of certain psychological states. By this I mean that a motive which seems to entail a specific behavior with respect to its goal determines this behavior with different probabilities under given conditions, or determines it in broader or narrower classes of different conditions. This means that these motives somehow differ in their *intensity*. When trying to define or to operationalize motives similar with respect to the goal but having different motivational potentials, we try to identify, for example, special features of the verbal reports which indicate high motivational potential (categorical statements of plans or desires), or to define in more general terms motives which are *more categorical* than others by virtue of occupying a higher position in the hierarchy of values of the individual or his group.

We may also build a general theory of the hierarchy of motives or values. Maslow's theory of the hierarchy of motives constitutes one such attempt. Any proposition about the hierarchy of values within a given culture attributes hypothetically different motivational potentials to these values within the given society.

Thus we may say that hypothetical models for theories of motivation which take certain psychological states as motives for behaviors constitute two-dimensional schemes — with different conditions facilitating the action of 'motives' as one classificatory dimension, and the different motivational potentials of the motives as the other. As I said earlier, the theoretical and observational meanings of both of these dimensions have to be specified independently of our specification of the behaviors resulting from these two classes of factors.

7. HUMAN RATIONALITY AS EXPLANATORY PRINCIPLE

A third approach to the construction of theories based upon 'understanding' of human behavior involves the notion of *human rationality*. Depending on the kind of situation dealt with, this approach may be roughly classified into three categories:

(1) The first category of hypotheses and theories deals with situations in which the individual uses his knowledge about the empirical relationships and especially certain causal connections which correspond to certain means-goal relations, so as to realize with certainty or with the highest possible probability *one specific goal*.

(2) The second deals with situations in which the individual takes into account various goals as well as attributes certain values both to these goals and to the various means so as to *optimize the overall utility* of his final decision.

(3) The third deals with situations involving *interaction* between two or more actors who either cooperate or compete in respect to one or more goals, each actor trying to optimize the overall final utility of his choice by taking into account the expected behavior of other participants in the *game*.

The notion of rationality may be involved — as Hempel[20] and others have pointed out — in either *normative* or *descriptive, explanatory* models of behavior. In its normative aspect each of these situations has a formal theory which *prescribes* how one *should* behave in the given situation in order to realize the given goal, to optimize the overall utility of one's decisions, and to find the most rational way in a game situation of the given kind. These normative theories can be used in the *guidance of our behavior*, if we accept the assumptions of the given formal model:

These assumptions are of three different kinds.

(1) First there are assumptions about the *values* of the different aspects of the situation in which the action takes place, both the values of the goal or various goals, and usually positive and negative values (gains or losses) attributed to different means.

(2) Second, there are assumption about some *empirical* (certain or

probabilistic) *relations* which connect certain means (usually related to our behavioral decisions) to alternative outcomes of the situation.

(3) Finally, in order to make a decision we have to accept some *criteria of rationality* which in a way expresses certain value judgments in itself. We may opt for a notion of rationality which maximizes the possible gain, or one which minimizes risks, etc.[21]

From these three sets of assumptions and *using the rules of logic of mathematics we can derive final conclusions as to how one should behave in order to be rational in the sense prescribed by the given model.* Such a model is, of course, as is any analytic construction, logically self-evident.

But the notion of rational self-evidence *has been transferred* from the area of prescription to the area of description and especially explanations of actual human behavior. The theories (or explanatory hypotheses) used in this context have two groups of independent variables: one includes the goal of the behavior (i.e. the actor's values and priorities), the other includes the actor's knowledge about the means-goal relation. But it should be made clear that the main interest of those theories lies in the second problem: they usually (although not always) take the goals for granted and concentrate upon explaining why certain *means* (e.g. certain behaviors) and not others are undertaken. Therefore to the degree that they leave unexplained or only partly explained the area of motivated behavior leading to goal satisfaction i.e. they do not explain sufficiently why the given goal is being pursued, they may be subject to the same criticism which I discussed in the previous section of this paper. And in this aspect of their structure they may (and should) also be improved in the way I suggested above. But let us assume we find that someone's needs or values (plus some additional conditions) explain in a satisfactory way why he realizes a certain goal and try to concentrate on the explanation of why he chose this course of action as a means directed to this goal. According to some writers (e.g., W. Dray[22]) for such explanation it is enough to assume that this person was in a certain situation, and that in this situation, it was rational to do what he did.

This kind of reasoning has been subjected to a deep criticism by Hempel in the study mentioned above. It is not enough to assume that it was the rational thing to do, Hempel states; we have also to assume that the actor was a rational agent himself. The logical explanation then looks as follows:

> *A* was in a situation of the type *C*.
> *A* was a rational agent.

In a situation of type C, any rational agent will do x.
Therefore, *A* did *x*. [23]

Moreover, as Hempel states, in order to make our explanation valid we must specify both the theoretical and the empirical meanings of its antecedents. This applies both to situation *C* and to the notion of rationality itself:

The notion of rational agent involved in scheme R above must, of course, be conceived as a descriptive psychological concept governed by objective criteria of application; any normative and evaluative constructions it may carry with it are inessential for this explanatory use. [24]

Let us look at the methodological character of the proposition in the above scheme which states: 'In a situation of type *C* any rational agent will do *x*'. Can we say that it is a 'regular' empirical generalization appearing in a scheme when we explain an event by a general law and its antecedent conditions? The situation is complicated and interesting, because once we assume that *A* is rational, we can *logically derive* predictions with respect to his behavior, and we do not need an empirical study to test it provided that our assumption is valid (and of course, provided that we know his values and the knowledge he uses in planning his behavior). Does this mean that those who insisted upon self-evidence of the rationality principle were right afterall?

In order to answer this question, let us specify two different functions the propositions of logic (and mathematics) may play in the formulation of social and behavioral theories. First, the rules of these formal disciplines *prescribe* how a theory should be built independently of the kind of variables and processes to which it applies. They permit us to describe the relations between the variables in the most unambiguous way possible. If our theory is of the deductive- axiomatic kind, these rules enable us to derive conclusions consistent with the postulates of our theory. If our axioms specify the basic laws of Durkheim's theory of division of labor, then if their consequences are empirically false, we have to modify the axioms. From this point of view, the psychoanalytic theory of 'irrational' defense mechanisms — if it would be axiomatized — should be as logically consistent as classical mechanics. With respect to this function there is no difference between the social and natural sciences in the use of logic and mathematics. In this function, the rules of formal discipline *prescribe for the scientist only the standards of rationality*.

But when we use the notion of rationality as an explanatory construct in the social and behavioral sciences, we build certain syndromatic constructs *which denote the psychological processes of the behaving individuals.* Then, of course, as with any theoretical models of specified processes, these constructs may be adequate or inadequate depending upon whether these processes work in reality as our model describes them. If our construct postulates that people in a certain situation reason correctly (e.g., according to the rules of elementary logic) whereas actually they don't, then our model is false, and therefore useless for explanation and prediction.

But suppose that certain empirically specified people do act in accordance with a certain notion of rationality. We still face the problem that under this assumption should be able to predict 'a priori' what they do without resorting to looking at empirical reality. Does this mean that this proposition is then somehow analytically valid?

Let us look at the problem in the following way. Suppose we have an axiomatic theory in the natural sciences the postulate of which specifies the basic relations among a set of variables. It is obvious that with respect to the postulates, the consequences are analytically self-evident. In the same way the behavior of a rational person is inherent in specific model of rationality we attribute to him. If he really thinks in the way described by the rules of formal logic, he will have to come to certain conclusions in exactly the same way a computer with a certain program has to come to certain conclusions when fed with certain data. In general, if our assumptions are correct then the consequences are logically derivable both in Newtonian mechanics and in rational models of human behavior. The postulate of rationality as an assumption of theoretical models in the behavioral sciences specifies, as I have already mentioned, certain structural dynamic relationships between variables and processes. If they specify these relationships correctly, it implies certain consequences in terms of the conclusions that people will have to make.

Let us quote here Hempel again:

Now, to be sure, the psychological concepts that serve to indicate a person's beliefs, objectives, moral standards, rationality and so forth, do not function in a theoretical network comparable in scope or explicitness to that of electromagnetic theory. Nevertheless we use those psychological concepts in a manner that clearly presupposes certain similar connections – we might call them quasitheoretical connections.[25]

But here again we encounter a special feature of the social sciences which does not exist in the natural sciences. When we apply the conclusions from

Newton's postulates to the behavior of a particular stone, we do not think that the stone 'consciously must' apply them in order to behave in the predicted way. In the case of explanations by the notion of rationality, we have to assume that both the assumptions of our model and the rules for deriving consequences from these postulates more or less adequately reflect the subjective psychological reality of the actor and his thinking. If we know the rules of his thinking its outcome is more or less predictable in the same way the outcome of behavior of any machine is predictable, on the basis of knowledge of the rules of functioning. In both cases the rules are self-evident for the scientist. But, the rules of human thinking are more or less selfevident for the actor himself as well.

If treated as empirical propositions about general patterns of interrelations among psychological variables, the rules of rationality connect a *potentially infinite number of antecedent variables with their decisional consequences, and therefore the postulate of rationality constitutes an extremely flexible principle of theory building with an almost unlimited range of application.* The limits of its applicability are defined by the limits of flexibility of the human brain in *efficiently processing* the initial data of psychological experience according to the rules of rationality which this brain more or less obeys.

The use of the word 'obey' is not accidental in this context, because whatever our assumptions about the structure of the human brain and its potential capacities, we have to assume that most of the rules of its functioning are 'imprinted' into it both by the individual experiences of the actor and by the culture in which he participates and which he assimilates in the course of his socialization.

This seems quite obvious in a respect to goals and values, and to the knowledge of the actor whose behavior we want to explain by the rules of rationality. We may be less aware of the social origins of the rules themselves and, what is more important, of the limits the results of individual and social learning impose on these rules. We know, of course, that our capacity to deal with a situation which requires the use of arithmetic and mathematics is determined by the extent of our formal education in these skills. We know that people who have been trained in formal logic or mathematical theory of decision *think differently* than do those who have not been. We also know that people who have had considerable personal experience with situations of a certain kind usually deal with them much better than those who have not had this experience. Therefore, we know that both individual experience and social learning can increase the scope of our rationality.

But when speaking of the applicability of the notion of rationality as an explanatory construct in behavioral theories, we should be quite sensitive to its limitations. One of these limitations has often been discussed in the literature, namely the fact that our behavior may be influenced by certain conscious emotions decreasing our rationality, or it may be governed by certain unconscious motives and impulses, and the basic role of our conscious thinking may then be for example, to 'justify' our unconsciously motivated behavior. What I would like to stress here are some more conscious limits to our rationality due to our perceived norms and values.

One obvious limitation is involved in models of rationality of the second type mentioned at the beginning of this paper, i.e. those which deal with maximization of the utility of decisions in the presence of a multiplicity of (often competing) values. Some actions which would be optimal from the standpoint of one particular goal might have to be completely excluded, because they would violate certain basic norms the individual accepts (or at least obeys in order to avoid the social sanctions). Categorical norms of our societies can from the standpoint of any particular goal be quite similar to a labyrinth which makes certain 'short-cuts' to this goal practically impossible. Other values make the 'cost' of certain alternatives so great that even if not completely prohibitive they cannot be 'afforded'.

A limitation of a different kind is the cultural articulation of knowledge people possess about goal- means relations. This knowledge may come from their own unique experience, but most of it is learned from the culture. Knowledge may be 'stored' in the culture in a variety of forms. It may be stored as descriptive propositions stating the relationships between different events or properties, without any implicit or explicit reference to goals which might thereby be realized, or it may be stored in the form of norms of behavior. These may take the form of *instrumental norms* – i.e., propositions stating that 'in order to obtain B you should bring about A,' which in its empirical context corresponds to the causal proposition 'A produces B'. But besides their strictly empirical content these norms may differ in the *degree of imperativeness* with which they are formulated (and usually sanctioned) in the given culture. This imperativeness may refer to the choice of means only, and the norms 'high' on this scale may be read:

'Whenever you want B, you must do A'. But they may be equally categorical with respect to the goal they prescribe, and then they will have the form: 'Because you ought to realize B, you must do A'. Finally, it may happen that even if there is an objective causal relation between doing A and realizing the state B, the rules of the given culture will not reflect this relation in the form of an instrumental norm, but will imperatively prescribe A and B indepently of each other, making both the goal and the behavior leading to it ultimate values of the given culture.

Now it is needless to say that the more categorically a given goal—means relation is stated in an instrumental norm, the fewer options for choice the person has.

Therefore when we want to use the notion of rationality as an explanatory principle, it is important to know to what degree the knowledge people possess has the form of categorical instrumental norms which actually leave little if any choice for alternative behavior. The notion of rationality should be applied only for the prediction of decisions made within the 'area of freedom' left within the normative context. Thus it seems that the proper explanatory construct of the behavioral sciences at the level of reality seen with a 'humanistic coefficient' should be of a 'kind of definition of the situation' by the actors, with a broader or narrow area left for individual rational choice. Only to this subarea can we apply our own models of rational behavior.

On the other hand, the notion of rationality seems to have great importance for the development of certain *comparative standards*. These standards may have either of two different functions in our research — i.e. *diagnostic or evaluative*. For diagnostic purposes, we should build (for the given social situation) an empirically valid model of rational behavior which — we assume — will be valid for at least sufficient number of members of the studied population. By comparing the actual behavior of all the members with its 'ideal' rational course, we can distinguish those persons who behave rationally from those who do not, and then ask an additional important theoretical question: Which factors are responsible for the behavioral deviations from the pure 'ideal' type? In other words, what can we say about the impact of 'irrational' factors upon this population's behavior? The reference to 'ideal type' is not accidental here — it was one of the important functions of the models of rational behavior in Max Weber's sociology.

The notion of *idealization* has many different meanings but one might be as follows: By idealization we mean a law or theory which intentionally takes into account only a limited number of variables whose relationships can be described by a fairly simple formula, and which is empirically valid provided other 'contaminating' variables do not occur e.g. which is valid in some extreme conditions. When their contaminating factors do occur, we still expect that the mechanism described by our ideal law will operate in the reality, but that now *it will act jointly* with other factors within a complex vector of forces; the final outcome will be the 'sum' of the action of the mechanism described by the ideal law and of actions of other factors which this theory does not take into account. Once we know the outcome predicted by the ideal law we can assess the impact of the other factors by comparing the overall result with the prediction, based on idealization only.

But this assumes that the 'ideal law' is at least partly valid, i.e. that the processes described by it actually occur in the studied reality, *together with other processes.* As long as we can assume this, we can use the ideal model of a rational course of behavior, assuming certain values on the part of the actors and certain knowledge they possess, as *a standard for estimating the degree of their rationality*. But what happens if the course of the observed action does not satisfy our ideal model? Unfortunately, this can be attributable to one of two very different causes.

First, it may be that our model was wrong. The values, assumption, the assessment of the population's knowledge, the area of freedom left for individual choice by the culture prescription, the criteria of rationality assumed in our model – any or all of these may not have reflected the sociopsychological reality they were supposed to reflect. It would not be easy to say which of our assumptions were wrong, since such a model is a complex logical structure with a whole set of antecedent variables, and only one dependent variable, the decision-making behavior of the studied population. If any one of these factors is wrongly assumed, the result will be a wrong prediction.

The alternative explanation is that the actor behaved irrationally i.e. he was acting under the influence of some conscious emotions or of more or less unconscious drives or impulses resulting in some kind of 'defense mechanisms'. When we discover the impact of any of these factors upon the kind of behavior they should be included in our multivariable theory of social behavior as additional variables within a multi-dimensional vector which can cancel or modify the course of the process based on the operation of human rationality. Unfortunately, this use of a standard of rationality to reveal the action of 'irrational' processes is based upon the previous assumption of an empirically valid model of rationality postulated for the given population. Since at this stage of social science, as I said above, we are rarely sure that our model of rationality is empirically valid, we seldom are able to derive our assessments about the action of irrational mechanisms by comparison of the course of action with a certain normative empirical model of its rational course. We have much more often to rely on a more direct model of assessment of irrationality, either by observing the action of a certain powerful emotions as the motivations of behavior, or in other ways well know both to clinical psychologists and to sociologists studying 'mass behavior'.

The second use of rationality as a comparative standard is of a quite different character from the diagnostic – namely as a tool for evaluating both the degree of rationality of given behaviors and the degree of rationality of the culture as a whole, from the point of view of certain _ *a priori accepted* normative models of rationality. Here we may apply a quite different approach than when using rationality as an explanatory construct, or when using it for the diagnostic purposes. When we use it as a *critical evaluative standard* we may define it as we wish, provided that we make explicit our evaluative standards (goals), our empirical assumptions about goal-method relations, and our criteria of rationality. By using rationality as a critical evaluative standard we may demonstrate that people often do not behave rationally with respect to their goals because their rationality is often limited by the stereotyped beliefs and prejudices of their own culture. We may also draw their attention for alternative goals which might become attractive to them once they are aware of their possibility. We might be able to reveal inconsistencies in the hierarchy of values they believe in, and propose certain other hierarchies free of such inconsistencies.

But when we do this, we are not *describing* the society and explaining the behavior of its members. We are indirectly trying to change the patterns of behavior of the society and to increase. the degree of rationality in the decision making of its members. I do not think that 'prescriptive sociology' is an 'illegitimate' task for the social scientist, provided he is aware of the distinction between describing social behavior and prescribing it. The more we succeed in attempts toward the rationalization of our society, the more the notion of rationality will be applicable also as an explanatory construct in social theories. At this moment, its uses are rather limited.

8. DEFINING AND ASSESSMENT – TWO OPERATIONS CALLED VERSTEHEN

Whatever the principle according to which we build our 'understanding theory', the theory should be formulated with the clearest concepts possible, the meaning of which specify both the hypothetical and the observable aspects of the phenomena denoted by them. Then the theory must be tested in a verificational study, which assumes that we are able to recognize in practical research situations the phenomena to which the concepts of our theory refer so as to see whether they really occur in the way described or implied.

Defining or 'grasping' the meaning of a certain 'understanding concept' is by no means identical with the correct assessment of referents of the given concept. These two 'operations of Verstehen', although related, are clearly different. The best illustration of this may be that even if we have a clear notion of what 'hidden hate' means, we may have difficulties in empirically assessing which of the persons we are studying are characterized by 'hidden hate'. In other cases, even when we have a rather clear introspectionist notion of what is going on in other people's minds (e.g. we assume that they feel the same as we do) we may still have difficulties in verbalizing and therefore in defining those feelings in clear conceptual language.

Whereas understanding in the first meaning belongs to the *area of theory formulation* and more specifically to the area of defining theoretical concepts, understanding as assessment of the state of mind of actual persons belongs to the area of *diagnostic and verificational social studies*, or to the area of theory testing and the application of theretical knowledge to the explanation or prediction of behaviors and social processes. The first operation consists in specifying what properties a person, group, culture or society *should* possess in order to be classified as referent of the given concept. The second consists in formulating diagnostic, singular or general, statement that persons belonging to the given group or category or being in certain specific conditions *really do have* those properties which the concept specifies.

On the other hand, the two 'operations of Verstehen' may be strongly interdependent in actual research, just as any concept formation may be dependent upon the observation of reality itself. It may happen of course that the concepts we need for a particular study already exist in the sociological vocabulary and that we are able to apply them in their existing form for the diagnostic assessment and in the formation or testing of our generalization. In most cases nevertheless, due to the great variability of social reality, we have to modify existing concepts and define new ones if our conceptual tools are to be flexible enough for our research purposes. But in order to do that we have first to 'understand' the studied reality itself — i.e. to relate in as correct a way as possible its hypothetical and its observable aspects. In such case the understanding constructs are in a way important *results of our study*, which at the same time permit us to present our results to others in an intelligible way. The more unfamiliar and complex the new phenomena we take into account, the more likely it is that we will not find proper concepts for denoting them in all the complexity of their behavioral-meaningful associations, and that we will have to define some new concepts.

How can we do that? In order to define a new understanding concept, we have to refer its content to certain other concepts which will be included in its definition. The most conscious and monumental effort in this direction was undertaken by Max Weber, who, starting with the most elementary notion of 'action', tried to define step by step his general concepts — *'allgemeine Gattungsbegriffe'* — and later on to use this conceptual scheme for the specification of meanings of some concepts so complex that they could refer to one specific case, one culture or society only.[26]

Actually the conceptual apparatus developed by Max Weber deserves some comment here. When we read the first chapters of 'Wirtschaft und Gesellschaft' our first impression is that we find there *only concepts* which are defined for the purpose of their eventual use in theories. But we should remember that even at their most elementary stage these concepts refer to 'motivational-behavioral syndromes'; therefore, according to the terminology proposed here, they already constitute certain elementary models of social reality. But since they do not take into account the conditions necessary for the operation of particular motives, they can be treated as partial models only. However, as Weber's conceptual scheme develops, the syndromes become more and more complex and involve increasingly more theoretical propositions about the interrelations of the phenomena involved in the meanings of particular syndromes. When we come to such concepts as 'bureaucracy' or 'charismatic power', they actually constitute complex theoretical models of the phenomena so 'defined', and to treat them as 'concepts only' would be to overlook the great theoretical richness of Weber's work even at the level of 'concept formation'.

Let us return to the problem of defining the understanding concepts. In doing this we have to assume that the terms in our definiens have, as nearly as possible, uniform meanings among their users, that they have *inter-personal reliability*. Since they refer not only to observable but also to meaningful sides of social reality, we have to assume that: (a) there is a kind of 'analogy of feeling and thinking' among the users of certain concepts at least with respect to some basic notions; (b) this is correlated with the use of certain *terms* denoting these feelings and thoughts.

If we agree that human thinking is determined by the social and cultural conditions in which people live, and especially by the verbal milieu of cultures, then we have to conclude that there must be a certain correspond ence between the terms people use and the way they think and feel when using them, because their thinking has been to a high degree determined by the words they use, and were taught to use.

The question is, How far does this unification go? This is an open, empirical problem. What I would like to stress here is that, even if this assumption has certain cultural limits as to its validity, *we could do much more toward the exact specification of meanings of theoretical concepts in the social sciences than has been done up to now.*

On the other hand, we know that the more distant the culture of the scientist from that of the studied society, the more likely it is he will not find in his own language certain elementary notions necessary for the specification of concepts with which to communicate the results of his (presumably successful) understanding of this culture to his readers. It may happen that basically different social conditions and basically different linguistic cultures will make the exact translation of certain meanings practically impossible. But here we should remember that the definitions we find in sociological dictionaries are not the only way of learning the meanings of new concepts (otherwise we couldn't explain how a child acquires linguistic skills). And there is always a way to convey the meaning of a certain concept by using it in many social and behavioral contexts, so that at least the *rules of its usage* will become clear for those who seek to understand a new notion. Anthropological field reports from 'exotic' cultures reveal how much can be done toward the specifications of concepts referring to cultural meanings which are very distant from our own. Here again instead of philosophical reflection on whether we can relate in a perfect manner some basic notions of these cultures to our own, I would suggest a more practical directive: Try to understand their basic notions as much as possible by specific definitional reference to our own conceptual scheme and at the same time by locating them in the indigenous social context in which they function. *It is better to understand at least something than to understand nothing.*

In the diagnostic sense Verstehen refers to the *actual interpretation* of certain behaviors and observable situations in terms of the social meanings the people involved attribute to them. The question of whether we correctly understand these attributed meanings and therefore assess properly the occurrence of entire meaningful-behavioral syndromes is therefore crucial both in diagnostic studies and in the verification of our understanding-models and theories. It is equally important to the application of such theories, to explanation and prediction, since these procedures presuppose that we have correctly identified the antecedents and consequents of our generalizations. Let us see now how we can understand the behavior of others, how we can validly *assess* that the phenomena denoted by our understanding constructs do really occur.

G. Homans, in discussing the problem of how to assess the occurrence of 'sentiment', asked the following pointed question:

We can see activities and interactions. But if sentiments are internal states of the body, can we see them in the same way? ...Science is perfectly ready to take leave of common sense, but only for a clear and present gain. Lacking more precise methods for observing sentiments, since the biological methods can only be used in special circumstances, have we anything to gain by giving up everyday practice? Have we not rather a good deal to lose? And what is everyday practice? In deciding what sentiments a person is feeling, we take notice of slight, evanescent tones of his voice, expressions of his face, movements of his hands, ways of carrying his body, and we take notice of these things as parts of a whole in which the context of any sign is furnished by all the others. The signs may be slight in that the physical change from one whole to another is not great, but they are not so slight so long as we have learned to discriminate between wholes and assign them different meanings. And that is what we do. From these wholes we infer the existence of internal states of the human body and call them anger, irritation, sympathy, respect, pride, and so forth. Above all, we infer the existence of sentiments from what men say about what they feel and from the echo that their words find in our own feelings. We can recognize in ourselves what they are talking about. All those who have probed the secrets of the human heart have known how misleading and ambiguous these indications can sometimes be, how a man can talk love and mean hate, or mean both together, without being aware of what he is doing. Yet we act on our inferences, on our diagnoses of the sentiments of other people, and we do not always act ineffectively.[27]

Other writers on 'Verstehen' stress the role of *empathy*[28] i.e. of psycho-logical insight, or the role of 'simulating experience' (German: 'nacher-leben') in making correct guesses about other people's contents of mind.

The only problem with empathy is that it is in its course and its results far from the methodological ideal of controllability of our conclusions. Intuition and empathy will for a long period continue to play a great role in our understanding assessment, but they cannot be presented as evidence. In order to present evidence we need more or less standard methods. The question is, Can we develop the *methods* as well as the *art* of Verstehen?

I think we can. And I think that in the methodology of social research a great number of rules refer implicitly to methods for the *standardization of Verstehen*. The broad area of methodology of attitude studies, for example, prescribes many rules which can give us relatively accurate descriptions of beliefs, value orientations, knowledge, etc. We find there rules for con-ducting interviews so as to obtain the most informative answers about people's goals, for recognizing respondents' frames of reference, etc. Some methodological rules of field techniques or laboratory observations tell us how we can make relatively valid guesses about the state of mind of the observed person, using his behavior as indicators. And so on.

From the behaviors we observe and the actors' utterances, in the sub-jective sincerity of which we believe we can construct with greater or lesser success the world of social meanings attributed to these behaviors: the structural dynamic syndromes of psychological-behavioral variables inter-related in the way the actors perceive them, with the goals and the means they accept, with the areas of freedom left to their rational choice, with the behavioral consequence as both the outcome of the meaningful sequence and the beginning of another one.

When doing this we do not have to rely only on the observation of particular individuals. Fortunately for the social sciences we can use also the *cultural patterns* involved in determining the behavior of individuals. These cultural patterns are often 'objectivized' in the form of certain documents which may be studied independently from individual behavior. We can use these documents as indicators of the subjective reality of the members of the given society, on the assumption that the cultural patterns (ideological beliefs and values, knowledge, social norms, etc.) expressed in them, are transformed more or less into the contents of minds of persons being socialized to that culture. If we have an access to a verbalized presentation of the system of values of a given group (e.g. its ethical code), and on the basis of other premises assume that these values have been internalized sufficiently well as far as the average group member is concerned, then instead of undertaking an empirical study of the value systems of the singular group members we can, with proper caution of course, use those values which are represented in the group's ethical code as indicators of the real values of the group members.

We can learn quite a lot about the typical values of the Nazi mentality by reading the official ideological texts of the Third Reich. We can learn a lot about the system of norms existing 'inside the heads' of members of a given population by reading a textbook of good manners accepted as the legitimate source of such norms by these people. The instances are innumerable.

The same refers as well to the *knowledge* that people poses the correct assessment of which is so essential for the application of models of rational behavior. Thus we can assume that the average person in Western civilization will know at least something about the role of viruses and bacteria in producing disease. On the basis of this assumption we will predict that most people will obey the appeal to have a protective vaccination at a time of dangerous epidemic.

It is obvious also that the scientist's knowledge of the content of the culture will permit him to make relatively correct guesses about the degree of rationality he may expect from the members of the given group and about what areas of issues are most likely to be treated in a rational way, what criteria of rationality they will use, etc.

Therefore we can say that relatively thorough knowledge of the content of cultural patterns shaping the contents of minds of the members of a given group can also be used be used in our attempts to standardize the understanding assessment.

Of course our assessment becomes here more correct if we can locate the cultural patterns in the context of a theory of socialization, which will tell us which categories of the members of the given society are most likely to internalize the values, norms, and knowledge of their culture most completely, and those who will treat the norms only as external pressures on their own behaviors. The understanding models of action we develop for the behaviors of those, who, due to our theory internalize these norms will be, of course, quite different from the models we develop for those who do not accept society's values as their own and treat them only as an element of the environment they have to deal with.

And here we come to the next section of this paper. If cultural patterns or any other variables are used as indicators of certain psychological processes, and predispositions, then we have to specify or describe as we can the nature of the *theoretical, logical and empirical relations* connecting the indicator and the indicated variable.

9. TYPES OF INDICATORS OF MEANINGFUL SOCIOLOGICAL VARIABLES

The notion of 'indicator' was (as far as I know) used for the first time in sociology by Charles S. Dodd[29] — one of the fathers of the 'empiricist' orientation — to denote observable correspondents of theoretical concepts, which for him had to be defined in the operational way in any case. Since that time the notion of indicator has been discusses by many methodologists, and especially by P.F. Lazarsfeld.[30]

Let me say here that *by indicator I of the concept C, I mean any event or property I whose existence or occurrence permits us to conclude with certainty, or with a given probability or at least with a probability higher than average, that a property or event denoted by concept C exists or has occurred.* When our indicator is described by a quantitative variable, assessment of the given intensity (e.g. score) of this variable I permits us often to make a definite or probabilistic assessment of the intensity of the variable C.

Assessment of the occurrence or existence of the indicator I may be either direct or indirect. It is (or at least it may be) direct observational assessment when the indicator I is a directly observable event or property. Most of the practical users of the term indicator, and researchers who construct the indicators of different characteristics of human beings and social aggregates and systems, seem to take for granted that indicators have to be composed of directly observable properties. We shall consider this problem later. But even at this stage of the discussion, we have to admit the possibility that some more or less indirectly assessed phenomena can be used as indicators of other events or properties related to them. When a psychoanalyst takes a patient's strongly expressed conviction about his love for his father as the indicator of 'repressed hate' for the father, he is not able to observe this 'love' directly but has to infer this conviction from the utterances of the patient. When we say that someone's strong moral values indicate his 'rigorism', and by moral values we mean only certain convictions of this person the existence of strong moral values cannot be directly observed. Nevertheless we can agree that in most cases the indicators we use in empirical research are composed of *at least partly observable phenomena and properties*. If we assume this limitation in our discussion we can say that an indicator is a phenomenon or trait (or a complex set of traits or phenomena) *I the observation of which* permits us to classify the objects of our interest according to a definite value of a given variable *C*. Thus we say, for example, that the make of his car is an indicator of the owner's income, that participation in political demonstrations indicates the political attitudes of the participant, that an I.Q. score indicates the intelligence of the respondent, and that a certain response pattern in a questionnaire study is an indicator of high work morale. By saying this we mean that observation of the given indicator permits us to classify the objects of our study from the standpoint of their income, political attitudes, intelligence or work morale.

But we have to relate the 'observability' of the indicators to what was said earlier. It may happen that we take as an indicator of something an event or property which is *strictly observable* in its entirety, i.e. in order to assess its appearance or existence we do not have to interpret it in the context of any social, cultural or individual meanings. This is a typical situation of course in the natural sciences – as, for example when the reading of the thermometer is for a doctor the indicator of the patient's temperature, and later on together with other symptoms, an indicator of the given illness. It may also happen in the social sciences that we take as indicators of sociological variables some events in their strictly physical form. Thus we may take someaone's good physical condition as an indicator of rather important sociological information: that he has enough money for nourishing food. Or we may take the physical presence of a large crowd in a city square as an indicator that someone must have organized this gathering, because its random occurrence seems rather unlikely.

Nevertheless, in typical cases of the 'use of indicators' in sociological research, the 'observable' events we use are already 'loaded with social meanings' – it is *precisely due to these meanings that we can use them as indicators of the variables* of interest to us. If we take the amount of money a person has an indicator of his belonging to the economic elite in a given country, it is needless to say that the indicator itself is a 'meaningful property'. If we take the fact of reading a Proust novel as an indicator of interest in 'high-brow' literature, the situation is the same.

In such case the event or property we take as our indicator can function in this way *only when our interpretation of the social meanings* associated with the given 'physical' event *is correct*. If an illiterate person holds Proust's book open in his hand he may, from a strictly physical point of view – at least at a given moment – be behaving identically to those who enjoy literature. But for such a person this would be a spurious indicator of high-brow taste, because the proper subjective meanings are absent. Even if we take the number of machines as the indicator of the productive capacity of a given society, we have to assume that there are people who are capable of using them and that the social structure permits effective coordination of their productive work.

But, since most of the concepts we use in sociology even at the most descriptive level belong to the category of those concepts that are 'loaded with social meanings', *we cannot classify our indicators as belonging to the 'language of observation'* in the sense proposed by Carnap. At least this does not apply to most of them.

Suppose now that we take as an indicator *I* of the concept *C* an event or property the occurence or existence of which can be assessed either completely or in part by observation. Then we may ask, *Why* is *I* the indicator of *C*, what is its relation to *C*?

When we analyze the indicator from the standpoint of its relationship to the indicated variable, we can distinguish two large classes of indicators, namely *definitional* and *empirical* indicators.

When we use, for example, the number of sociometric choices received by a person as the measure of his sociometric status, our indicator *defines* for us the measure of the indicated property. The notion of status in sociometry *means* the number of positive choices scored by the person. The relation between the indicator and the indicated property has been established by a *terminological convention*. When we say that certain kinds of reactions in interpersonal relations are the indicator of 'aggressive behavior', we simply specify the meaning of aggressive behavior in terms of certain kinds of reactions. We do not have to assume any empirical relation between the indicator and the indicated phenomena; the relation between the indicator and the indicated phenomenon is 'secured' by the meaning we give to the given term.

Definitional indicators may also be called *internal* indicators because they either constitute completely the indicated phenomenon or are at least its partial *definitional components*.

Empirical indicators are those which are related to the indicated phenomenon or variable by a non-definitional, i.e. empirical, relation. They may also be called *external* indicators, because the concepts denoting them do not belong to the meaning of the concept referring to the indicated phenomenon.

When we classify people into two categories of self-esteem — those with low and those with high self-estreem — taking as the indicator of that variable the sociometric ratings scored by them, we assume that there is a high empirical correlation between self-esteem and the position of the person in his primary group. However, we do not define self-esteem in such a way as to suggest that a person's sociometric status belongs to the definiens of this term, but treat it as an external factor of the defined phenomenon. When we construct an index of political party preferences on the basis of people's socioeconomic status, ethnic group, religion, etc., we treat these indicators as external determinants of political preference and not as elements of this attitude. An example of an internal indicator of political preferences might be voting behavior, or an answer in a public opinion poll, etc., in the same way an internal indicator of self-esteem might be a man's own opinion of his abilities, his perception of his popularity, etc.

When we are not able to say whether an indicator belongs to the meaning of the given concept or is related to the referent of the concept by empirical relation only, it is obvious that we have not defined the meaning of the corresponding concept clearly enough. Unfortunately, this is the case with many sociological concepts currently in use.

One could ask why we put such emphasis on this distinction. The reason is that as long as we don't know which phenomena are *denoted by the concept we use* and which are *conceptually external* to it, *we don't know what our theory is about*, and by the same token, which observations — and for what reasons — are confirming or rejecting it. Thus, for example, when we know that certain alternative sets of indicators are definitionally external to the indicated phenomenon, and that each indicates it by the force of postulated empirical relations, we can use each indicator to check the validity of others, and additionally to *integrate the findings through use of different indicators into a common social theory*. In this way we can integrate the results of studies conducted *under different research conditions, or through use of different research techniques*, within a theory formulated in a more general conceptual scheme. Moreover, the negative results of verificational studies using definitional indicators by definition reject our theory, whereas negative results with the use of empirical indicators always leave open the question of what is wrong: the tested theory or the auxiliary propositions on which we based our choice of indicators.

When we look more closely at the *empirical indicators*, we can classify them again into two categories, depending on the nature of the empirical relation involved.

The assumption that a certain completely or partly observable variable I is an empirical indicator of the variable (concept) C may be based upon knowledge of some relationship which is accessible to a more or less *direct observational test*. Thus, for example, after having classified the inhabitants of a given town into certain 'income categories' on the basis of make of car they own, we can check the real income of the car owners to see how correct our classification was. In this situation the indicator I stands for some other 'observable' characteristic which also might be observed independently (with all the references to its meaningful interpretation mentioned above). The only reason for using the indicator I in our research instead of the characteristic C indicated by it is the fact that it is usually *easier* to observe I than C. Let us call such indicators *observational indicators*. It should be remembered that to call these indicators observational does not refer to the fact that the indicator is here an observable property, but to the fact that the *relation* between I and C is accessible to a direct observational test — which presupposes the observational character of both I and C.

The other category of empirical indicators is of a different character. In preceding sections of this essay I discussed the importance of latent properties in general and psychological concepts in particular in the conceptual apparatus of the social and behavioral sciences. Whether defined in purely introspectionist terms or in the theoretical terms of any psychological theory, the referents of these concepts *cannot be directly observed by an external observer*. Some, like the phenomena denoted by concepts of psychoanalytic theory, are inaccessible even to our introspection and have to be assessed in an even more indirect way. In all these cases the scientist must *infer* the existence of these latent, hypothetical traits from the occurrence and pattern of relationships of some observable traits. If the observable traits which permit us to infer the existence of the hypothetical traits are taken as their indicators, I would call them *inferential indicators*.

As I mentioned above, when we are using a sociological variable I as a definitional, observational, or inferential indicator of another variable, we usually use a variable which already has a specific social meaning in itself. But then, this implies that in order to understand our indicator correctly we have to treat its strictly observable components as inferential indicators of something else, i.e. we have to relate proper social meanings to it. Later on, assuming that our interpretation of the meaning of the indicator itself is correct, we make other uses of this variable in our inferences to other variables.

So we see that in most cases where we use a sociological variable as an indicator of any type, the assessment of proper meaning of this variable itself requires first that we use certain strictly observable events or properties as inferential indicators of some other hypothetical events, and then we may use such syndromes as the indicators of other kinds of social phenomena.

One can summarize the above comments on indicators in the following way. If the term referring to the indicator I belongs to the connotation of the concept C, I is a definitional indicator of C; otherwise it is empirical. When it is empirical (and at least partly observable), and a phenomenon denoted by the concept C is at least partly observable *in a way that does not involve the observation of I*, I is an observational indicator of C. Otherwise it is an inferential indicator of C.

But there are many cases which violate this simple classificational scheme. These are the cases where C in itself constitutes a complex syndrome, the meaning of which postulates that a certain object in order to be C must have a set of characteristics A, B, D, and I, and that these characteristics are interrelated in the syndromatic way characterized above where I gave the most simple examples of 'meaningful-behavioral' syndromes. Let us assume that we define a concept C (say, 'panic escape') in the following way: C = df ($I.P.$), where I refers to a strictly behavioral sequence and P to psychological processes that correlate strongly with certain kinds of behavior. I may then be of course the indicator of the whole syndrome C. But is it an empirical or a definitional indicator? In a way, both.

I is a *partical definition* of the concept C, therefore it is a definitional indicator of C. But since the meaning of the concept C assumes a strong correlation between P and I (let us for simplicity's sake assume their equivalence: $P=I$), then *due to the meaning of our syndromatic concept* (and under the assumption that the phenomena I and P are really correlated in this way) I is also an *empirical (inferential) indicator* for the other component P of the syndrome C.

The make of a car may be a clearly empirical (observational) indicator of the income of the owner − or (if it is a Ferrari) an inferential indicator of his liking for fast driving. But if we take the make of an expensive car as an indicator of a 'high standard of living' and understand the notion of standard of living in such a way that having a luxury car is relevant to it, then of course our indicator is both empirical and definitional. It partly defines the whole concept C and it empirically indicates other components of the syndrome C, i.e. the entire syndrome of conditions of life denoted by 'high standard of living'.

As we know many concepts in social theory and in social research belong to this category, i.e. they denote composite clusters of syndromatic traits. 'Conservative ideology', 'high social status', 'authoritarian behavior' are examples of a broad category of terms denoting such composite syndromes of traits, which can be represented by one or a few well chosen indicators. These indicators are, as I have already noted, empirical indicators for other elements of the syndrome, as well as definitional ones − i.e. partial definitions of the concept of the whole syndromatic variable.

These syndromes are usually more complex than the one described earlier, because not only do they include phenomena from both the observational and hypothetical levels, but at each level they assume a whole cluster of phenomena. In a way they constitute fairly complex theoretical models of social or behavioral reality in which the latent factors and forces and their observable manifestations are related to each other. When we say that we understand a certain behavior, we mean by this that we think we have located its strictly observable components within the context of the proper meaningful syndrome. *The operation of Verstehen – in its diagnotic sense – is nothing else than the interpretation of observable aspects of human behaviors as indicators of the whole meaningful syndromes to which they belong.*

10. BEHAVIORAL INDICATORS AND 'SURPLUS MEANINGS' IN BEHAVIORAL THEORIES

As we have seen from the above analysis, an observable indicator may belong to a meaningful syndromatic construct both constituting its partial definition and indicating its other components by an assumed empirical relation. The most important class of such indicators are the overt behaviors. These behaviors may be related by different kinds of causal relations to their psychological counterparts (to be discussed in the next section of this chapter).

Let us consider here a category of situations in which the behavior is understood as an indicator of a certain *disposition D*. D thus constitutes an element (or the only element) of the postulated latent phenomenon.

The word 'disposition' requires some comment. If we say that a given disposition D is recognized by a certain behavioral indicator I, the problem may be posed as to what kind of relationship connects D and I, or, in other words, whether this relationship is an inferential or a definitional one. As we know, there are two different possibilities of understanding the relation between such an indicator and an unobservable dispositional property: (1) We may say that the disposition D *means* a tendency to the occurrence of the given behavior I and that in this situation I is a definitional indicator for D, being its complete or at least partial (conditional) definiens. (2) We may say that the disposition D is *empirically* related to the behavior I which, therefore, constitutes its *inferential* indicator.

We may treat these two ways of understanding the term 'disposition' as two incompatible philosophies of theory construction, one of them strictly 'empiricist', according to which all scientific terms should be finally reducible to strictly observational statements, when the other admits theorizing in terms of 'latent properties' and processes and, additionally (in the social sciences), may accept that the processes to which such theories refer 'actually occur' in the minds of the acting persons, and may be experienced introspectively by them.

But even when we accept the second approach to theory formation, the two meanings of the term 'disposition' may constitute *two steps in our analysis of human behavior.* In the first step, we observe certain regularities of behavior, and therefore we assume that people have a 'disposition' to behave in the given way; in the second step we try to understand the character of the mechanisms responsible for this disposition and to build, as Hempel termed it, the notion of disposition as a broadly dispositional concept.[31] In such case, even if at the first stage of our analysis word 'disposition' means practically only a behavioral regularity, we have to treat this behavior as the empirical consequence of an *unknown* but, at least potentially, independently definable explanatory causal mechanism, the nature of which we would like to specify in the second stage of our analysis.

In the context of 'understanding theories', which is the primary focus of this paper, the independent meaning of the term 'disposition' refers to such phenomena as 'willingness to do something', 'decision', 'emotional compulsion', 'conviction that one ought to do something', etc. These terms also refer, of course, to independently identifiable phenomena, their relation to subsequent behavior being causal, (and obviously conditional), depending upon the additional occurrence of many factors and counterfactors.

Within such theorizing the *occurrence of a disposition is itself explainable within a broader theoretical model,* and in previous sections of this chapter we considered various kinds of models which can be used for this purpose. What is characteristic of these models is that they often not only permit us to explain the behavior we are interested in, but *also have additional empirical implications which can be used for testing their validity. We say, that the concepts of our theory have 'surplus meanings' – they imply more than is necessary for the purpose of a specific explanation.* Due to this they permit us later on to explain, for example, the correlation between two different kinds of behaviors.

But how do we arrive at concepts which have 'surplus meanings'? One of the most fruitful ways in the social sciences is to try to look 'inside human minds', and to assess by our Verstehen the existence of certain 'meaningful syndromes', Such a syndrome is a whole cluster of dispositions along with the psychological mechanisms leading to them.

The better we 'understand' an individual or a population, the greater the number of predictions we are able to derive from the richness of 'surplus meanings' which occur in the concepts we use for denoting the results of our Verstehen. On the other hand, the validity of our 'understanding concepts' and the hypotheses in which they occur is always subject to empirical test by the observation of the behavioral consequences they imply.

If it happens that the empirical relationship between behavioral indicators I_1 and I_2 does not correspond to the pattern of relationships derived from the meaning of a certain theoretical model, the variables of our psychological model or the relationships between them must be changed so that the relationships derived from the model correspond to the correlations of observable phenomena. This means that we should reformulate our theoretical model so that it implies either the disposition to I_1 or the disposition to I_2 (or permits the derivation of either I_1 or I_2 under certain additional conditions.) To recall a well-known example: It was assumed by the authors of *The Authoritarian Personality* that political conservatism constitutes one of the interrelated characteristics of the 'authoritarian syndrome'. But later on further discussions and studies revealed that one can be authoritarian in the context of various ideological orientations, which led to the redefinition of the 'authoritarian syndrome' so that political views do not belong to it.

Obviously, we can also use behavioral consequences in the *choice* of one from several alternative models. Let me give a most simple example. Suppose that we find in a certain society a large category of people who both do not steal and do not act aggressively under any circumstances. Then we introduce another common-sense notion by saying that these are 'good persons'. This disposition to 'be good' is supposed to explain this correlation between nong-stealing and non-aggressiveness. But later on we try to find a deeper theoretical mechanism as our 'explanatory construct'. Here we face two alternative models. One is based on the fact that we have discovered that theft and aggression are evaluated highly negatively by the norms of the given culture: therefore we assume that the 'good' persons are such because they have internalized the values of their culture very deeply. The other explanation, a more 'psychological' one, might be that these persons are especially sensitive not to do any harm to others. Then we start to look for the *'Experimentum crucis'* which will permit us to choose the proper model. In this case we try to find any behaviors which are sanctioned by the values of the given culture but imply some harm to others. If it turns out that such behavior is correlated positively with the first two indicators – non-stealing and non-aggression – then the 'cultural' explanation is valid. If it is correlated negatively with them then we accept the 'psychological' model.

The procedure of choosing one from among alternative theories by 'experimentum crucis' is too well-known from the natural sciences and from psychological experimentations to be elaborated here. I mention it for one reason only: In order to test a theory or to choose one from among several alternative explanations, the chosen theory itself should be so logically consistent that its relation under certain conditions to subsequent behaviors should be unambiguous. The main problem with 'understanding' analysis of social reality is that we rarely obtain a degree of logical consistency that permits such tests with a certainty comparable to that in other sciences.

11. THE VALIDITY OF INDICATORS IN EMPIRICAL SOCIAL STUDIES

Whenever we use in our research an indicator I for the concept C, and this indicator does not constitute a complete operational definition postulating identity of the indicator in this concept, we face the problem of *what is the relation between the extension of the term I and the extension of the term C.* In other words, what is the *validity* of the phenomenon I as the indicator of the concept C? The importance of the correct answer to this question for proper theoretical interpretation of the empirical findings is too obvious to be discussed at length here.

We can distinguish *three different measures of the validity* of the indicator I for the concept C.

First, we can ask what is the *screening power* of I for C. Let us say that I has a perfect screening power for C if all the objects which belong to the extension of the term I belong also to the extension of the term C – i.e. I is able to reject all non-C's. And in general by the screening power of I for C, we mean the probability (relative frequency) that the objects of the class I are also of C. If we denote the screening power of I for C by SP_{IC}, then the formula for it is:

$$SP_{IC} = \frac{nIC}{nI} = pC/I.$$

In cases where we would like to have in our study only those objects (or as high a proportion of them as possible) which are C, then we will have to look for an indicator or a set of indicators having maximally high screening power for C, even if this means that some C's will not be covered by the extension of I.

Second, in some cases we may wish our indicator to have as high, or nearly perfect, an *inclusion power* as possible. By this I mean that our indicator will include as high a proportion of C's as possible, even if at the same time it might also include many non-C's,. Thus the formula for inclusion power is:

$$IP_{IC} = \frac{nIC}{nC} = p(I/C).$$

Third, it may happen that we would like somehow to 'balance' these two functions together. We can then use the *discriminative power* of I for C, meaning by this of course the correlation between I and C:

$$DP_{IC} = R_{IC}.$$

Let me now present graphically several situations with different relations between C and I, in order to illustrate the different measures of the validity of I for C:

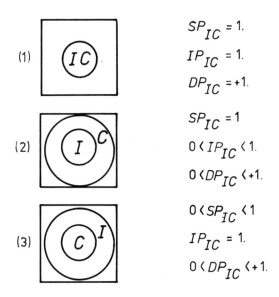

(1) $SP_{IC} = 1.$
$IP_{IC} = 1.$
$DP_{IC} = +1.$

(2) $SP_{IC} = 1$
$0 < IP_{IC} < 1.$
$0 < DP_{IC} < +1.$

(3) $0 < SP_{IC} < 1$
$IP_{IC} = 1.$
$0 < DP_{IC} < +1.$

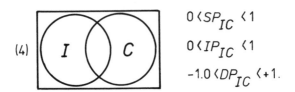

$$0 < SP_{IC} < 1$$
$$0 < IP_{IC} < 1$$
$$-1.0 < DP_{IC} < +1.$$

In different research situations we may be interested in different aspects (measures) of the validity of our indicator. It would, of course, be optimal to have our indicator with the discriminative power as close as possible to 1, but since this is rarely the case we face the problem, whether in our choice of an indicator *we should rather try to maximize its inclusion power or its screening power*.

The choice may be obvious in some practical life situations. When during an epidemic the authorities try to put under quarantine everybody who might have had any contact with the infection, they use indicators that have maximal inclusion power, not caring too much how many people are included in their extension 'innocently'. When officials look for candidates for the crew of a space-ship they use indicators of physical and mental health that have maximal screening power, being aware that many candidates will be rejected unjustifiably.

But the need for enhancing one or another of these functions (when perfect discriminatory power cannot be obtained) becomes obvious also in the case of *verificational studies*. Let us say that we have a simple theory which combines two theretical concepts into a *general relationship* of the type $C_1 \rightarrow C_2$ (C_1 is always followed by C_2). Each of these concepts is empirically identifiable by one indicator so that I_1 is the indicator of C_1 and I_2 of C_2. Now let us assume that for each pair: concept-indicator there may be three different relationships between the extension of the concept and its indicator which we denote as follows:

$I = C$ when the indicator has a discriminative power equal to 1.

$I \subset C$ when the screening power of the indicator for the concept is equal to 1, but its inclusion power is lower than 1.

$C \subset I$ when the inclusion power of the indicator is equal to 1, but its screening power is lower than 1.

In our empirical research we of course test the relationship $I_1 \rightarrow I_2$ from the confirmation of which we would like to derive the conclusion that the *theoretical hypothesis* $C_1 \rightarrow C_2$ is also true. Suppose we are 'lucky' and find in our research that I_1 actually is always followed by I_2. What are the implications of this for our theoretical hypothesis?

The answer depends of course upon the kind of validity of each of these indicators separately. We represent this in the form of Table I, showing the cases in which we can accept our hypothesis $C_1 \rightarrow C_2$ and those in which we cannot. The last category is marked throughout by (?), meaning 'uncertain'.

$I_1 \rightarrow I_2$	$I_1 = C_1$	$I_1 \subset C_1$	$C_1 \subset I_1$
$I_2 = C_2$	$C_1 \rightarrow C_2$	?	$C_1 \rightarrow C_2$
$I_2 \subset C_2$	$C_1 \rightarrow C_2$	?	$C_1 \rightarrow C_2$
$C_2 \subset I_2$	?	?	?

Thus we see that in the case when $I_1 \rightarrow I_2$ our theoretical hypothesis is confirmed:[32] (a) when the indicator of the *antecendent* has either perfect discriminative power or perfect inclusion power, (b) provided that the indicator of the *consequent* has either perfect discriminative power or perfect screening power.

Suppose now we are less lucky in our research and do not find that I_1 is always followed by I_2 but may be followed by both I_2 and by non-I_2 ($\bar{I}_2$). Does this necessarily mean that our theoretical hypothesis connecting the variables C_1 and C_2 has to be rejected? The answer depends again upon the configuration of the measures of validity of the indicators we use in our study. Let us denote the case of rejection of our theoretical hypothesis by $(C_1 \rightarrow C_2)'$, and the situation in which the conclusion is still uncertain by (?).

The corresponding Table II then looks as follows:

$I_1 \rightarrow (I_2 \text{ or } \bar{I}_2)$	$I_1 = C_1$	$I_1 \subset C_1$	$C_1 \subset I_1$
$I_2 = C_2$	$(C_1 \rightarrow C_2)'$	$(C_1 \rightarrow C_2)'$	?
$I_2 \subset C_2$	?	?	?
$C_2 \subset I_2$	$(C_1 \rightarrow C_2)'$	$(C_1 \rightarrow C_2)'$	?

We can summarize this by saying that in the case of negative results of our attempts to confirm a general theoretical hypothesis at the operational level, we *must* reject the theoretical hypothesis when: (a) the indicator of the *antecedent* has either perfect discriminative power or perfect screening power, and (b) the indicator of the *consequent* has either perfect discriminative power or perfect inclusion power.

Otherwise, we may be still 'hopeful' and start to think about a better set of indicators which might 'save' our theory in a future study.

Thus we may see that the measures of validity specified above have important implications both in practical research situations and in verificational studies, and that in verificational studies *different measures of validity should be taken into account for dependent and independent variables of our research design*, depending upon the positive or negative results of our study.

I have discussed above some extreme types of verificational studies in which at least one measure of validity of our indicator for the corresponding concept was equal to 1. We know of course that in most cases in actual social research none of these measures of validity is equal to 1, that they will have at best *probabilistic* values only. One could extend that above reasoning to probabilistic indicators in seeking the *degree of confirmation* of the theoretical hypothesis when the indicators of the antecendent and the consequent of the hypothesis have a given probabilistic inclusion or rejection power, but an analysis of these problems in their technical details would go beyond the task of this paper. Let us consider, rather, some more general problems related to the question of *how we can estimate the different measures of validity of our indicators*. To the degree that these indicators refer to understanding concepts, our analysis deals here again with the problem of 'standardization of Verstehen',

With respect to this problem let us look first at the situation in which the relation between the indicator and the phenomenon denoted by the corresponding concept is of an empirical character, and therefore in itself constitutes a part of (or is based on) a social theory or generalization which has somehow been tested before. It seems that it is not enough to state the empirical character of the relationship but we should try to interpret it in terms of causality, looking at the nature of the causal connection between the indicator and the indicated variable. From this standpoint we can distinguish six types of situations.

(1) Cases where we are unable to do more than point out the empirical character of the relationship between indicator and indicated variable, and cannot tell whether the observed relation between these variables (either general or statistical) reflects a causal connection between them or not.

Thus, after having established that a certain kind and level of education is highly correlated with a certain pattern of values, we can use educational attainment as the empirical indicator of this value system. But at the same time we may be unable to say whether this level and kind of education is the cause or the effect of the given types of values — or whether their relation is spurious due to the fact that both are dependent on one's social origin and family background.

(2) Cases where we are able to say that *there is no causal connection* between the indicator and the indicated variable, or in other words, that the relationship between indicator and indicated variable is a spurious one. If we say, for example, that a liking for abstract paintings is a good indicator of other elements of a 'sophisticated cultural taste', we do not usually suppose that a liking for abstract paintings is either the cause or the effect of reading good novels, liking symphony or chamber music, etc. We suppose, rather, that all these habits, including our indicator, are elements of a more general cultural pattern of thinking and behavior, and that they were all learned (or somehow acquired) in a way that excludes a causal relationship between them and makes some other factors external to them — such as level of education, certain types of social contacts, etc. — 'responsible' for their existence and correlation.

(3) Cases where we are inclined to say that the indicator and indicated variable are causally related in such a way that the indicator is the *cause* of the indicated variable. The use of cultural patterns as indicators of the values of those who will later on become socialized into that culture belongs of course to this category. Another example is when we take a certain situation in which our subject finds himself as the cause of his internal reaction to it, and try, using both our empathy and all available generalizations from previous research, to intuit how he feels in this situation.

(4) Cases where the indicator is the *effect* of a postulated latent property. This refers to situations where we take behavior as the indicator of the motive. We have, of course, the same situation when the behavior is verbal (e.g. the answer to a question), and we try to derive from the 'verbal behavior' the real beliefs and opinions of the respondent.

(5) Cases where, on the basis of our theoretical knowledge, we assume that the indicator and the indicated hypothetical property are related to each other in a *causal feedback* — each of the variables influencing the other. Thus, if we have a man occupying a top position in an organization,

we assume that he accepts to a high degree the goals of this organization, and we can use his position as the inferential indicator of his attitudes. But we also know that these two elements of our syndrome influence each other causally: the fact that he accepts the goals of the organization is relevant to his having this position – but at the same time his being at the top increases his acceptance of the organizational goals.

(6) The sixth category is of a special kind, because it involves use of at least two indicators. One is assumed to be the cause of a certain psychological process, the other its effect. Then the latent variable identified in this way constitutes a *mediating link* in the causal chain. Thus a clinician may infer from an interview in which the patient says only the best things about his father that he, in fact, subconsciously hates his father. But his inferences are much stronger when he additionally and independently learns that the father was and still is a very authoritarian person.

I mentioned above that in the course of understanding, we have to build psychological-behavioral syndromes of interrelated variables, denoted by our understanding-constructs. Here I would like to add that *we should try to specify as much as we can the internal causal structure of these syndromes, and their causal relations to the variables definitionally external to them.* This is of course an important theoretical goal in itself, but if it has additional implications for the assessment of the validity of observable components of the syndromes (or their external causes or consequences) as indicators of the whole syndromes. By taking these causal relation into account we can establish the character of the general relationships between each of the components of these syndromes and additionally the probabilistic consequences of different causal assumptions, the latter permitting us to estimate the statistical validity of indicators in the process of Verstehen.

When looking for causal connections and feedbacks in our theoretical models of meaningful behaviors, we should be aware of course that most of the relation they involved are of a *conditional character*. This means that indicator I may be the cause (or effect) of C under the assumption that some other conditions K are also present, which may be presented as follows:

$$(I.K) \to C \text{ or } (C.K) \to I.$$

The conditions which codetermine and modify the internal causal relations within the syndrome may be of course either other external variables or variables postulated at the same level as the inferred ones i.e. the psychological states and dispositions of the social actor. In the latter case the variables should be included in the structure of the syndrome itself as internal modifiers of its functioning.

If the relation between I and C is conditional with respect to K as stated above $[(I.K) \rightarrow C]$, then our *indicator is valid* only for these conditions K, and the same applies to situations where the relation is the effect of the indicated variable C in conditions K. Needless to say, almost all the empirical indicators we use in sociological studies are of a conditional character only. But then the *nature of their conditionality* and the character of the conditions under which they may be used constitutes an important element of the social theory, and the theoretical premises for each indicator should be specified in each sociological study as clearly as possible.

In general we can say that the indicator I which is the cause of C, may constitute either a necessary but not sufficient or a sufficient but not necessary or both a necesary and a sufficient condition of C, or it may be only a necessary component of one of several alternative sufficient conditons of C in practically possible (i.e. existing in the studied population) causal chains. As we will see below, this typology has certain consequences for its validity for C. The same holds when I constitutes – according to any of the causal patterns mentioned above – the *effect* of the indicated variable C.

As I demonstrate in Chapter V of this volume, we can use the notion of conditional causal relations also as a *'bridge' between the general causal relation and probabilistic relations*. The models of reasoning presented there could be used as tools for estimating the degree to which indicators with only a probabilistic validity maximize each of the two measures of their validity and therefore the degree to which one could say that our theoretical hypothesis seems to be confirmed, too.

Also we can say in general that if, according to the assumed causal mechanism, the indicator I – being the *cause* of (a manifest or latent) variable C – constitutes the *necessary condition* for the variable C, then it has a *perfect inclusion power* for it. The greater the number of alternative causes of C, and the more negatively they are correlated with I, the weaker the inclusion power of I for C. If according to our theory I constitutes a *sufficient condition* for C, then it has perfect screening power; if it is only one of the components of such a sufficient condition, its screening power is determined by the probability of I occurring together with all the other components of the sufficient condition for C.

In the case where I is the effect of C (e.g. as in the case of the motive-behavior sequence), the situation is reversed. If C is a sufficient condition for I, then I has perfect screening power for C. If C is only an element of such a sufficient condition, the screening power of I for C will be the function of statistical relations between C and other concepts comprising the sufficient conditions for I. If C is a necessary condition for I, then I will have perfect inclusion power for C. The greater number of alternative causes of I, the weaker the inclusion power of I for C.

As we know, in the choice of indicators we rarely use a single indicator for the indicated variable and usually use whole 'batteries' of them. But here again one can formulate certain general rules for the use of such 'batteries' of indicators I_1, I_2... I_n in the identification of the latent variable C. Suppose that due to the poor shape of our theory, all we know about any of these indicators is that they are positively correlated with C, i.e. the existence or occurrence of any of them increases the probability of C. In this situation it is obvious that if we were to use all of them in the form of an additive index we would increase the screening power of our highest score of our index. The objects (persons) who reveal all the behaviors $I_1...I_n$ (or other indicators of C) are most likely to have the property C. The smaller the number of items characterizing the given person the smaller the screening power of the given score of index I for the property C. On the other hand, if we want to enhance the inclusion power of our battery, possession of only one of these indicators should be enough to include our subject in class C.

With respect to the foregoing, it is clear that we should apply different types of batteries of indicators and indices for the operationalization of independent and of dependent variables in our theories.

In empirical social research, we often treat the *positive correlations* between a set of indicators as the criterion of their relation *to the same latent variable*, and lack of such correlation as evidence that the latent attributes indicated by them must be different. Nevertheless, direct and unreflective inference from positive correlations of observable indicators to the hypothetical property they indicate may be erroneous, especially if we take into account that the underlying psychological property may lead to different behavioral consequences under different conditions. Suppose we have a 'latent variable' C which — as we assume — leads to different behavioral indicators I_1 and I_2. If these indicators are not perfectly correlated with $(r = 1.0)$, this means that there are some additional situational conditions (let us say K_1 and K_2) such that when K_1 occurs with C, C leads to I_1, and when C occurs with K_2, I_2 occurs. Suppose now that K_1 and K_2 are empirically mutually exclusive. It is obvious that in this case I_1 and I_2 will also be mutually exclusive. If K_1 and K_2 are negatively correlated, the same will occur with I_1 and I_2. This doesn't mean that I_1 and I_2 do not indicate the same latent property C. They do, but they are its valid indicators for different conditions which are negatively correlated with each other. Therefore the practice of choosing a battery of indicators for a 'latent attitude' in order, for example, to construct a scale on the basis of analysis of the correlations between them, without interpretation of these correlations in the context of a causal and usually conditional theories, is valid only under the assumption that all these possible conditions which may determine the occurrence of each of our indicators are randomly related to each other, which is not necessarily true in more or less structuralized social conditions.

It is obvious that the notion of necessary and sufficient conditions should be taken into account also in situations where we try to estimate the validity of *definitional indicators*, although here these relations do not refer to the character of causal connections but to relations between the extension of indicator and the extension of the indicated concept which are established on the basis of a terminological convention. If our indicator defines a concept completely, which means that I constitutes a *necessary and sufficient condition for C*, then of course I has, by definition, perfect discriminative power for C, and by the same token all its measures of validity are — due to our terminological convention — equal to 1. But many of the definitions we use in the social sciences do not postulate such strong definitional relations. First, the definitions may be of a clearly conditional character — i.e. we assume the relation of definitional identity of I and C but only under some specific conditions. For example, when we define social status in a small group by the number of sociometric choices scored by the person from the members of the group, we can say additionally that this definition is valid only for groups which have already existed for some time before the sociometric measurement was made, and is not valid for groups which are artificially constructed under laboratory conditions and therefore may have no stable internal structure with the consequence that all choices must be made more or less at random. But then of course we should be aware that any theory formulated with such concepts applies by definition only under the conditions (generally or historically) specified for the validity of the concepts.

The conditionality of concepts may also occur in another less obvious form. Sometimes we may say that I constitutes a *partial definition* of C, meaning by that that I constitutes a necessary but not a sufficient condition for C. Thus we may say that holding a certain kind of belief is necessary in order for a person to be classified as 'authoritarian' but by no means sufficient. *If according to the meaning of a given concept,* indicator I constitutes a *necessary but not a sufficient condition* for C, then of course its inclusion power is perfect, but its screening power is probabilistic only and depends upon the empirical relations between I and other phenomena which, due to the meaning of this concept, belong to C and constitute its *others definitionally necessary components. Of course C occurs only when all these components are present together.*

In other cases I may define C partially in the sense that it constitutes a sufficient but not a necessary condition for C. In this sense 'food' constitutes a partial definition of 'reward' in some systems of the theory of learning. 'Being a writer' may definitionally imply that one is a member of the 'intellectual elite' of a given country, but it is by no means a necessary condition for belonging to the intellectual elite; being a scientist or an actor may constitute another sufficient condition for elite status. If I constitutes a definitionally sufficient condition for C but not a necessary one, then, of course, it has a perfect screeding power, but its inclusion power is probabilistic and depends upon the empirical relations between the extension of I and the extensions of other classes of phenomena also denoted by C.

In both these situations we face a serious problem in verificational studies where I *stands for the whole concept C.* In the first case, where I is necessary but not sufficient for C, we must decide whether in empirical relations I 'behaves' in the same way when occurring with other components of the concept C as when without them. Only in cases where it does can we extend the conclusions based on the observation of the 'behavior' of phenomena denoted by a partial definition of the concept C to phenemona denoted by the whole concept C. In the second case where I is a sufficient but not necessary condition for C, we have to decide whether the phenomena defined by other partial definitions of *C* obey the same regularity as *I* does. Only in such cases where they do can we apply the conclusions from our study to the whole area to which the concept *C* applies.

Finally, it may happen that the definitional relation between *I* and *C* is even weaker than specified above: *I* constitutes a sufficient condition for *C* only when some other necessary components of *C* are also present, and at the same time *C* may also be defined by some other definitions which may also constitute sufficient conditions for *C*. In such cases the validity of *I* for *C* is probabilistic in both senses of the term, with all the implications for the theoretical conclusions from the results of our studies mentioned above.

I have discussed above the notion of the conditionality of definitional relations with special reference to definitions which the scientist would like to construct himself for use in his own explanatory theories. But it seems that a similar procedure can be applied when we try to reconstruct certain understanding concepts in the way they are understood by the members of a given community, i.e. when we look at social reality with the use of humanistic coefficient only. Analysis of the internal logical structure of the subjective 'social meaning' of certain objects and behavior and of the 'definition of situations' as they exist in the given culture, could permit us to grasp more clearly the relations between two aspects of social reality – as defined by the scientist and as seen with the humanistic coefficient – than we usually do in our studies. This would enable us to see much more clearly the structure of the meaningful side of 'social phenomena' and eventually to introduce new 'surplus meanings' into our understanding-theory. Whether we finally accept these cultural notions in our own theoretical explanations and theories of social behavior, or decide to redefine them for our theoretical purposes, or finally to introduce quite new concepts, is of course quite another problem.

NOTES

* This paper is an elaboration of ideas presented in other papers by the author, and especially in 'Obserwacja i rozumienie ludzkich zachowań a problemy budowy teorii' (Observation and Understanding of Human Behaviors and the Problems of Theory Construction) in S. Nowak, *Studia z metodologii nauk społecznych (Studies in the Methodology of the Social Sciences)*, Warsaw, 1965; in 'Terminy oznaczajace stany i przezycia psychiczne w teoriach socjologicznych' ('The Terms Denoting Psychical States and Processes in Sociological Theories'), *Studia Filozoficzne*, 1 (68), 1971 in

'Functions of "Verstehen" in Assessment, Explanation and Prediction of Social Phenomena', paper presented at the Polish-Danish Philosophical Conference in May 1973.
Published in Danish Yearbook of Philosophy, Vol. 10, 1973; and in 'Correlational,
Definitional and Inferential Indicators in Social Research and Theory', *Polish Sociological Bulletin,* Vol. 3-4 (5-6) 1962.

[1] See A. Schutz, *The phenomenology of the Social World,* Evanston, 1967.

[2] Max Weber, *Economy and Society,* edited by G. Roth and C. Willich, New York,
1968, p. 4.

[3] Quoted from G. Lundberg, *Foundations of Sociology,* New York, 1939, p. 12.

[4] See P.W. Bridgman, *The Logic of Modern Physics,* New York, 1932.

[5] G. Lundberg, *Foundations of Sociology,* p. 14.

[6] *Ibid.,* p. 11.

[7] See R. Carnap, 'Methodological Character of Theoretical Concepts', in H. Feigl, M.
Scriven (eds.), *Minnesota Studies in the Philosophy of Science,* Vol. I, 1956.

[8] K. McCorquodale and P. E. Meehl, 'Hypothetical Constructs and Intervening
Variables', in H. Feigl, M. Brodbeck (eds.), *Readings in the Philosophy of Science,* New
York, 1953.

[9] P.F. Lazarsfeld, 'Evidence and Inference in Social Research', in M. Brodbeck (ed.),
Readings in the Philosophy of the Social Sciences, New York, 1968, p. 610.

[10] For a detailed discussion of this problem see Chapter XI.

[11] F. Znaniecki, 'Social Actions', quoted from L. Coser, B. Rosenberg (eds.), *Sociological Theory,* New York, 1957, p. 223.

[12] B. Malinowski, *The Diary in the Strict Sense of the Term,* New York, 1967, p. 217.

[13] See R.K. Merton, 'The Self-Fulfilling Prophecy', in: *Social Theory and Social
Structure,* New York, 1968.

[14] See B.F. Skinner, *Beyond Freedom and Dignity,* New York, 1971.

[15] G. Lundberg, *Foundations of Sociology,* pp. 9, 24, 25, 28.

[16] Th. Abel, 'The Operation Called Verstehen', in H. Feigl and M. Brodbeck (eds.),
Readings in the Philosophy of Science, New York 1953, p. 683-684.

[17] F. Alexander, *The Logic of Emotion and Its Dynamic Background,* quoted from
Abel, *The Operation Called Verstehen.*

[18] Th. Abel, *The Operation Called Verstehen,* p. 685.

[19] For an analysis of the methodological problems of theories in which culturally
determined motives function as causes of human behaviors, see Chapter IX.

[20] See C.G. Hempel, 'The Concept of Rationality and the Logic of Explanation by
Reasons', in C.G. Hempel, *Aspects of Scientific Explanations,* New York, 1965,
pp. 463-487.

[21] See *ibid.,* p. 467.

[22] W. Dray, *Laws and Explanations in History,* Oxford, 1957.

[23] C.G. Hempel, *op. cit.* p. 471.

[24] *Ibid.,* p. 472.

[25] *Ibid.,* 474.

[26] Max Weber, *Economy and Society,* Vol. I.

[27] G.C. Homans, *The Human Group,* New York, 1950, pp. 37-39.

[28] See: S. Ossowski, 'Experimental Sociology and Inner Experience', *Polish Sociological Bulletin,* Vol. 3-4, 1962.

[29] C.S. Dodd, *Dimensions of Society.*

[30] See P.F. Lazarsfeld, *Evidence and Inference in Social Research*. See also the 'introduction to the Part I, Concepts and Indices', *The Language of Social Research*, P.F. Lazarsfeld, M. Rosenberg (eds.), New York, 1955.

[31] See C.G. Hempel, *Aspects of Scientific Explanations*, pp. 457-463.

[32] It should be clear that the term 'confirmation' refers here to one aspect of verificational process only, namely: to what degree the observed relationship between the indicators permits us to derive the conclusion, that *within the studied sample* the phenomena denoted by our theoretical concepts are related in the way specified by our hypothesis. This is quite another problem than the question to what degree we can *generalize* the relationship established in our sample to a broader (possibly universal) population. This question is discussed in Chapter VII.

BIBLIOGRAPHY

Abel, Th., 'The Operation Called Verstehen', in H. Feigl and M. Brodbeck (eds.), *Readings in the Philosophy of Science,* New York, 1953.

Bridgman, P.W., *The Logic of Modern Physics*, New York, 1932.

Carnap, R., 'Methodological Character of Theoretical Concepts', in H. Feigl and M. Scriven (eds.), *Minnesota Studies in the Philosophy of Science*, Vol. I, 1956.

McCorquodale, K., and P. Meehl, 'Hypothetical Constructs and Intervening Variables', in H. Feigl and M. Brodbeck (eds.), *Readings in the Philosophy of Science,* New York, 1953.

Dodd, C.S., *Dimensions of Society*, New York, 1940.

Dray, W., *Law and Explanation in History*, Oxford, 1957.

Hempel, C.G., *Aspects of Scientific Explanations*, New York, 1965.

Homans, G.C., *The Human Group*, New York, 1950.

Lazarsfeld, P.F., 'Evidence and Inference in Social Research', in M. Brodbeck (ed.), *Readings in the Philosophy of the Social Sciences*, New York, 1968.

Lazarsfeld, P.F., and M. Rosenberg (eds.), *The Language of Social Research*, New York, 1955.

Lundberg, G., *Foundations of Sociology*, New York, 1939.

Malinowski, B., *The Diary in the Strict Sense of the Term,* New York, 1967.

Merton, R.K., 'Self-Fulfilling Prophecy', in: *Social Theory and Social Structure*, New York, 1968.

Nowak, S., 'Causal Interpretation of Statistical Relationships in Social Research', in this volume, p. 165.

Nowak, S., 'Correlational, Definitional, and Inferential Indicators' in *Social Research and Theory, The Polish Sociological Bulletin*, 3-4 (5-6), 1962. Nowak, S., 'Cultural Norms as Explanatory Constructs in the Theories of Social Behavior', in this volume, p. 319.

Nowak, S., 'The Logic of Reductive Systematizations of Social and Behavioral Theories', in this volume, p. 376.

Nowak, S., 'Functions of "Verstehen" in Assessment, Explanation and Prediction of Social Phenomena,' in *Danials Yearbook of Philosophy*, Vol. 10, 1973.

Nowak, S., 'Obserwacja i rozumienie ludzkich zachowan a problemy budowy teorii' (Observation and Understanding of Human Behaviors and the Problems of Theory Construction') in S. Nowak, *Studia z metodologii nauk spotecznych* (Studies in the Methodology of the Social Sciences), Warsaw, 1965.

Nowak, S., 'Terminy oznaczajace stany i przezycia psychiczne w teoriach socjologicznych', ('The Terms Denoting Conscious Psychical States and Processes in Sociological Theories), *Studia Filozoficzne* I (68), 1971.

Ossowski, S., 'Experimental Sociology and Inner Experience', *Polish Sociological Bulletin*, 3-4, 1962.

Schutz, A., *The Phenomenology of the Social World*, Evanston, 1967.

Skinner, B.F., *Beyond Freedom and Dignity*, New York, 1971.

Weber, M., *Economy and Society*, ed. by G. Roth and C. Willich, New York, 1968.

Znaniecki, F. *Social Actions*, New York, 1936.

CHAPTER II

VERBAL COMMUNICATIONS
AS INDICATORS OF SOCIOLOGICAL VARIABLES*

1. THE STRUCTURE OF THE 'COMMUNICATION CHAIN'

It is probably not an exaggeration when we say that in social investigations the most important source of information about the facts and processes we are interested in are the verbal communications we 'receive' from other human beings, either directly or through the various 'media of communication': survey questionnaires, autobiographies, books, the press, radio and TV. This is certainly obvious to any specialist in public opinion polls and attitude studies for whom the questionnaire and interview have become standard tools of investigation of 'verbal behaviors' and who sometimes might even be inclined to identify the field of his study as the study of 'verbal behavior'. The same is true for those sociologists for whom the personal document is the basic source of sociological data. It may be less obvious to a researcher for whom the observational technique applied to artificial (experimental) and natural social situations is a basic tool of investigation, and for whom 'non-verbal behavior' seems to be the basic object of scientific interest. But even in this kind of studies a great number of our inferences are based upon verbal comments people affix to their 'non-verbal' behaviors, i.e., on verbal communications from which we often infer the 'meanings' of their non-verbal behaviors. The anthropologist may be interested in the study of religious rites and customs and therefore be inclined to treat the observation of 'religious behaviors' of the members of the given community as his main research technique, but when we look closer at the nature of his primary data we readily discover how much of this knowledge he owes to the fact that he is able to converse with the members of the studied community, to listen as they talk spontaneously in various situations, or if the community is literate to read 'communications' of either a private or a public character which reveal to him the individual and social meanings of the studied phenomena. Even in the study of history, where a written account of events is the basic source of information, the document is nothing more than a verbal communication 'materialized' in a

written form, although in other areas of historical study, (e.g., in archaeo‐
logical studies of illiterate cultures) verbal sources may be practically nil.

In a broader perspective even a secondary analysis of results of earlier
sociological studies may be classified as the use of very complex verbal
communications, our 'communicators' here being the researchers who con‐
ducted the original studies and thus delivered the data for a new problem.
We may, by analogy, say that a historian who is working on a 'synthesis' of
a certain period or of a historical process treats the various monographs on
the subject of his interest as verbal communications which are for him a
source of indirect information about the studied phenomena.

Thus we can see how extensive is the use of verbal communications as
sources of information in any social study. This implies the importance of
the study of the process of 'verbal communication' within the methodology
of the social sciences, and especially the study of what might be called the
logical structure of a 'one-way communication chain'. The structure of this
chain and the substantive nature of assumptions we have to make in the
interpretation of the 'message' going along the chain may vary in all the
different situations, but it seems there are certain general features of all
these situations in which a verbal communication (or written message) is the
source of our inferences about various categories of social data. In other
words, *verbalized messages are for us the indicators of different categories
of sociological variables* or of variables in any other humanistic discipline.
The aim of this paper is to present a general paradigm which might be useful
for the interpretation of verbal communications in their various indicative
functions.

Let us assume for simplicity's sake that we are going to use the *content*
of a certain communication *as it was 'received' and understood by us* as an
indicator for different kinds of phenomena we are interested in, in our
investigation. It seems reasonable to assume that the 'point of departure' in
the whole process of interpretation of indicative value of the indicator is *our
own reception, perception and interpretation* of the message and not the
message in its original form and content. On the basis of the content of such
a message as understood by the investigator he may make three different
kinds of inferences:

(a) He may reach a certain conclusion on the basis of the content of the
'received' message as to the content of the original message; that is, as to the
content of the communication as formulated by the communicator or

informant. Let us say that in this case he is interpreting the message according to its *communicative function*.

(b) He may try to make inferences on the basis of the content of the received communication as to the content of the speaker's (or writer's) opinions, convictions and beliefs with respect to the subject of communication. In this case he interprets the communication according to *expressive function*.

(c) Finally, on the basis of the content of the received communication he may try to make some inferences as to the existence, nature or features of the things, states, events or processes to which the communication refers. In this case he interprets the communication according to its *descriptive function*.

In all these cases the content of the received communication is for the scientist an *empirical indicator*[1] for the phenomena of the given category. An empirical indicator is any trait or event that is related to the indicated phenomena not by a terminological convention but by a certain empirical relationship, from which the scientist may conclude that the indicated phenomena do occur, when he states the occurrence of his indicator. But – as in the case of any indicator – he faces the problem of its *validity* for the phenomena for which he would like to make his inferences; in other words, he faces the problems of *degree of correspondence* between the message as *received* and understood by him, and all the other 'links' of the 'communication chain'. In short, in order to make valid inferences about the phenomena we are interested in we have to know to what degree our reception, perception, and understanding of the communication corresponds to: (a) what has been actually communicated to us; (b) what the author of the communication actually thinks about its subject; (c) what are the actual features of the reality the communication refers to.

But the empirical relations that interest us in the case of the communication chain are of a special character, because we are interested in *degree of correspondence* between its different links.

This fact has certain additional implications because we have to make our inferences about the indicative implications of verbal communication in a certain order. I introduced the notion of the communication chain in order to stress that we have to 'make our inferences step by step, in a certain order'. As I said, we have to start from *our own* understanding of the content of the communication and then make inferences about the content of the original communication under the assumption that the relation between these two links of the communication chain is a *relation of sufficient correspondence*. Let us call the relation between these two links the *communicative relation*. Then — assuming the correctness of our interpretation of the 'sent message' — we may start to surmise what was 'inside the speaker's mind' with respect to the subject of the communication, under assumption that the relation between words and thoughts is that of high correspondence. Let us call the connection between these two links of the communication chain the *expressive relation*. Finally — under the assumption that the author of the communication really meant what he said or wrote -- we may start to make inferences about the things or events he is talking about, again assuming sufficient correspondence between his knowledge and the features of reality he is reporting about. Let us call this connection the *cognitive relation*.

Therefore we may 'reach' to any link of the communication chain only under the assumption that all relations between this link and the last one (i.e. the content of the received communication) are characterized by sufficiently high correspondence. If we cannot assume or assess empirically a sufficient degree of correspondence for a certain relation, our inference has to stop at this point of the communication chain as long as the content of the communication is for us the only indicator, i.e., *if we are not introducing other kinds of empirical relations or premises which permit us to use the verbal communications as indicators of sociological variables*. The structure of the communication chain is presented by Figure 1.

In most cases the content of the received communication is not the only indicator for our inferences; we use many other cues which either tell us something about the possible degree of validity of verbal communication or serve as the source of additional informations. It would be useless to recount here the vast number of experimental studies in the area of 'non-verbal communication[2]', it is a fact of our everyday experience that the tone of voice or a certain gesture of the posture of a speaker may be much more informative as to both his general state of mind and his specific attitudes and opinions than a long speech. These non-verbal cues may strengthen or weaken our credibility regarding the content of the communication intentionally directed to us, or they may tell us something about the speaker's views and attitudes which are unrelated to the content of his verbal message. Moreover, the content of the communication itself may be

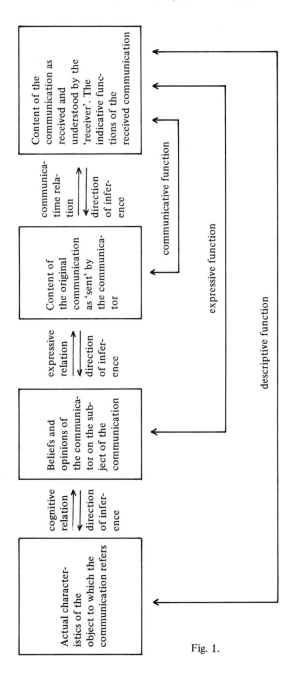

Fig. 1.

used as an indicator of certain other variables about which the speaker does not mean to inform us, at least in the given communication. From the richness of his vocabulary we may derive conclusions about the speaker's education; from the logic of his reasoning, about his intelligence; from the extremity of the position he takes, about his emotionality; etc.

Nonetheless, even when we use all possible cues (both the psychological and the non-psychological ones) in making our inferences about variables of interest to us, it is still true that the *content* of what people tell us constitutes for us one of the main sources of information both about the characteristics of human individuals and about the social phenomena in which they play a part. In the present chapter I would like to concentrate especially on this source of sociological data, looking at all the other cues and indicators primarily from this perspective. The question may be asked, How is the problem area covered by this paper related to the more general frame of reference of communication theory? To quote a classic text on this subject:

A completed act of communication has often been described in these terms: *Someone says something, somehow, to someone, with some effect.* The fundamental questions, therefore, are: *Who says what, how, to whom, with what effect?* When the studies are focused upon 'who', we speak of control analysis; when they deal primarily with 'what', it is content analysis; if the subject is 'how' we call it media analysis; if 'whom', it is audience analysis; and if 'effect', it is of course effect analysis.[3]

In a way this chapter covers the questions: 'who, what, how and to whom', leaving the question of effect outside its problem area. Nevertheless these questions are asked from a special perspective. Since we are interested here in the use of 'verbal behaviors' as indicators of sociological variables of various kinds, the substantive propositions about the relations between different links of the communication chain are for us primarily of instrumental value: they permit us to estimate the degree to which our inferences from verbal communications are valid. On the other hand, as we see from the above diagram, the simple paradigm 'who says what, how and to whom' under closer look appears to refer to a much more complex phenomenon than is usually believed.

2. TWO CATEGORIES OF FACTORS DISTORTING THE COMMUNICATIVE RELATION

The correctness of our inferences along the communication chain depends on the degree of correspondence between each of the two neighboring links and on the different aspects of validity of one of these links as empirical indicator of the other. The central problem seems here to be identification of possible factors of distortion operating on each of the relations distinguished above. Let us begin our analysis with the communicative relation.

It is an elementary fact of human interpersonal relations that a given message may be 'sent' with one meaning and 'received' with another. This applies to all possible channels of communication, both verbal and non-verbal, but I think that the various causes of distortion of the communicative relation going by the verbal channel can be classified into two distinguishable categories.

The first group of factors has to do with change or distortion of the physical 'shape', the material form, of the message itself. Each message is transmitted through a certain 'medium' — whether this be a piece of paper with ink lines of a given shape, a design on a slab of marble, a series of sound waves, or magnetic traces produced on the tape of a tape-recorder. This physical shape of the message may be more or less distorted at the moment it is 'received' by the person who would like to interpret its content as the indicator of various phenomena he is interested in. The extreme case of distortion is complete disappearance of the material shape of the message. The paper with the written text may be destroyed by fire, the voice may be inaudible from a certain distance, the traces of ink on papyrus may fade with time, etc. Needless to say, this means a complete breakdown of the communication.

Even when the message does not disappear completely its physical shape may be considerably distorted, which means that we will receive it in a form that is somewhat different from its original. The sound of a human voice may scarcely be audible over the 'noise' (whether this noice is produced by the voices of other persons speaking at the same moment of is the effect of static in a radio transmission). The original inscription on a papyrus may be visible only in fragments, etc.

Distortions in the physical shape of verbal 'messages' are obvious causes of trouble for historians. The sociologist, especially if he is studying human beings by direct contact with them, might be inclined to believe that he can disregard this category of factors of distortion of communicative relation. This is true to a fairly high degree. But, it seems that if we extend our frame of analysis and treat, for example, the interviewer as our 'medium' (on the assumption that he is only supposed to write down the exact text of the respondent's answers), then all deviations from the original answers due to the fact that they were written 'freely' _ i.e., 'incorrectly' — by the interviewer might be viewed as similar to those distortions caused by deterioration in the physical shape of written communications or the 'noise' in radio transmission.

The analogy is nevertheless not complete. Changes in the physical shape of communication if they are due to some natural process (like the function of distance in time or space) are usually *random in relation to the content*. This does not mean of course that they will be random with respect to the *physical properties* of the communication. It is quite likely that words spoken more loudly will be better heard over noise than words spoken in a softer tone. Characters of a certain shape written on a piece of marble will be more likely to get wiped out with the flow of the time than some others, etc. But, certain words written on a metal plate will not be more likely to corrode *because of their meaning*, nor are certain formulations more likely than others to reach the ear of the 'receiver' in their correct physical form due to their content. On the other hand, we can expect that distortions of the text introduced by the inaccurate or freely transcribing interviewer will not be random in relation to content of the communication; and we know that the interviewers opinions about the subject of the study may not be irrelevant here, as we know that the changes in classical philosophical texts introduced by medieval copyists were not random in relation to content but were organized in meaningful patterns.[4]

For these reasons it seems more proper to interpret the interviewer or copyist and other such factors, as an additional link in an *extended communication chain* than as a 'medium' which *transports* the content according to some laws of physics or chemistry as it would transport any other physical traces independently of their meaning or lack of any such meaning for human beings. The problem posed by extension of the communication chain will not be discussed in this chapter.

Returning to distortions in the physical shape of communications, the scientist who tries to restore their original shape has two kinds of methods at his disposal. The first consists of applying certain methods which are characteristic for technology or for natural sciences. The Assyrian inscriptions may be reconstructed by putting together all the fragments which can be made to fit each other. The illegible ink on papyrus may be read after the application of chemicals or ultra-violet rays. And so on. The other method consists in the reconstruction of *missing fragments* by assuming that they are necessary for a certain 'meaningful whole'. Even when we can hardly hear someone who is speaking in a noisy, crowded room we can reconstruct what was said, especially if the subject of the communication or the speaker's pattern of expressing his views is familiar to us. The more the historian knows about the history, structure, and culture of the studied society the more easy it is for him to interpolate the missing fragments of certain inscriptions, etc.

The reason we are able to apply such 'humanistic' means in the restoration of the physical shape of a communication is that each communication was built (shaped, formulated, etc.) according to certain rules of communication, which permit people to use certain objects, events or processes as 'transporters' of the meaningful contents between themselves.

There are probably few words that have a greater range of possible meanings than the word 'meaning' itself. For the moment, it is enough for us to distinguish the *linguistic meaning* of a certain word or sentence, or of the whole communication, from all its other possible meanings. We say that someone understands the linguistic meaning of the given 'message' when he *knows the rules of its usage* or at least is able to apply them succesfully in everyday practice, i.e., he knows in what kinds of situations the given word, phrase or whole message should be used, what kinds of feelings and opinions it is 'entitled' to express, what kinds of objects or events it refers to according to the rules of the given communications system. Thus an English or Polish speaker can say that he understands the meaning of the sentence 'Il pleut' when he knows it may be used in French for describing the fact it is raining, or understands the meaning of the German sentence 'Ich liebe dich' when he is aware it is used, by German speaking people, for expressing the feeling of love.

This brings us to the second category of factors distorting the communicative relation, and decreasing the degree of correspondence in it. Here the author of the communication and its 'receiver' differ in their understanding of the linguistic meaning of the communication. The extreme case is when we don't understand the content of the communication simply because we don't know the language or system of symbols in which it has been formulated. But al least, in this kind of situation, no incorrect inferences will be made. Much more insidious sources of distortion are the more or less strongly articulated variations of meaning attributed by different persons (convinced that they are using 'the same language') to the same term, or the same expression composed of many terms, or even to the whole content of a given communication. Any elementary textbook of techniques of social research warns about the errors which may occur when the investigator interprets the answers of his respondents *in his own frame of reference* instead of theirs, and attributes to the terms occuring in their answers his own meanings. The more vague or ambiguous the meaning of the given term, the less clear its expressive connotation or empirical descriptive denotation, the greater is the danger that it may be understood in different ways by the 'sender' and by the 'receiver' of the communication.

Distortion of the communicative relation may stem from the fact that one of the partners in the act of one-way communication (usually but not necessarily the 'receiver') knows less than the other about the rules of the language, or is at least less able to apply them in decoding the message. It may also be due to their belonging to different subcultures of the same general culture, thus differing in their way of interpreting the meanings of some terms, given ways of expression, or the rules of contextual usage. In such case, the communication chain is distorted in its initial communicative relation, which has, as we will recall, obvious consequences for all the other functions of the communication.

3. VERBAL BEHAVIORS AND THEIR BEHAVIORAL CORRELATES AS 'LEGITIMATE' OBJECTS OF SOCIAL STUDIES

It is obvious that in most cases evaluation of the degree of correspondence between the meanings attributed to the message by its 'sender' and its 'receiver' is only the first step for more 'distant' inferences along the communication chain. Assuming that we understand what is being said in the communication, we usually try to make further inferences about what the speakers really think in regard to the subject of their communication.

Further on we often try to estimate the degree to which what they think may be treated as a source of information as to *how things really are*. For the social investigator, understanding the content of the interview is usually the first step toward assessing the existence of certain *attitudes* in the speaker's mind or of the facts he refers to. For the historian, understanding the content of a written source is usually the first step toward the assessment of the occurrence of the historical facts mentioned in the source, although some contemporary historians are interested in the expressive function of the 'sources' as well. While in most cases assessment of the content of the original communication is only a first step in the process of inferences, it should nevertheless be mentioned here that in certain classes of situations the inferences simply end at this link in the communication chain. This means that there may be formulations of research problems in which we use verbal indicators on the assumption of correspondence only between the sent and the received content of communication, without necessarily assuming a high correspondence on any other relation of the communication chain.

The reason for this is that for a humanist the *form or content of verbalized human communications may in itself be an important research object*. The most obvious case is, of course, the whole area of linguistic studies, for which the general rules governing the verbal behaviors are the main object of study independently of what they really express or refer to in any particular case of the use of the language. But, there is a whole range of 'valid' subjects of sociological investigation in which we want to assess *what people say* – or used to say – in a certain group and in certain situations, without trying to go further in our inferences and to guess the degree to which what is being said by the group members correctly expresses the contents of their minds, or describes certain facts.

We can distinguish two kinds of situations when our interest is focused on the message itself, and we would like to 'grasp' its content as accurately as possible. In the first category, we do not exclude a priori any expressive or descriptive functions that the message may also have; it is just that we are not interested in these functions, at least at this stage of our investigation. Many problems in quantitative 'content analysis' of mass media belong to this category of research problems.[5] This does not mean, of course, that we are not entitled later on to ask other questions, e.g. about the intentions of the communicator, or more broadly about the value system expressed in the given content; it only means that the content itself is a 'valid' subject of study.

Another type of situation arises when we know (or assume) that the given verbal behavior *does not have* (at least in general) an *expressive value* corresponding to its literal meaning. When I ask someone "How are you?" This does not express my interest in the matter of how he really is at the time of my asking. At best it plays the role of a greeting ritual. This may, of course, express something else – my tendency to sustain social contacts with the given person, my social conformism in general, etc. Nevertheless, such rituals themselves may be a fascinating subject of investigation. Another example of this kind of interest would be the study of verbal behaviors on public occasions[6] e.g. in social groups or in social systems which have achieved a high degree of strictly external conformity in respect to such behaviors and occasions. Even if we know that they most likely do not express the speakers' thoughts, these verbal behaviors constitute in themselves essential components of more general behavioral patterns of these societies, and the sociologist interested in studying these cultures will probably give his attention as well to these *non-expressive verbal behaviors*. Another area of cultural phenomena in which our interpretation may 'legitimately' stop without going to the next link of the communication chain is the *world of fiction* – in the literal sense of the term – which constitutes a fairly large part of our cultural environment. When we read a novel or a short story our interpretation stops – or at least may stop, unless we are interested in the personality of the author himself – with grasping the content of the fictional communication. An average reader usually does not try to guess to what degree Hamlet was really interested in the basic dilemma of human life, or to what degree the story reported by Shakespeare corresponds to reality, because he knows that he is dealing with the world of literary fiction. This is the typical approach to the world of fiction by a 'naive' receiver, for whom such communication does not express anything real, brings no information about actual facts, but still is an important element of his cultural world. Even a scientist may study the content of a communication without referring it to anything else. A lot of theoretical analyses both in traditional and in modern theories of literature do not go beyond the literary work itself and look for certain intrinsic 'structural' regularities governing the phenomena of the world of literary fiction. In all these cases correct assessment of the content of the 'message' is a sufficient condition for the further analysis of the phenomena being the object of scientific interest.

Needless to say, our research problem usually becomes more interesting when we are able to put the results of analysis of the content of the 'message' into a broader motivational and functional context. Thus after identifying the intrinsic rules of Shakespearean drama we can try to explain at least some aspects of Shakespeare's fictitious world by relating them to the author's personality and to the configuration of social processes of the world he was living in. On the other hand, we may consider the way Shakespeare's world has functioned in the culture of mankind for the last three centuries, shaping certain literary trends and being responsible for some essential features even of our everyday thinking. But this is quite another problem; while it may be important for some of us, others will be happy to enjoy only the story content of events which befell a Danish prince.

When we say that a great part of our social environment consists of the culture of our society or the cultures of various groups we belong to, we should also remember that a great part of these cultures is composed of norms and values more or less established on an inter-generational or interpersonal basis. Those established for a given group are often verbalized in the form of certain written communications – moral codes for the group, normative prescriptions for the proper kinds of behaviors in certain situations, etc. They thus become materialized verbalizations of – to use the Durkheimian concept _ the 'collective representations' of the given group. These components of the group culture may of course be analyzed in relation to the intentions of their authors (in which case would take the next step along the communication chain) or in relation to their functions in the society so that we would evaluate the degree to which they are able to control actual human behaviors. But the analysis of verbalized 'collective representations' may in itself be an important and interesting goal of social study. We can study the internal logical relations among the norms of a penal code, regardless of the question how many persons conform to these norms or accept them. We can try to set up some general classification schemes for these normative phenomena or establish some correlations between different norms or values on the basis of determining the occurren_ ce of *various* 'collective representations' in various societies. 'Structural anthropology' has discovered many interesting correlations in exactly this area of variables.

As we can see, studying the contents of verbal (or verbalized) messages may lead to the discovery of certain regularities, or certain correlations between different variables which may *characterize this strictly verbal level of social phenomena*. Once these correlations have been established, they can be used as indicators of other variables at the same level (i.e. at the content level of the communications themselves, independent of what they eventually might express). This means that having established certain correlations for phenomena at the strictly verbal level, we can say that persons who have said (C) (the content of the communication as *sent to us*) are also likely under other circumstance to say (C_1), because we know that the elements or contents of communications (C_1) and (C) tend to be the interrelated components of a broader syndrome of verbal behaviors, whatever these verbal behaviors might express or describe. The assumptions of our inferences can be shown as follows:

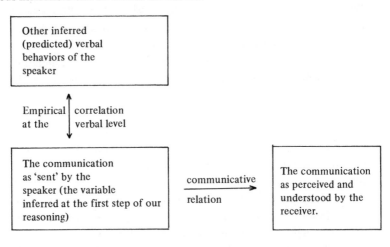

As an example of this kind of inferences; we know already that people who express an extreme opinion on one specific ideological issue are likely to express extreme opinions on other issues dealt with by this ideology. We also know that this may be for one of two reasons — either they are strongly committed to each of the various components of the ideology, or they are merely 'behavioral conformists' to it — and investigators usually try to distinguish the two. But even without distinguishing them we are entitled to use the correlations in assessing the variables of the 'second step'. Knowing that the correlations at the level of verbal behavior covers both classes of opinionatedness, we can say that if someone says something strongly pro Catholic he is likely to endorse other pro-Catholic opinions, even if we don't know to what degree his answers reflect a sincere commitment.

The inferences from the first- to the second-step variables at the same level of verbal behavior can go in any direction and to any area of verbal communications or verbalized messages, where either a correlation between two simple patterns of verbal behavior, or a correlation within a whole complex syndrome at the verbal level, has been discovered in previous investigations.

The same kind of inferences can be made *from verbal to non-verbal, 'real' behaviors*, whenever we know that certain verbal behaviors constitute the elements of certain broader behavioral syndromes. Thus, for instance, knowing someone's strong religious declarations, we can infer (whatever their expressive value might be) that this person is likely to regularly attend church services. Knowing that someone is a 'high-scorer' on the verbal scale of ethnocentrism, we can predict that, at least in situations where his behavior is under the social control of other members of his social group who have the same attitudes toward the given ethnic minority, he will behave in a way more or less congruent with his declared opinions, whatever their psychological, expressive, meaning might be.

4. INFERENCES BASED ON ASSUMPTION OF CORRESPONDENCE AT THE EXPRESSIVE LEVEL

In most of studies, the determination of what people *actually say* is only the starting point for another question: What are their real thoughts and opinions on the subject of their communication? In other words, after analyzing the linguistic meaning of the communication we start to think about its psychological meaning. The boy who hears the statement 'I love you' will hardly be satisfied by the mere linguistic meaning of this com munication but will usually ask himself: 'Does she really love me?' The sociologist who during an interview hears the person declare that he is going to vote for the Democrats will likely ask himself 'Does he really intend to do that?' This means that *we usually treat verbal communications also as indicators of certain mental states and predispositions of the communi-*

cators; as manifestations of their knowledge, intentions, and opinions; as reflections of their preferences; etc. Thus, both the sociologists and the layman are interested not only in the external aspects of human behavior but also in its expressive aspects.

This is of course quite another type of 'understanding' — the understanding of the subjective, psychological meaning of the communication. We are faced here by the problem of the relation between the content of verbal declarations and what it is they express. I have discussed the problems of 'understanding' in social science in another essay of this volume[7] so I do not intend to duplicate it here. May I only say briefly that human beings live in a 'meaningful world' and discovery of the subjective meanings people attribute to their own behavior and to the social situations they act in is therefore an important task of sociological investigations. These meanings may be assessed in many different ways but one important way we can learn about them is to listen to what people have to say about them.

Such communications usually come to us from individuals and occasionally from a group which has collectively prepared a formal message or collectively adopted a verbalized set of norms of behavior. Our interests may also focus on particular meanings — motives of certain individual behaviors, values certain people hold, expectations they have regarding the behavior of others, etc. On the basis of such data we can try to understand — in a more or less 'Weberian' way — the behavior of individual persons to explain such behavior and to predict its future course. But our inferences from such data are by no means limited to these individuals only. Having information about the characteristic values and attitudes of *members* of a certain group, we can come to certain conclusions about the value system of this *group as a whole*. Having information about the mutual sympathies of members of a given small group, we can try to establish its 'sociometric structure'. Having information about the role expectations within a certain internally differentiated group, we can try to determine the system of roles and other structural features of this group. To the degree that all social phenomena are 'loaded' with social meanings, the discovery of these meanings requires to a greater or lesser extent our using the verbal communications of members of the studied populations in their expressive functions.

This does not mean, of course, that the contents of verbal communications (even if we assume their subjective credibility) are the only source of our inferences here. We usually try to put the opinions and convictions of the persons studied into the context of the 'real behavior' in order to see, among other things, the degree to which they really do what they say they ought to do, how they react to those who do not behave according to their expectations, etc. In order to obtain such information we either observe their behavior directly, or make our inferences from their communications which we use now according to the descriptive function. In any case, knowledge of the psychological context of 'real behavior' is extremely important for any social or behavioral scientist. To give an example: Suppose we know that in a certain society people do not behave in a certain way; it makes a lot difference whether we learn that a majority of them think they nevertheless ought to behave in this way (and therefore feel more or less guilty) or that in their opinion this kind of behavior is from a normative point of view irrelevant or even undesirable.

5. EXPRESSIVE AND OTHER INSTRUMENTAL FUNCTIONS OF VERBAL BEHAVIOR

In order to make such inferences as those described above we have first to make certain that the expressive relation in the communications chain is that of high correspondence. Therefore we face the problem of identification of possible factors which might disturb this relationship. It is no secret to anybody that there are factors which function to produce insincerity — from concealing our true opinions to open lies. Besides, hundreds of experimental studies aimed at controlling the research validity of interviews and questionnaires have revealed various kinds of factors responsible for distortions of the expressive relation. From the standpoint of our problem in this chapter we are interested only in the impact of the interviewing situation on the expressive relation in the communication chain. We know that the ways people perceive the interviewing situation or the goal of the research, the social context in which the communication is expressed, and many other factors co-determine in various ways the degree of correspondence between what people really think and what they actually say at the given moment. But I think that we can look at this problem in a broader frame of reference and ask about the motives behind any verbal behavior, assuming that these behaviors are usually instrumental for the satisfaction of motives of certain kinds.

The first category might be called *expressive motivations* of verbal behavior. I mean by this the kind of motives which can be exemplified when, in an intimate conversation, one's best friend is confiding a serious personal problem. Also governed by expressive motivation would be the man who answers a questionnaire because he wants the scientist to know what he thinks about the subject, and believes he has nothing to hide. Such motivations, especially when they are loaded with emotion, are sometimes so strong that people have to tell someone what is worrying them or what is making them happy at the given moment.

In general, we can say that expressive motivation, or the need to 'open one's mind' to another person, usually increases the correspondence of the expressive relation. Therefore in our studies, when we want to use the content of what people are saying as the source of sociological information we should try to stimulate their expressive motivations. In order to be able to do this we will need to better understand the nature of such motivations and the conditions which contribute to their enhancement. Undoubtedly there are special kinds of needs to 'communicate' which probably are rooted partly in certain personality features and partly in the specific situation of the individual at a given moment. Other factors are related to the social context in which the communication is being made, and here confidence in the sympathy and loyalty of the 'receiver' of the communication is probably the key factor in determining the arousal of expressive motivation.

But the verbal behaviors may be also (subjectively) instrumental towards other kinds of motivations — those which are oriented toward *speaker's adaptation to the external world or his tendency to influence, to control external world.*

Thus for example, when a candidate gives a speech at a political convention, he usually wants it to have certain effects beyond his being properly understood. He would like to bring about certain political changes in his country and therefore is presenting some proposals for these changes in his speech, as he would like to be elected to office. When one constantly tells his boss that he is the best possible boss one could have in the world, his goal is probably not limited to his wanting the boss to know the content of this message, but he also hopes that this will have some positive effects for his future promotion.

Neither of these examples implies that this kind of instrumental motivation necessarily distorts the expressive relation. The politicians proposals may express values. It may happen that someone really admires his boss. When a patient tells his doctor about his health problems he is obviously instrumentally motivated, but we may at the same time reasonably expect that he will try to be as accurate as possible in his report. When I say to the waiter in a restaurant, after ordering my meal, "Please hurry, because I am so hungry!" I express my needs with the obvious intent to influence the water's behavior, but this kind of instrumental motivation does not decrease the degree of correspondence between the content of my communication and the content of my mind. These two motivations – may nicely coincide and work in the same direction. Confiding in our friend may be motivated *both* by the need to communicate with someone who is sympathetic and is able to understand the problem *and* by the expectation that this friend might be able to help us in one way or another – even if only by giving us some sound advice. Rioting students are both expressing their strong negative emotions against the 'Establishment' and trying to influence the course of political events.

On the other hand, when people express views that are approved in their society only *because* they want to be approved too, this kind of instrumental motivation usually introduces serious distortion into the expressive relation of the communication chain.

The rule is simple. Sometimes expressing one's genuine views, opinions, and expectations is instrumental toward one's external goals; other times, *hiding* one's thoughts or even 'expressing' those which do not exist may be instrumental toward one's purposes. This is a rather obvious rule of interpretation of the expressive validity of communication, and all of us use it both in our everyday social contacts and when we are collecting or interpreting sociological data.

I said above that the 'need to communicate' is usually realted to a high degree of correspondence between the speaker's thoughts and words. But it may lead as well to a distortion of the communication chain at this link, particularly when the need to communicate is an element of a broader social syndrome: 'the need to have strong social ties'. In such a situation it may happen that the act of communication will have adaptative rather than expressive functions, in as much as it will be oriented (more or less consciously) toward gaining the approval or interest of the listener. This may lead to the speaker's conforming the content of the communication toward the real or imagined expectations of the receiver or to the expression of certain non-existing thoughts if the speaker thinks this will make him 'more interesting' to the listener.

Here again one should stress that once having established what the speaker really feels or believes in or knows, we can further on infer many other characteristics of the content of his mind as well as many other psychological properties, using all the correlations and all the laws and theories which might be useful for the given purpose. Some of these correlations will be described in terms of general pshychological or socio-logical theories which tell us, for example, that people who hold certain ethnocentric opinions are also likely to have conservative views in regard to the socio-political system, or that people who are superstitious in certain areas probably have little scientific knowledge about these areas. We also know that people who accept one opinion of a socially established belief system of a group will be likely to accept the other ideological beliefs of this group, and many social theories and research reports describe such atti-tudinal syndromes in a quite detailed manner.

Nevertheless, we should be aware that when we speak of inferences we can make from one psychological state to another psychological state on the basis of 'known correlations' existing between them, these correlations are in themselves of a strictly *inferential character*: they are theoretical con-structions derived form analysis of the correlations between the phenomena at the observable level, under the assumption that these observable pheno-mena are inferential indicators for the corresponding attitudes or other psychological states and predispositions. But – for the reasons I discussed in the first chapter of this volume – the relations between the observable indicators and their hypothetical correspondents are usually rather vague in social theories, and the validity of behaviors (and utterances) as indicators of the inferred states is far from being perfect. This has, of course, direct implications for the eventual relationships between phenomena at the hy-pothetical level which we would like to derive from the correlations between their observable counterparts. At this stage of the development of social theory we can seldom say that a certain relationship at the hypotheti-cal level is *general*, or that it has a *definite probabilistic value*; we usually say that a relationship at the hypothetical level seems to be positive and therefore we can say at the best that the occurrence of a certain attitude or other psychological characteristic somehow increases the probability of occurrence of the other, 'hypothetical property' we would like to infer.

Suppose we come to the conclusion that the existence of one attitude somehow increases the probability the existence of another attitude we are interested in, and we are inferring the existence of the other attitude (with all the reservations as to the degree of uncertainty of our inferences). Needless to say, on the basis of such indirectly established psychological states or processes in the speaker's mind (either those expressed in the communication or those inferred) we can make predictions at the behavioral level of the person — as regards both his verbal and his non-verbal behavior. We then assume that what people think, know or intend to do may be relevant for what they will say or do in the future. For this purpose we should use all the knowledge at our disposal. Thus, from the indirectly inferred state of the speaker's mind, we can infer the content of other verbal communications he is more or less likely to make at times when we are not observing him. We can also use the inferred states of the speaker's mind as indicators (predictors) of his non-verbal behaviors causally related to these states. The scheme of these inferences may be diagrammed as follows:

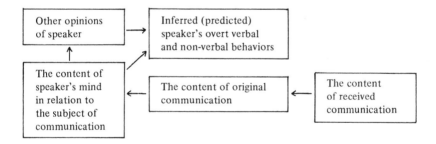

6. PROBLEMS OF THE VALIDITY OF SUBJECTIVELY SINCERE COMMUNICATIONS

There is one problem that should be discussed when we speak of the expressive function of communications. Suppose we have come to the conclusion that someone is completely sincere in his answers and that therefore the correspondence between what he thinks and what he says is practically perfect.[8] Does this mean that we can take the content of the communication as an 'adequate presentation' of the given fragment of his 'interior' to which the communication refers? This would be the case only if we were to assume that the man is a perfectly self-conscious being. But, at least since the time of Freud's discovery of unconscious states and processes, we have known whole areas of psychological processes may be completely inaccessible to the communicator's introspection, while at the same time he may perceive incorrectly the functions of other states which he 'observes', in his mind.

With respect to the unconscious processes and states, the communicator is in a situation similar to that of a person who does not know about certain *external* facts, events, or relations. It is obvious that he is unable then to report them. When we ask a woman brought up in a rigid, traditional value system whether she has strong inclination for extra-marital sex she may sincerely (and probably with a certain indignation) answer, 'Of course not'. If we say that her answer is not true, we are treating her verbal report in terms of its *descriptive* rather than its *expressive* function. Independently of such an answer, we may infer the existence of such 'repressed' states and attitudes from a non-literal interpretation of the content of her verbal reports (which would be a typical case of the psychoanalytic interview) and from other kinds of cues such as the existence of certain neurotic symptoms.

Our procedure here is similar to that discussed in the previous section; from the conscious states revealed by our communicator we make inferences to his other, unreported conscious states, using for this purpose all known correlations and theories — except that now we are making inferences on the basis of a specific theory which describes the relationships between the conscious and the unconscious, repressed psychological states and mechanisms.

On the other hand, communication which are not reliable descriptive indicators as valid descriptions of respect, mental states, processes and predispositions, are not irrelevant as long as we treat them in their strictly expressive function, i.e., as the speaker's conviction about his thoughts and motivations. Rationalizations do not play a basic role in motivational mechanisms but it may be essential for a social psychologist to distinguish candid cynics from those who need rationalizations in certain actions which they would otherwise perceive as morally improper. To say nothing of the fact that the study of the conscious contents of human minds may be in itself a legitimate object of scientific interest. Therefore the main problem is, *What kinds of inferences do we want to make from subjectively sincere verbal communication*? If we distinguish the descriptive function of such reports from the expressive one, we have to conclude that the subjectively sincere but descriptively unreliable communication reveal correctly those areas of mental states which... they express.

There is another, quite different category of problems related to the unreliability of subjectively sincere communications, which is often mentioned in methodological discussions about the validity of questionnaire answers in survey of experimental studies. As is known, the situation in which the communication is made and factors immediately preceding the expression of the respondent's views and opinions may play an important role in creating in his mind views which never existed there before. The wording of the question in a guestionnaire may strongly influence the answer, inasmuch as it may 'create' by its mere formulation certain inner reactions of the respondent. We can inculence the respondent's *attitudes* by formulating the question in suggestive, strongly emotional terms. Sometimes by asking a question pertaining to a problem which until then had not existed for the respondent, we can compel him to 'solve' the problem for himself at this moment and *thus* to answer the question. In either case the answer is subjectively sincere, but we sometimes have doubts about its validity as an indicator of 'genuine' attitudes of the respondent. This doubt may sometimes go so far that we call such attitudes 'fictitious'. In other cases, the changes produced by a research instrument are much more subtle, since by asking a question formulated in terms from another frame of reference than that of the respondent, we compel him to 'reformulate' his thoughts in a way that does not correspond to his own usual way of thinking. Some writers even say that asking a question in a questionnaire or interview should be treated in the same way as any experimental situation, i.e., the question is no more than a 'stimulus' and the answer is no more than the 'response' to it. We shouldn't worry – they say – about what we may infer legitimately from the stimulus to the inside of the 'black box', i.e., the speaker's mind, because it is a 'meaningless' problem; we should only worry what kind of predictions to other 'responses' can be made knowing this or that 'response' to the given 'stimulus'. As I tried to show in the first essay of this volume, making inferences about the contents of other people's minds is not a useless undertaking for a sociologist, even if he tries to do so regardless of the possible use of these conclusions for the future predictions of 'overt behavior'. So if one accepts that some situations, including that of being interviewed with a long battery of items of a questionnaire, may more or less change the content of the respondents' mind, before answering the question of the validity of our verbal indicators for the speaker's content of mind *we should ask about the exact purpose of our investigation*. In all our inferences we start from the fact that at the moment of the research our respondents really believed in what they said to us, and then we ask the question whether the assessment of respondent's views at the time of interviewing or answering our question in other conditions is interesting for us, either in itself or as the starting point for other kinds of inferences.

We can distinguish several typical patterns of inferences from utterances in situations when we suspect that the respondent's views were influenced by them. Assuming that a person in a 'stimulus situation' (be it questionnaire study or a more visible kind of external influence, upon the respondent's views or perceptions, such as, e.g., in Ash's well-known experiment on perception under group pressure)[9] really believes in what he reports to us, the options are as follows.

(A) We may say that in these specific conditions people really think and feel in the given way, and we don't care about anything else (1) because *we are interested* in their feelings in this kind of 'experimental' situation and in nothing more, or (2) because we think that the experimental situation is a sufficient approximation to a certain (usually rather narrow) class of 'real' situations for which we would like to make our inferences.

(B) We may try to reach conclusions about the person's opinions in his 'normal' situation outside the research conditions. We might then say that the opinions we discover in such conditions are *not representative* either of the person in his normal states or for a broader population for which the studied person may be the sample (i.e. one of the members of our representative sample). The question of the direction and strength of the possible impact of the research conditions upon respondents' attitudes, feelings, and opinions then becomes the crucial one.

(C) We may have a situation in which, as the result of the research stimuli, the attitudes and opinions of the studied persons become changed for a longer period − e.g. the well-known 'panel effect'. The situation here is different than in (B) because the expressed views are *now* not representative for the rest of the population which was not included in our study, although they are representative for all the persons participating in the study.

(D) We may want to determine the existence of certain features which are interesting to us for strictly correlational purposes (e.g. for the purpose of prediction). In such case the only think that matters is the correlation of the verbal behaviors (which we then use as definitional indicators of these variables) with the other phenomena — with the 'criterion variables' we are interested in. The problem of 'experimental bias' is here completely irrelevant, if the correlation is empirically valid.

In some situations the nature of our inferences may be unclear, even to ourselves. Let me give an example. Suppose we would like to use the *questionnaire answers in the prediction of future human behavior. We face* here two possible research strategies. One would be of a strictly 'psychometric' kind, whereby we take any 'battery of indicators ẁhether they are, with respect to their content, meaningfully related to the given behavior or not and select our items from the standpoint of the maximal correlation with the 'criterion variable', i.e. the predicted behavior. Here we do not have to care about such things as the possible impact of the research situation on the respondents' views. The only thing we are interested in is the construction of a battery of tests with maximal predictive power. Since we want to use *this battery of tests* in actual predictions, the amount of bias or influence it introduces into the respondents' minds is irrelevant for us. We assume only that it will act in the same way in all cases where it is used for prediction.

But we may apply the other strategy. If we would like to base our prediction upon the 'understanding' of the motivations of the given person, upon our assessment of his real values and his perceptions of the world — and assuming that this person will be guided in his 'rational behavior' by his knowledge of the world and his value system — we try to get the most unbiased picture possible of his views, feelings, opinions, values, goals, etc. In such research strategy the amount of change introduced temporarily in these feelings and opinions by the research conditions and research instruments becomes a problem of great importance.

7. INFERENCES FROM AND OF DISTORTIONS OF THE COGNITIVE RELATION

When we know the beliefs and opinions of our respondents on a given subject, we often stop our inferences at this link of the communication chain. The studies of attitudes or public opinion polls are the best examples of research of this kind. But even in this kind of research at least some answers are treated not according to their expressive but to their *descriptive* function. When we ask the respondents about their age or education we are interested in 'what they believe, their age or education is' but in how old or educated they really are. The same is true for all the 'social background characteristics ꞓwe need for explaining the attitudes and opinions of the respondents. When our questionnaire includes certain questions about the respondents' behavior we seek to determine the occurrence and character of his 'real behavior ꞓand not his opinions about it — how he thinks or believes he did behave is in this kind of problem only a step toward the assessment of certain other phenomena. The respondent's knowledge and beliefs are here only empirical indicators of certain objective facts or events he reports about in his communication. His knowledge may correspond to these facts more or less correctly and this correspondence defines for us *the accuracy of the cognitive relation.* To say it in the most simple way, we can infer the correct shape of the facts reported in the communication if, and to the degree that, the speaker (or the author of the communication) knows them correctly.

Now we come to the next question: Under what conditions are people able to have correct knowledge of events, and what are the implications of various kinds of distortions of this relation for the techniques of social research?

It would be useless to write here a treatise on the gnoseology, psychology, or sociology of cognitive processes and human memory, if the report refers to past events. To recall the most obvious categories of factors which might co-determine the cognitive relation of the communication chain. Let me just briefly note what is necessary in order for it to have a relatively adequate knowledge of events or facts.

(a) People must be in a position to know about these events. This is why most people are able to describe certain events from their own life history and *their own* social characteristics, but very few could report accurately on the car accident that took place at a certain *intersection* on a given day. A valid description of this event can be given only by those who were eye-witnesses to the accident. The validity of a given report interpreted according to it descriptive function should be evaluated by checking whether its author had to be in a special situation in order to have the correct knowledge, and if so whether he *was* in this situation.

(b) People must possess the *means* necessary for correct assessment of the given phenomena, including special instruments (if they are needed), and training, capabilities, etc. Thus the average respondent may be able to describe correctly his age or the furnishings in his apartment, but not his blood pressure or certain of his personality traits, although he potentially may have 'direct access' to them. Therefore the variables we would like to infer from the verbal reports, treating them according to their descriptive functions, should be classified from the standpoint of what special capabilities and observational instruments are necessary for their correct assessment, and then the degree to which the given category of respondents have these capabilities and instruments at their disposal must be determined.

(c) People's cognitive capacities can be affected by *emotional factors*. Certain emotional pressures may, as we know, both disturb the perception of certain phenomena and transform mental pictures with the passage of time in such a way that the emotional needs of the individual are more or less met by the resulting distortion. The psychology of cognitive process knows very well the various mechanisms for distorting the picture of reality — from the projection of our own feelings upon the reality and transforming its picture to correspond more with our emotional needs than with features of the reality itself, to the various mechanisms for striving toward 'cognitive consonance' of the picture of the phenomena, the emotional balancing of different components of our opinions, and so on.

(d) Preconception and bias characteristic for certain social groups can disturb the cognitive relation. Obvious examples of this are the various social and cultural stereotypes of prejudices which predispose people to perceive some conditions or situations, the behavior of certain persons or their characteristics, in a more or less distorted way, congruent is the given stereotype.

When we interpret the content of the received communication according to its descriptive function we are interested in minimizing the possible impact of any or all of these factors upon the cognitive relation. Only if we can assume (or determine) that they did not operate in a distortive way (or that their operation was too weak to matter) can we make descriptive inferences about the events reported.

It would be useless to recount here the kinds of variables sociologis may infer from a respondent's valid observation, both events and characteristics which refer to the speaker's directly (characterizing their objective features of their immediate social environment) and the events which the speakers can reliably describe. After assessing the validity of these reports at the individual respondents' level, the researcher may later on' aggregate' them in order to construct different characteristics of the groups as wholes, beginning with the 'mean income' of the group members and ending with 'social mobility rates' characteristic for the given society. The same applies to different behavioral characteristics which we may infer from verbal reports. If we can assume that the individual reports describing the behaviors both of our respondents and of those whose behaviors they describe are valid, we can construct whole, more or less complex patterns both of individual behaviors on a mass scala and of the patterns of social interactions on a smaller or greater scale.

What is more important here is that we are seldom interested in the strictly 'overt' aspects of human behavior. We usually want to look at their occurrence in the *meaningful context of beliefs, motives and value orientations* governing them or at least accompanying them, and in the context of the perceptions of those to whom this behavior is directed and who, according to the meanings they themselves attribute to it, will react in some say to this behavior. This means that when analyzing the reported behaviors as certain meaningful wholes — i.e., *as both overt and psychological syndromes* — we have to treat the reports simultaneously in their expressive and their descriptive functions. Nevertheless we should be aware that the logic of these two inferences is based upon different premises and that assessment of the subjective meaning of behaviors may sometimes be more reliable than assessment of their strictly overt aspect.

In some cases we ask questions regarding certain events or facts our respondent might know about or have his spontaneous report on such events, nevertheless the degree of correspondence between the reality itself and its subjective picture in the eyes of our respondent is irrelevant for us. This means that our question only pretends to be a question about facts, in reality, we are interested in the responent's opinions about them. If we ask a representative sample of members of a certain society what are exactly the changes in the standard of living of their whole society during the last two years, we know that this is usually not a question which could be correctly answered by the average respondent in case we should wish to interpret the answers according to their descriptive function. Asking such questions in an attitude survey usually means we plan to interpret the answers according to *their expressive functions* — i.e. as indicators of attitudes and not as adequate descriptions of economic processes.

In such cases, whether the perceptions are correct or not they are always important social facts in the sense of their manifesting attitudinal characteristics of members of the society. The communications are thus interpreted not according to their descriptive but to their *pseudo-descriptive expressive function*. The degree of possible distortions of the communicative relation will then be for us irrelevant.

Sometimes it is just these distortions that are important for us, so that we want to maximize their operation in the cognitive process. Here, even the content of the respondent's opinions may be only an indicator of something else — namely, certain psychological or sociological predispositions, which can cause, among other things, distortions of cognitive processes. Textbooks of projective techniques in sociological or psychological research give many examples of such pseudo-descriptive questions or tests.

In some other cases we proceed in exactly the opposite way: We ask the respondents for their *evaluations* of certain things or events, which would usually imply that we intend to make inferences about their views and opinions, but here we infer from their answers certain conclusions regarding the things to which these evaluations refer. This is possible *when we know the standards or criteria of evaluation* which our speaker (or some identifiable group to which he belongs) uses in evaluating these things or events. It is not difficult to make inferences about a man who has been evaluated as a 'good husband' by his wife in a typical working class Polish family. We know at least that he is not an alcoholic, that he brings his salary home regularly, that he takes care of the children in his free time, that he does that share of the domestic chores which 'belongs' to the husband, etc.

The more knowledge we have about the standards used by certain people or certain groups for the evaluation of certain objects, events, or situations, the more we are aware of the *descriptive contents of these evaluations*, even if they were not explicitly stated in the utterances. In such cases we may use these utterances in their descriptive as well as their expressive function. But to the degree that we interpret them in their descriptive function — i.e., as a source of information about the actual characteristics of the evaluated events — we have to apply all the criteria discussed above in order to estimate the degree of possible distortion of the cognitive relation of the communication chain. This does not apply, of course, to interpretations of the evaluations when we interpret them in their expressive functions, as manifestations of the speaker's attitude toward the evaluated event: here we try to estimate only the possible disturbances in the expressive relation.

Whether we come to our conclusions about the characteristics of the 'objective' facts on the basis of their direct description in the communication, or indirectly by inferring them from certain evaluative statements, these facts can constitute again a starting point for other inferences; they can be used as empirical indicators of phenomena to which the communication did not refer. These correlations may refer to relationships of various kinds. First, they may refer to certain relationships at the observable level. If someone describes the kind of furnishing he has in his apartment we can make inferences about his income, even though he does not mention income in his communication. When someone says he is the owner of a large piece of land, we can infer a lot about his everyday life style if we know the culture of big land owners in this society.

Once we have established the non-reported objective correlates of the events and situations described in the communication or have derived them from respondents evaluations, we can use these inferred events as well as the events and situations directly reported and described to us as the starting point for other inferences, now going to the level of respondents' attitudes, predispositions, etc. provided that our theory is rich enough to include the relationships we need for that purpose. We can also take our inferences to the behavioral level. This we can do in two different ways. We can infer from the speakers' attitudes and opinions that they are likely to behave in such and such a manner. But we can also make our inferences on the basis of direct correlations between certain situations or certain objective social characteristics of people and their verbal or non-verbal behaviors. These are, of course, two different patterns of reasoning and require different kinds of theoretical assumptions.

NOTES

* Basically modified version of the paper: 'Funkcje wskaźnikowe odpowiedzi w badaniach ankietowych' ('Indicative Functions of Answers in Questionnaire Surveys'), published in S. Nowak, *Studia z metodolgii nauk spolecznych* (Studies in the Methodology of the Social Sciences), Warsaw, 1965.

[1] For the typology of indicators see 'Concepts and Indicators in Humanistic Sociology', in this volume, p. 1. It should be also added here that the different measures of validity discussed in the paper quoted above may apply also to the characterization of the degree of correspondence between the neighbouring links of the communication chain. Nevertheless, it seems reasonable to assume that in the study of communication chain we will be primarily interested in the *screening power* of our indicator for the next link of communication chain, i.e., in the probability that, e.g., the author of the communication really thinks what he said, but not that he said *all* what he thinks about the given subject (what would imply then maximization of *inclusion power*).

[2] See R.L. Birdwhistel, *Introduction to Kinetics*, Univ. of Louisville (Ky.) 1952. See also J. Ruesch and W. Kees, *Nonverbal Communication*, Univ. of California Press, Berkeley, 1956.

[3] H.D. Laswell, D. Lerner, and Ithiel de Sola Pool, *The Comparative Study of Symbols. An Introduction*, Stanford, 1952, p. 12.

[4] In the detailed analysis of the problem of interviewer's bias, see H. Hyman, *Interviewing in Social Research*, Chicago, 1954.

[5] See B. Berelson, *Content Analysis in the Communication Research*, Glencoe, Ill., 1952.

[6] See I. Goffman, *Behavior in Public Places*, Glencoe, 1969.

[7] Concepts and Indicators in Humanistic Sociology.

[8] In terms of perfect screening power, of course, see note 1.

[9] S.E. Ash, 'Effects of Group Pressure Upon the Modification and Distribution of Judgments , in H. Guztzkow (ed.), *Groups, Leadership and Men*, Pittsburgh, 1951.

BIBLIOGRAPHY

Ash, S.E., 'Effect of Group Pressure Upon the Modification and Distortion of Judgments', in H. Guetkow (ed.), *Groups, Leadership and Men*, Pittsburgh, 1951.

Berelson, B., *Content Analysis in the Communication Research*, Glencoe, Ill., 1952.

Birdwhistel, R.L., *Introduction to Kinetics*, Univ. of Louisville (Ky.), 1952.

Goffman, I., *Behavior in Public Places*, Glencoe, Ill., 1969.

Hyman, H., *Interviewing in Social Research*, Chicago, 1954.

Laswell, H.D., D. Lerner, and Ithiel de Sola Pool, *The Comparative Study of Symbols. An Introduction*, Stanford, 1957.

Nowak, S., 'Concepts and Indicators in Humanistic Sociology, in this volume, p. 1.

Nowak, S., 'Funkcje wskaznikowê odpowiedzi w badaniach ankietowych' (Indicative Functions of Answers in Questionnaire Surveys), in *Studia z Metodologii Nauk Spolecznych* (Studies in the Methodology of the Social Sciences, Warsaw, 1965.

Ruesch, J., and W. Kees, *Nonverbal Communication*, Univ. of California Press, 1956.

CHAPTER III

MEANING AND MEASUREMENT IN
COMPARATIVE STUDIES*

1. CONCEPTUAL AND OPERATIONAL ASPECTS OF
PHENOMENAL AND RELATIONAL COMPARABILITY

The development of comparative social research seems to be one of the most characteristic features of contemporary sociology. The idea of broad, cross-cultural, cross-regional or cross-historical comparisons is, of course, not a new one in the social sciences. In fact it is as old as the tendency to base theoretical generalizations about social phenomena upon broad inductive, empirical evidence. It can be traced in the history of social sciences from Aristotle and Thucydides, through Machiavelli, Ibn Khaldun and Montesquieu, to Marx, Comte, Spencer, Durkheim and Weber. There are nevertheless some special features in contemporary comparative sociology. One is that the studies called 'comparative' are usually those which collect and analyze data from *more than one society* — or perhaps better, from more than one nation-state — trying at the same time to apply the same degree of standardization of research techniques which until one or two decades ago could be found in the studies conducted within one national sample.[1]

The so-called comparative studies pose many methodological problems, and — as has been pointed out by many writers — most of these problems are by no means unique for this kind of study only. Nevertheless one has to admit that the problems which exist in any inductive study are *magnified* in 'comparative' studies by the fact that here the differences between the different settings in which our study is being made are usually significantly greater. Therefore it seems worthwhile to concentrate on the central methodological problems of comparative studies, both for their own sake and from the standpoint of what such conclusions imply for more general problems of the methodology of sociological investigations.

One of the central problems in cross-cultural or cross-national studies is the problem of *comparability of the studied phenomena with respect to the research variables* in the different populations in which we make our study and to which we would like to apply our theoretical conclusions. In the general logical scheme of induction, things look quite simple: we should state the existence of the events denoted by a certain concept (variable) (x) and then try to find out whether there is a general, cross-cultural relation between this variable and another variable (y). If the relation turns out to be not general but limited to some populations only, we should try to discover some general characteristics (z) which occur in those populations for which our regularity 'works', and which are absent in those populations where the relationship xy 'does not work', If we find them, we may conclude that the *relationship between x and y is conditional upon the occurrence of z.*

But when it comes to the application of this simple logical scheme in an actual empirical social study, things become less obvious and the problem we face often is: How do we know we are studying 'the same phenomena' in different contexts; how do we know that our observations and conclusions do not actually refer to 'quite different things', which we unjustifiably include into the same conceptual categories? Or if they seem to be different, are they really different with respect to the same (qualitatively or quantitatively understood) variable, or is our conclusion about the difference between them scientifically meaningless?

I think that when a comparative researcher is uncertain whether the phenomena of his interest are identical or different in different national settings, he may have one or more of three different questions in mind:

(a) The scientist may be uncertain as to the *meaning* of a theoretical term he is using in his comparative study. That is, he is not certain what attributes of the compared objects should be taken into account in order to decide whether the compared objects belong or do not belong to the same conceptual category. This problem is especially serious in the social sciences, because we often use certain theoretical terms the meaning of which is more or less vague and unclear. It is obvious that as long as we are unclear about the meaning of our term, we are unable to decide which phenomena in quite different social or cultural contexts belong *by definition* to its denotation if it is a qualitative (dichotomous) variable, or which are measured by that variable if it is a quantitative one. Uncertainty about 'comparability' at the level of our theoretical concepts unavoidably accompanies the vagueness of our concepts. It is impossible to say whether the objects are similar or

different as long as we do not know the properties with respect to which
they should be compared.

*It is obvious that from a strictly definitional i.e. conceptual point of
view, the phenomena are identical with respect to the given concept
defining the standard for their comparison if they possess the characteristics
which belong to the contents of the given concept, whatever other
characteristics they may possess also.*

*As Sidney Verba wrote in his study on the problem of con-national
comparability of research variables:*

The fact that we are searching for functional equivalence makes clear that we are not
looking for measures which are equivalent in all respects. What is important is that the
measures be equivalent for the problem at hand.[3]

If, for example, we define 'the elite of economic decision making' as a group
of people occupying the top positions in the system of economic decision
making of the given country, then this concept *applies by definition both to
owners of the means of production in a capitalist country and to the top
economic managers in a socialist country.* These two groups become
definitionally equivalent due to our terminological convention, whatever
other differences might exist between them. On the other hand, if they do
not possess the characteristics specified by the given qualitative concept or
do not have the same magnitude of a variable of a quantitative kind, then
with respect to this standard of comparison they are different, whatever
other similar features they might possess.

(b) Having a certain meaning of the concept in mind or trying to define
it for his study, the scientist may have doubts as to the *theoretical
fruitfulness* of the concept so defined, i.e. whether it will be useful for the
purpose of given comparative analysis of social phenomena within the
different systems. If the goal of his comparisons is to test a theoretical
proposition, the scientist may be uncertain whether he will be able to
formulate interesting theoretical generalizations about the phenomena so
defined.

The precise definition of the given concept should be therefore
distinguished from the *'legitimization' of theoretical usefulness* of the
concept defined. One cannot object to the conceptual equivalence of the
two groups of 'elite of economic decision making' on the basis of the above
definition. One could, on the other hand, question its theoretical fruitfulness if one were to say, e.g., that no interesting generalizations or theories
could be formulated about 'elites of economic decision making' if this

concept is understood in such a way. *The problem of the fruitfulness of the given concept is different from its precise meaning, and the two should be treated separately.* And both these problems should be distinguished from a third one, namely how we can *recognize* in the practice of social investigation the objects or phenomena which possess the characteristics specified by our concept and therefore how we can in practice perform the task of classification of the objects of our study by the *indicators* we would like to use in it. As we know the decision concerning the choice of an *indicator* for a research variable may sometimes be identical with the procedure of defining a certain concept. We are then constructing an 'operational definition' of the given concept in terms of some observable phenomena. If we additionally assume that such an operational definition exhausts the whole meaning of the given concept, the *question of the validity of the indicator for this concept becomes meaningless* – it must be valid by reason of our definitional convention. No research evidence can prove or disprove the validity of such an operational definition. The question of the validity of indicators for the indicated phenomena becomes empirically meaningful only when these indicators do not define the given concept completely. This occurs either when the concept is defined in such a way that the indicator does not belong to its meaning, or when it is defined in such a way that the indicator constitutes only its partial meaning.[4]

The answers to these three questions are, of course, not independent from each other. We usually try to begin with establishing such – possibly precise – meanings of our concepts as will be useful for the purpose of the study, and especially – since this is primarily what we are interested in in this chapter – fruitful for the formulation of theoretical hypotheses we intend to verify in our research. Nevertheless we should be aware that this poses two different, although related problems. The first is knowing the properties with respect to which we compare the studied objects. The second is knowing *why* we compare these objects with respect to these properties. The answer to the third question also depends – at least partly – on our definitional decisions. If some of the components of the meanings of our concepts denote fairly easily observable phenomena and these phenomena are related in a predictable manner to the phenomena described by other components of the given concepts, we are fully entitled to use these observable phenomena as the indicators in our research. On the other hand, if we come to the conclusion that none of the phenomena designated by our concept may be used as its observable indicator, i.e. that we have to use indicators which are *conceptually external* to the studied phenomena, then the question of what indicators should be used in the given national or cultural setting for the identification of the given theoretical variable is an open and strictly empirical one.

Let us forget for a moment the problems posed by the choice of indicators by assuming, for example, that in our study we are dealing with a domain of fairly easily observable phenomena which therefore do not require additional external indicators. One might think that in such a study, once we defined our concepts with sufficient precision, there would be no problem with the classification of objects with respect to them, with decisions as to whether the compared phenomena are identical or different with respect to the given standard of comparison, just as we do not have a problem with the identification of an object such as a 'stone', regardless of whether we find it in the woods or in a box in a museum of natural history. Unfortunately, things are not always so easy.

Let me first give an example. When describing the methodological problems of famous Seven Nations Attitude Study conducted in Western Europe in 1951, Eugene Jacobson wrote:

It was clear from the outset that it would be inappropriate toward our ideal of identical research in many details... because it seemed likely that the identical operations might in some cases result in measurements which were not equivalent.[5]

When reading this one would first have the impression that the authors of the study were looking for non-identical, external indicators of the variables which were anyhow sufficiently similar. But when we read the research reports in the volume quoted above, we can see that in many cases the authors were studying in different societies, phenomenally quite different variables, because they assumed their *relational* equivalence.

As we know, our concepts may denote certain 'absolute', i.e. nonrelational, characteristics of persons (such as age, sex, possessing a certain attitude, behaving in a given way) or non-relational properties of groups, institutions, etc. Then we might say that we are interested in *phenomenally the same* traits or events in different settings and classify them as identical with respect to the characteristic specified by our concept. The fact that they are the same with respect to one characteristic does not mean, of course, that they are indentical with respect to others to which our concept does not refer.

But the contents of our concept may also include a variety of *relational characteristics* of people or communities. In order to define such relational characteristics we must of course specify in phenomenal terms one of the elements of the relation in question (i.e. the criterion variable of this relation) as well as the relation itself, but the *relational counterpart of our criterion variable may be definitionally open* with respect to its phenomenal 'absolute' characteristics. Everything that satisfies the given relation with the criterion variable belongs then by definition to the denotion of this relational concept. To give an example: If a sociologist studies the 'motives' of a certain behavior and he understands by 'motive' the antecedent psychological process causally related to this behavior (when the given behavior is the criterion variable belongs then by definition to the denotation of this psychological event that is causally related to this behavior becomes by definition its 'motive'. From a strictly phenomenal point of view these motives may be quite different for different persons. They are identical (or if we prefer, equivalent) only with respect to the given relation with our 'criterion variable', which is specified by the meaning of our relational concept.

The denotation of a strictly relational concept is delimited by occurrence of the relation in question, all objects, phenomena or processes in the societies compared which are related in a given way to the criterion variable belong to it. Whether they are phenomenally (in terms of their 'absolute', i.e. non-relational characteristics) similar or different is conceptually irrelevant here. When we define the concept of 'poison' by its effects on an organism, the chemical composition of the 'poison' becomes irrelevant. The only thing that counts then is the relation between different 'poisons' and the organism, their harmful effects for its functioning.

Sometimes the phenomenal range of applicability of a relational concept is only *partly open*. Our concept refers then to a certain range of phenomenally different objects or events which are related in the given way to the criterion variable but not to all of them: the range, although fairly broad is nevertheless specified to some phenomenally defined limits. When we define the term 'motive' in the relational way, as the cause of the given behavior, we nevertheless specify additionally that this refers only to psychological causes. Someone might go further and limit the term 'motive' only to those psychological causes of behavior which are accessible to the person's introspection. When we define 'conformist behavior' as that corresponding to the behavior of the majority of members of the actor's group, we add that this applies only to the range of this behavior that is regulated by the norms of the given group. Some strictly physiological regularities and uniformities such as sleeping at night would not constitute for us examples of 'conformist behavior'.

It may happen that the relational counterparts of a certain variable are in several social contexts phenomenally quite similar. 'Educational institutions' are, at least within the industrial societies, sufficiently phenomenally similar: they consist of school, universities, special training centers, etc. But it may also happen – if we extend the range of our comparisons – that we will have to include into the denotation of the concept 'educational institutions' certain institutions in 'primitive societies' which are phenomenally quite different from European or American schools, but nevertheless perform the same functions.

If we know a certain society well enough, we know which objects or phenomena are related in the given way to the criterion variable of a relational concept. Then these (phenomenally specified) objects are included implicitly in the extension of our relational concept. When speaking about 'educational institutions' in our society we have at the same time in mind 'schools', 'universities', and various 'centers of vocational and professional training'. Even if the exact meaning of our term is specified in relational terms only, *we implicitly include in this meaning all the phenomenal features of the objects satisfying the given relation*. And when doing our research on educational institutions we know in advance exactly what we are going to study, and how to recognize the absolute features of objects which are denoted by our relational concept.

The situation is different if we don't know the relations among the social phenomena in the society well enough, because then we don't know what the relational counterparts of the given criterion are and – in a way – what we are looking for. We may think we do not know the meaning of our concept, when actually what we don't know are the phenomenal features of the counterparts of the relation in question. In order to find these counterparts we have to discover in empirical study which phenomena satisfy the relation specified by our concept. Once we discover this we may then 'redefine' our concept in terms of absolute characteristics of the phenomena which are — in the given social context — related in the given way to the criterion of our relation.

Let me give one example of such a situation. In the study presented in the book *Values and the Active Community* by Philip Jacob and others, the authors present a strictly relational definition of one of the central concepts of their research, that of 'activeness':

Activeness is the combination of individual and collective behavior directed to, or having consequences for, an increase of individual involvement in solving problems at a collective community level. [7]

And then we read:

The definition of activeness was influenced not only by consideration of data available at the collective level, but also by investigation of the kinds of behavior appropriate for local governmental units. *The components of the definition of activeness were continuously shaped by information gathered at the local level of the four countries.* [8]

I would be inclined to say that the authors were rather establishing in the first stage of their study the phenomenal extension of the relational concept which they defined with sufficient precision to know what they would like to observe. If the concept is defined with the use of absolute characteristics only, we know that in order to find its referents we have to observe certain phenomenally understood *features* of the studied objects; when it is relational we have to establish whether these phenomena satisfy the relation in question. The only problem is that to establish whether certain phenomena are actually related in the way our concept specifies may be more difficult than to establish whether they have certain absolute characteristics. And therefore the phenomenal classification of objects distinguished with respect to a relational property often may be done at a later stage of the research.

We do this sometimes at the stage of the pilot study, when we try to establish whether the same behaviors or institutions have 'the same meaning' in different cultures; and sometimes we have to do it at the stage of data analysis, when we look for correlations of different variables in different settings with the same criterion variable. *For the assessment of relational properties the correlational analysis of our data may play the same role as the observation of single 'items' for the assessment of non-relational properties.*

The relational concept with the counterpart of the given criterion phenomenally 'open' is of course different from the syndromatic concept which I discussed in another study of this volume.[9] In speaking of a syndromatic concept we have in mind a certain set of phenomenally specified characteristics which additionally are related to each other in the given way. Then in order to decide whether the phenomena denoted by the given syndrome occur, we have to assess whether certain a priori specified absolute characteristics of the individuals are interrelated in the syndromatic way, but contrary to the case of phenomenally open relational concepts we know in advance what we are looking for.

The problem of assessment of objects defined by absolute or relational concepts should not be mixed up with the problem of the choice of indicators for them, if these indicators are definitionally external to the indicated phenomena. Unfortunately we are not often able to say whether the given indicator belongs to the content of the indicated concept or is definitionally external to it; therefore we don't know whether we use the relationships established in our study for the specification of the phenomenal extension of our relational concepts or in order to indicate some phenomena we are interested in. These indicated phenomena may be understood again as absolute characteristics of certain specified objects or they may be the relational counterparts of certain phenomena. In the latter case, on the basis of correlation between our indicator and the criterion variable, we infer that a certain relation between the criterion and the phenomena theoretically interesting to us also occurs. In such event, we have to use a quite complex set of theoretical assumptions, which, unfortunately for the social studies, are seldom formulated in an explicit way.

2. TYPES OF RELATIONAL EQUIVALENCES

As I said, with respect to a concept defined in absolute terms the phenomena are 'identical' when they possess the characteristics specified by the given concept, whatever their other features might be. In exactly the same way, if we use a relational concept the compared phenomena are conceptually indentical if they satisfy the relation specified by our definition, whatever their absolute properties, provided these properties are within the permissible range of phenomenal variation defined by the content of the corresponding concept. When the researcher uses the notion of 'equivalence'[10]

·rather than of 'identity', he usually wants to stress that the relationally identical phenomena are not phenomenally identical; they differ with respect to their absolute properties, and this is the only way in which the notion of 'equivalence' should be understood in concept formation.

When looking for 'equivalence' of various kinds we have to specify first *the relational counterparts* to which the compared phenomena are equivalent, i.e. the criterion of the given relation. This criterion may be understood in absolute, phenomenal terms because we want to study the relational counterparts of some phenomenally identical objects; or our understanding of its phenomenal extension may be due to someone's having established previously, that the phenomena of the given kind satisfy certain theoretically important relations, and should therefore be included under the same conceptual category. In the latter case *the criterion itself may be phenomenally different in a different social context*, but be classified as conceptually equivalent in the cultures or societies compared on the basis of previous theoretical knowledge. Once we accept certain (identical or equivalent) criteria for our relational concept, we may then start looking for *their* relational counterparts. And here let me distinguish several of the most typical relations we usually use in defining relational concepts in the social sciences, and later on in establishing the 'equivalences' of phenomena in the compared groups or societies.

(1) The objects, phenomena or behaviors are perceived, understood and evaluated in a similar way in different cultures. One could refer to this as *cultural equivalence*.

This is an extremely important aspect of 'equivalence' or relational identities, because it refers to the similarity of the 'social meanings' of certain objects, actions or social situations, to the 'subjective definitions' given to these objects, actions or situations by the cultures of the compared societies. On the basis of similarity of social meanings we may then classify different behavioral sequences (understood in strictly observational way) as culturally equivalent, or phenomenally the same overt behaviors as nonequivalent with respect to the cultural meanings related to them.

These meanings will never be identical in all respects. If we take into account only the fairly simple aspects of 'subjective definitions' it may turn out that the given phenomena are 'identical' in the two or several cultures. The more complex meanings are taken into account, the more likely it will be that the compared phenomena are not culturally equivalent.

But here again we should ask what aspects of the cultural meanings are relevant from the standpoint of our own research problem. If it turns out that from this standpoint it is enough to take into account certain simple aspects of the subjective meanings of certain objects and situations, and that in terms of these meanings the 'objects' in the compared societies are similar, then we can say that the compared phenomena are culturally equivalent whatever other differences in meanings attributed to the same social objects and situations should exist.

At a more individual psychological level we may classify as equivalent certain behavioral sequences of certain observable facts taking into account the meanings attributed to them by given individuals. Then instead of cultural equivalence we should speak of *psychological equivalence*, although the nature of the relation is basically the same and we differ only in the focus of our attention.

In both of these cases we classify the observable sequences as identical if they are given similar meanings by the behaving persons, or are perceived in a similar way by other participants in the given social interaction. It is obvious that since ones individual's perception of a certain behavioral sequence does not have to be identical with the perceptions of those with whom he interacts, the behavioral sequences identical with respect to the one relation may be different with respect to the others.

(2) But we may also change the criteria of our relational concept and take as the reference for the estimation of equivalence the behavior itself, treating as equivalent all the 'subjective meanings' of certain identical behavioral sequences. Then we can speak of the *behavioral equivalence* of various meanings. On the basis of a concept so defined we might classify as equivalent different motivational patterns which accompany the fact that people work hard or that they conform to the norms of the model behavioral patterns in their societies. One interesting subcategory of behavioral equivalences might be *motivational equivalence:* we could then treat as equivalent all possible motives which might lead to the same behavior.

Of course, instead of a simple behavioral sequence we might take as our criterion mass behavior on a societal scale or some more or less complex pattern of behavioral interactions, but in doing this we should be aware again that the reference for the estimation of the equivalence is observational behavior and not subjective meanings as in the previous relational category.

(3) It may also happen that the criterion behavior itself will be classified as identical on the basis of the social meanings we attribute to it. Then as a second step we would try to discover *other elements of social meanings* which might be associated – as either 'identical' or 'equivalent' – with the behavior so defined. To give an example: we might classify as conceptually identical various patterns of interaction that are behaviorally quite different saying, for instance, that all of them constitute examples of 'cooperation', and then looking for different psychological states which might be moticationally equivalent in inducing people to cooperate with each other. I think this category of 'behavioral equivalences' is more typical in social studies than situations where the criterion behavior is understood in a strictly 'behavioristic' way.

We might also view the problem of 'equivalences' from quite another perspective, looking at analogies of relations between certain elements (persons, groups, institutions, etc.) and the more 'inclusive' social system to which they *belong*. Here again the prior assumption is that the compared societies – constituting the criteria for our relations – are analogous with respect to certain of their characteristics. Once we assume this, we may distinguish different kinds of relations between a social system and certain of its components.

(4) We may say then that the (phenomenally more or less different) objects, groups or persons are *contextually equivalent*, meaning by this that they are the elements of or *belong to some 'higher level' groups* or systems so defined. Thus from this point of view all members of the same school class are 'contextually equivalent', independently of their individual school achievement, and two students from different schools may also be contextually equivalent if they are members of school classes characterized as analogous, e.g. with respect to their average academic achievement. Persons belonging to quite different cultures may then on the basis of certain definitions be contextually equivalent if we find, for example, that their cultures are similar with respect to certain features important to us.

(5) Once we decide to include two groups or systems under a common conceptual category we may later on try to distinguish certain elements (whatever the phenomenal differences between them) which have the same function which we might call *functional equivalence*. Durkheim's definition

of religion is a typical case of definition in functional terms only: according to it, religion is any system of beliefs which is able to play the role of integrating the society around it, whatever its content might be. The concept of 'educational institution' mentioned above belongs, of course, to this category.

I would like to stress that the notion of functional equivalence as defined here is narrower than it sometimes is in discussions of the methodology of comparative research, where the notion of functional equivalence covers various kinds of relational equivalences discussed here.

(6) Another category of equivalence is that of *structural equivalences*. To say that two persons or groups in different societies are structurally equivalent presupposes that we are able to specify certain structural analogies between the two societies, and later on to demonstrate that *within the structures so defined* the compared objects occupy the same position. The character of the structural analogy and the degree of preciseness of our description of the structures compared depends entirely upon the goal of our study. If someone were to compare a poor Indian village with a rich suburban community in the United States, to look at the income distributions in these two communities, and finally to say that the richest peasant in the village is 'structurally equivalent' to an American millionaire living in the U.S. community, he would be, of course, formally correct. The only thing is that we doubt whether the concept so defined would have any scientific use. The concept of the 'elite of economic decision making' discussed previously is a case in point of structural equivalence.

We can look at the problem of relational equivalence from quite another perspective altogether by inquiring into the nature of the *causal relations* between the criterion variable and its relational correspondent. And here we can distinguish three different situations:

(7) In the first kind of situation we state only that phenomenally more or less similar characteristics are correlated (statistically or in a more loose sense) to certain criterion variables which either are phenomenally identical or have been found to be analogous in previous research, while at the same time we are unable to say anything else about the nature of the causal relations underlying these correlations. Here we can speak of the *correlational equivalence*[11] of the compared phenomena.

(8) The phenomena may be termed *causally equivalent* if they have *similar effects* in the systems compared. Needless to say, the *functional*

equivalences are a subclass of causal equivalences: from the totality of effects of certain variables we concentrate in the case of functional equivalence on those which are relevant for the stabilization of the system or contribute in an essential way to its 'normal' functioning. If we concentrate on the *dysfunctions* we look at those effects which undermine the system or contribute to its more or less rapid transformation.

(9) Finally, we can say that the phenomena are *etiologically equivalent* if we come to the conclusion that − although phenomenally different − they are the effects of identical causes in different social contexts. In some situations the identification of phenomenally different effects of the same cause can be scientifically important.

It should be remembered that the foregoing typology of relational equivalence does not assume a disjunctive classification. The phenomena compared may have more than one relation in common and the concept may involve equivalence in the respect to several relations at the same time. The Marxist notion of 'social class' stresses etiological equivalence, structural and contextual equivalence, various functional equivalences, and sometimes also the cultural equivalence of groups which are called 'classes'.

With respect to different relations to the same criteria and with respect to different criteria, it seems unnecessary to point out *different* groups, persons or cultural and institutional patterns *will be found equivalent* among the societies compared. It is also obvious that unless we specify as clearly as we can the goal of our comparison, it is impossible to decide which criteria and which relations should be primarily taken into account.

Let us now look at the concept defined in absolute, phenomenal terms as compared with relational ones. When we look more closely at this problem it turns out that the differences between these two categories of concepts are less profound than one might believe. As I said above, once we find out which phenomena are related in the given way to the given criterion in the given society, the phenomenal aspects of the given relational concept are clear to us; whether we redefine this concept in phenomenal terms, or have the phenomenal aspects of its meaning implicitly in mind, is more or less irrelevant − we know in any case what it refers to. On the other hand, the concept defined primarily in terms of its absolute, phenomenal properties is defined thus in order to be useful for certain scientific purposes. If such concepts are useful, this is often because their referents are related in a specific way to certain other phenomena, and the goal of the study is to discover and describe these relations. As I said above, the precise meaning of a concept is not identical with its usefulness, and these two aspects should be treated separately. But when we ask *why* we decided to define (in absolute terms) a certain concept in the given way, why we decided to include in its meanings certain characteristics and to disregard certain others, it may turn out that this is simply because the referents of the concept are related in a given way to some other phenomena — which assures our concept theoretical fruitfulness. To give a most simple case: We decide that we should define a certain concept in terms of a set of phenomenal properties because its referents are causes (or effects) of phenomena of some other category; by defining our concept in the given way we are able later on to formulate a 'nice' causal theory. *The fruitfulness of the given concept in this case* (and in the case of all theoretical concepts) means the number of generalizations in which our concept may validly occur. These generalizations refer of course to certain relations. Therefore we may say that theoretical concepts defined in absolute terms have additionally certain implicit relational meanings; this relational meaning is equivalent to the concepts' theoretical fruitfulness. In a certain way the relational and phenomenally defined concepts constitute rather different starting points for our work on the theory of a given science: in a more or less complete theoretical structure the absolute phenomenal aspects of meanings of theoretical concepts and the relations in which such concepts occur (or which are involved in their meanings) are closely interconnected with each other.

3. DECLARED VS RECONSTRUCTED MEANING OF ATTITUDES

The first sections of this chapter deal with the general logical problems of the comparability of research variables. I distinguished first the absolute and the relative properties, and discussed various types of relational equivalences. One of these was 'correlational equivalence'; the phenomena are relationally identical if they are correlated in the same way with certain 'criterion variables', This notion may be useful in the comparative studies of attitudes.

In one sense attitude is an 'absolute property' of a person. By this I mean that in order to determine whether a person has this or that kind of attitude (when 'attitude' means a certain state of mind) we do not have to assess anything else, but only certain aspects of the state of mind of this person defined in phenomenal terms. Depending on the definition of the given attitude, we may take into account a great many or only a few aspects of this psychological phenomenon.

But as with any other absolute property, attitude can be redefined in a relational way. We can say, for example, that by a positive attitude toward a certain object we mean a set of predispositions 'responsible' for the 'positive behaviors' toward that object. Here we interpret attitude as a relational concept, having in mind the motivational relations between any psychologically understood attitude and positive behavior toward a certain object whatever the 'absolute' characteristics of this attitude might be. We can also look for the *correlational equivalences* of certain attitudes. This method is especially useful when we are dealing with 'general attitudes' usually expressed verbally by some vague declarations of 'positiveness' (or 'negativeness') toward something and which are often for us (and sometimes for the speaker himself) free of more concrete connotations. We may then try to relate this attitude to more specific and less ambiguous opinions, evaluations and predispositions existing in the speaker's mind and to see whether in the minds of different persons this 'general attitude' as expressed by certain general and vague declarations coexist with the same criterion variables indicated by certain 'verbal behaviors' of the studied persons in which more concrete opinions and predispositions are expressed.

Let us now concentrate on these verbal indicators, which we usually obtain in a questionnaire or interview. Suppose that someone expresses in an interview a certain attitude the content of which is not clear to us, and the only thing we know is his 'verbal behavior' which is for us the indicator of this vague attitude. In order to grasp the possible differences or identity of attitudes corresponding to the same verbal indicator by two or more persons we usually proceed in one of two ways:

(A) We simply ask the respondent, 'What do you mean by saying this?' and on the basis of his answer to an open-ended question in our questionnaire or on the basis of a detailed report of an intensive interview we have the *declared psychological meaning* of the term denoting our behavioral indicator.

This can be done in many different ways. We may use a certain probing question, relying on the reflectiveness of our respondents. Or we can add some more or less strict rules for the control of the declared meaning. The most developed technique of such control is the Osgood-Tannenbaum technique of Semantic Differential[12] where the respondents are asked to locate their attitude toward a certain object at a certain number of 'semantic continua', thus specifying as precisely as possible the content of this attitude.

But there are reasons in some research situations for trying to avoid any type of probing. There are also studies in which the probing (or the use of Semantic Differential) was not applied in the interview and the question of the meaning of certain attitudinal items is an open one. In this situation we may apply another technique.

(B) We can put into our questionnaire or interview schedule many items revealing (expressing) in much more concrete way various perceptual, emotional and 'behavioral' aspects of the subject's attitude toward x. On the basis of answers to these questions we can locate our 'initial indicator' of this vague attitude in the context of other indicators expressing very detailed components of the subject's attitude toward x. In this way we can reconstruct the psychological meaning of the term denoting our initial behavioral indicator of the attitude, even without asking the subject, 'What do you mean by saying that?'

The differences between the declared (by the respondent) and the reconstructed (by the scientist) psychological meanings of the term denoting our initial attitudinal indicator seem to be rather important, and there are several possible relations between these two constructs.

(1) There may be full equivalence between the psychological connotation of the given 'attitude' as declared by the respondent and the connotation of this term as reconstructed by the scientist on the basis of analysis of the respondent's state of mind.

(2) The scientist and the respondent may agree as to the content of these two constructs but at the same time differ as to the relative importance of the components. e.g. their motivational force. They may agree as to the existence in the respondent's mind of a given set of feelings, convictions or dispositions, but disagree as to their mutual connections and their connection with the initial indicator, as well as to the general meaning of their attitude. What the respondent is inclined to declare as the basic motive

of his attitude toward x the scientist may be inclined to treat as only a rationalization, discerning the 'real motives' in components which seem to the respondent to be purely marginal and non-essential.

(3) In some cases the respondent might accept the whole structural construction of the scientist, but object to the *term* he uses to denote this construction. Thus, for example, many subjects of the study, *The Authoritarian Personality* would probably oppose such constructs as the F-Scale and would also oppose the statement that they are 'Fascists' or 'Ethnocentric', even while not denying that they possess all the components of the phenomenon which the authors term Fascist or Ethnocentric orientation.

(4) If the respondent's answer are used as indirect indicators of deeper traits, hidden motives, repressed drives, etc., the difference between the declared and the constructed meanings of the given attitude may be that of the existence or non-existence of the given traits in the respondent's mind.

It is also obvious that two people who express their attitudes toward x by the use of different verbal or non-verbal indicators may later attest to (or the scientist may later attribute to them) the same psychological reality related to different initial indicators.

Thus we see that: (a) the same expression may have either the same or a different meaning from the perspective of the subject and the perspective of the scientist who tries to assess its meaning by his techniques; and (b) at the same time different verbal indicators may have identical or different connotations from the standpoint of these two perspectives.

When the number of subjects in a study is more than a handful, it is practically impossible to construct a picture of the psychological reality corresponding to the given indicator separately for every case. We are therefore compelled to apply *statistical analysis* in order to discover what kind of psychological meaning should be attributed to the given indicator for the members of the given population. This can be done in order to obtain both the *declared* meaning based on the relationship between the initial indicator and the answers to the probing questions and the *constructed* meaning based on analysis of the relationships between the indicator and other items of the questionnaire concerning attitudes toward x.

The statistical approach is more than just an enforced necessity; it has some advantages of its own as well. For after we have classified our respondents from the standpoint of the items expressing a certain attitude, our next step is to cross-tabulate the variable initially denoting the general attitude with the other variables representing specifications of the meaning of this initial indicator. This procedure usually shows that some of the variables representing the 'components of meaning' of the 'general indicator of the attitude' are on a mass scale significantly related statistically to this indicator and that their relationship is a positive one. But other ones which might be attributed by the scientist to some individuals as the psychological correlates of their general attitude may turn out to be either statistically independent or even negatively correlated with the initial indicator on a mass scale. It may happen as well that some psychological characteristics of the respondents which in a study of separate individuals might not be related by the scientist to the given indicator – their coexistence with the general attitude having been for him purely 'accidental' – are shown by mass-scale analysis to be strongly correlated both with the initial indicator of the attitude and with other clustered phenomena.

Sometimes there may be low or no correlation between components of the meaning of the attitude as declared by the respondents and the initial verbal indicator of this attitude, because people classified by the initial indicator as belonging to different 'attitudinal groups' do not differ on these more specific items. It may happen too that some important traits not related by the respondents to their general attitude will later on be discovered as correlated with its general indicator.

In the case of such a divergence between the results of correlational analysis and the scientist's assessment of individual declarations, the scientist may still insist that for his case the real psychological meaning of the variable denoting the initial indicator is as stated previously, either by him or by the individual respondents. He is then obliged however, to explain why he treats his individual cases as exceptions from the general pattern of relationship between variables existing in the given population.

For the population as a whole it seems reasonable to distinguish, among all the items concerning x, those and only those items which are positively correlated with the initial indicator, and then to put them together as the inferential indicator of psychological connotation of the term denoting the of general attitude indicator for this population.

What follows in this chapter is an example of such a procedure.

4. IDEOLOGICAL CONNOTATION OF ATTITUDE TOWARD 'SOCIALISM' AMONG WARSAW STUDENTS – A CASE STUDY OF 'MEASUREMENT OF MEANING' OF POLITICAL ATTITUDES

In 1958 and 1961 a representative consecutive survey was carried out among the students of Warsaw schools of University level.[13] Most of the items of the questionnaire were concerned with students' attitudes toward diverse aspects of socialist ideology. The authors of the questionnaire were concerned not only with getting very specific and detailed views of the respondents on some concrete aspects of socialism but also with constructing an indicator which might measure a *general attitude toward socialism*. After long deliberations and extensive testing of this item in the pilot study, this general indicator was formulated as follows: 'Would you like the world to move toward some form of socialism?'

In 1958 we received the following answers: Definitely yes – 24%; qualifiedly yes – 44%; no opinion – 19%; qualifiedly no – 9%; definitely no – 2%; N.A. – 2%. On the basis of these answers we were able to say only that the respondents had a distinct tendency to react favorably toward the term 'socialism'. But what did they mean by that term?

Fortunately we had many much more specific items in our questionnaire measuring various concrete attitudes of our respondents toward social and political issues. In order to discover the meaning of 'socialism' for Warsaw students, all of these ideological items of our questionnaire were cross-tabulated with our indicator of 'pro-socialist' orientation, and the level of significance and a coefficient of contingency (Tschuprov's *T*) was calculated for every relationship. The results of these calculations and cross-tabulations are given in Table I.

A general picture of the social ideology of Warsaw students may be seen from the lower marginals of this table. I do not propose to analyze its main features. Instead, let us see in what specific ideological characteristics the 'definite socialists' differed from the 'moderate socialists' and the 'socialists' as a whole different from the 'non-socialists'. In other words, let us see what was the *ideological meaning* of this indicator, as revealed by its statistical relationships.

When we distinguish those items which are significantly related to our indicator and rank them according to the strength of the relationship (as

Relationships between the indicator of general 'attitude t‹

The wording of the item	Categories of answers
Do you consider yourself to be a Marxist?	definitely yes + rather yes (%) definitely no (%)
Do you think one should risk one's life for social ideology?	yes (%)
Do you feel the need for social and political activity?	definitely yes + rather yes (%)
Do you agree with those who say that in the second half of the 20th century 'patriotism' is a worn out idea?	definitely yes + rather yes (%) definitely no + rather no (%)
How do you think historians in forty or fifty years will judge the period 1945-1955 in Poland's history?	definitely favorably + rather favorably (%) definitely unfavorably + rather unfavorably (%)
Do you think it permissible for the State to limit civil liberties for the attainment of the important social goals?	yes, even for a longer tim‹ yes, but the time of lim‹ should not be too long
Are you in favor of permitting unlimited 'free enterprise' in the following branches of the national economy? (definitely no + rather no (%)	handcrafts small-scale trade small industrial enterpris‹ wholesale trade medium industrial enterp‹ foreign trade heavy industry large land states
Index of acceptance of nationalized economy	mean score of index (range of index score 0‹
What role should the Workers' Councils play in our industry?	they should be the real m‹ gers of the factories (%)‹
Do you think it is important to try to eliminate the exploitation of some people by others?	very important (%)
Do you agree with those who say that the differentiation of salaries is an indirect form of exploitation of some people by others?	definitely yes + rather yes (%)
Do you think it right that everyone in the country should have more or less the same income?	yes, and should be put in‹ practice at an early date‹ yes, but should be put i‹ practice slowly and care‹
Do you think that the present range of wages and incomes in Poland should be now limited?	definitely yes + rather yes (%)
What should be the upper limit of highest monthly income of one working person in Poland½	arithmetic mean of the p‹ posed upper limit (zloty‹ per month)
No. of persons = 100%	

...alism' and other ideological items in Warsaw student's ideology

	Would you like to see the world to move towards some form of socialism?						
total sample	definitely yes	rather yes	no opinion	rather no	definitely no	Level of significance	Tchuprov's T
3	35	9	2	3	–	0.001	0.25
4	22	31	43	61	86		
4	49	30	29	15	7	0.001	0.17
0	66	47	45	32	21	0.001	0.16
5	9	15	20	21	14	0.05	0.09
6	85	70	70	66	78		
8	47	24	24	10	–	0.001	0.225
1	52	73	72	89	100		
3	34	21	17	11	21	0.01	0.10
4	1	1	1	2	–	not calculated	
0	28	20	15	13	7	for separated	
5	36	25	20	8	–	items	
4	76	60	57	35	21		
4	78	62	54	37	50		
0	83	72	66	43	43		
5	93	87	73	73	64		
5	87	73	71	67	29		
5	5.30	3.79	3.25	2.85	0.001	0.001	0.17
5	63	54	46	45	36	0.05	0.09
2	97	86	76	66	77	0.001	0.18
2	37	31	23	31	9	0.05	0.09
4	36	22	25	13	7	–	0.07
5	49	45	42	46	35	–	0.09
5,847	6,262	6,882	6,192	7,222	6,000	0.05	0.10
733	181	325	140	63	14	–	–

measured by T), we see that for the population of Warsaw students in 1958 'to want the world to move toward some form of socialism' meant the following:

(1) They evaluated less negatively than the majority of the students all the political and social changes which had taken place in Poland during 1945-55 (T=.55).

(2) They were less non-Marxist than the average student (T=0.25).

(3) They had a stronger conviction that the 'abolition of the exploitation of some people by others is very important' (T=0.18).

(4) They were more ready than the average to accept the system of nationalized economy (T=0.17).

(5) They were more convinced that it is proper to risk one's own life in defense of social ideology (T=0.17).

(6) They felt more than the others a need for political and social activity (T=0.16).

These first six items are strongly related to our indicator and constitute the *nuclear correlational meaning* of the term 'socialism' in the population of Warsaw students. But another group of items should be added to them as constituting more *peripherical areas of the meaning* of the term, and as having much weaker relationships with our indicator. Thus 'the socialists' were also to a lesser degree characterized by the following attitudes.

(7) They were a little less sensitive[14] than the rest of the students about the limitations of civil liberties by the State in order to obtain important social goals (T=0.10).

(8) They were rather more than the others in favor of full authority for Workers' Councils in factories (T=0.09).

(9) They were a little more patriotic — although the same is true of the definite anti-socialists, which would mean that we have two kinds of 'patriotism' in our population (T=0.09).

(10) There were a little less opposed to the opinion that the differentiation of wages constitutes an indirect form of economic exploitation — although this view was not too popular among the students overall (T=0.09).

As we can see from Table I the last three items, concerning economic equality among individuals, cannot be included in the constructed meaning of the pro-socialist attitude as indicated by the question: 'Would you like to see the world move toward some form of socialism?' The 'prosocialist' students did not significantly differ on these items from the 'non-socialist' ones.

5. 'MARXISM' AND 'SOCIALISM' – TWO VARIABLES WITH MEANING AND DIFFERENT INTENSITY

Table I shows that the population described here had a rather strong pro-socialist orientation, and that their 'socialism' might be characterized mainly by the acceptance of some traditional elements of socialist ideology (especially a nationalized economy) and by their attitudes toward some political aspects of post-war Poland.

In contrast to 'Socialism', the term 'Marxism' was not too popular among Warsaw students. When asked: 'Do you consider your self to be a Marxist?' the students answered as follows: Definitely yes – 2%; qualifiedly yes – 11%; no opinion – 17%; qualifiedly no – 34%; definitely no – 34%.

This difference in the responses to 'Marxism' and to 'Socialism' may be attributed partly to the philosophical aspects of Marxist ideology (which the religious majority of the students found difficult to accept) and partly to the very dogmatic form of Marxism which the subjects knew from their own personal experience prior to 1956.

The study then tried to find out (in the same way as described above) what the students 'meant' when they said they were definite or moderate Marxists, or definite or moderate non-Marxists.

Curiously enough the general pattern of correlation between Marxist items and the rest of the ideological items of our questionnaire was almost identical to the 'Socialist' pattern of relationships. This was true both for the content of the related variables and for the rank order of their coefficients with 'Marxism', So the obvious conclusion was that the correlational meanings of 'Marxism' and 'Socialism' where very similar. Upon studying the correlational equivalences of these two items one is tempted to say that *they expressed a very similar ideological continuum.* But were they expressing also *the same intensity* of the 'underlying continuum'? To say it more precisely: Did the identically worded answers to these two items (e.g. 'positively yes', etc.) express the same values of the indicated variable – or did the analogically worded answers to correlationally equivalent items indicate different points of the underlying dimension in the same way that

100° on the Fahrenheit scale indicates a lower temperature than 100° centigrade when both of them measure 'temperature'?

The first indication that these two items might measure different intensities of the same variable was their mutual relationships: we found that the items referring to 'socialism' and 'Marxism' were related to each other in a Guttman-scale continuum. Warsaw students showed the following distribution on this scale (in percentages).

Marxist Socialists	12
Non-Marxist Socialists	56
Non-Marxist Non-Socialists	31
'Errors of the scale'	
i.e. Non-Socialist Marxists	1

The conclusion that these two items refer to *different points of intensity of the same latent variable* was confirmed even more strongly when we compared the relative frequencies of positive answers to our controlling questions (constituting correlational criteria of meaning) within the groups of subjects distinguished by their positive answers to 'socialism' and 'Marxism'.

As we can see from Table II, the 'definite Marxists' are more likely to accept many of the items of our questionnaire than are the 'definite socialists'. The same is also true for 'moderate Marxists' and 'socialists'. Analysis of Table II also shows that in their responses to most of the items of our questionnaire the 'definite socialists' were more similar to the 'qualified Marxists' than to the 'definite Marxists'; from the standpoint of correlational meaning the answer 'definitely yes' to the question concerning 'socialism' was equivalent to the answer 'rather yes' to the question concerning 'Marxism'; the definitely positive answer to the question concerning 'Marxism'; the definitely positive answer to the 'Marxist' items had no correspondence on the 'socialism' scale.

Thus we see that we are able not only to discover the similarity (or differences) of general correlational meaning of attitudes corresponding to two different indicators, but also to evaluate the *relative equivalence of answers* to both indicating questions from the standpoint of the degree of 'loading' of these answers with convictions, emotions or dispositions to act, which constitute the given attitude.

We have seen from the above that on the basis of analysis of the pattern of relationships between our initial indicator of an attitude and other

TABLE II

Comparison of relative equivalence of corresponding
answers to 'Marxist' and 'Socialist'
items, among Warsaw students

Answers among (%)

The item and the category of answers	'definite Marxists'	'definite Socialists'	'rather Marxists'	'rather Socialists'
In your opinion is it proper to risk one's life for social ideology? (% – yes)	62	49	49	30
Do you feel the need for social and political activity? (% – definitely yes + rather yes)	87	66	70	47
Do you agree with those who say that in the second half of the 20th century 'patriotism' is a worn out idea (% – definitely no + rather no)	92	85	82	70
How do you think historians in forty or fifty years will judge the period 1945-55 in Poland's history½ (%– definitely favourably + rather favourably)	69	47	48	24
General index of acceptance of the nationalized economy (% – total sum of scores 4–7, i.e. more approving nationalization)	85	83	80	67
What role should Workers' Councils play in our industry? (% – they should be the real managers of factories).	69	63	58	54
Do you think it is important to try to eliminate the exploitation of some people by others? (% – very important)	100	97	89	86
Do you agree with those who say that the differentiation of salaries in an indirect form of exploitation of some people by others (% – definitely yes + rather yes)	46	37	37	31
Do you think it right that everyone in the country should have more or less the same income? (% – categories of answers, see Table I)	38	36	27	22
Do you think that the present range of wages and incomes in Poland should be now limited? (% – definitely yes + rather yes)	62	49	52	45

attitudinal items of the questionnaire we can at least partly solve three rather important problems of cross-cultural surveys:

(1) Whether and to what degree an *identically worded* question really has the same meaning for members of the different populations.

(2) Whether and to what degree two *differently worded* questions have the same meaning for members of the different populations.

(3) Whether and to what degree we can say that *corresponding answers* to (identical or differently worded) questions for which the similarity of correlational meaning in both populations has been confirmed also have an equivalent 'intensity', and to what degree the intensity of attitudes expressed by identically worded answers to questions with similar correlational meanings is different.

Thus we see that the use of correlational analysis may help us to control the meaning of indicators of some more or less vague attitudes used in cross-cultural surveys independently of whether these indiators are worded in the same or in a different way. This is true, of course, only under the assumption that the items which are used for the control of meaning of the other more vague items possess relatively the same (or sufficiently similar) meaning in the compared cultures, i.e. that they themselves are equivalent. This is not always the case, of course, but the assumption seems to be more valid in application to items which refer to more concrete specific problems than to those which are indicators of more general (and usually more vague) orientations and attitudes.

NOTES

* The first three sections of this chapter constitute the continuation of ideas from another paper: 'The Strategy of Cross-National Survey Research for the Development of Social Theory' presented at the international conference on cross-national survey research organized in 1973 in Budapest by Centre Europeen de Coordination de Recherche et de Documentation en Sciences Sociales in Vienna. The last two sections presenting a case study of 'correlational control of meaning' are based on the paper 'Correlational Control of Meaning of Attitudinal Variables in Cross-Cultural Surveys' presented at the Conference on Cross-National Survey Research organized by International Social Science Council in La Napoule (France) and published in *The Polish Sociological Bulletin*, 5–6, 1962.
[1] See: R.L. Merritt and S. Rokkan (eds), *Comparing Nations; The Use of Quantitative Data in Cross-National Research*, New Haven, 1966. See also: A. Przeworski and H. Teune, *The Logic of Comparative Social Enquiry*, New York, 1970.

[2] For the analysis of problems of cross-national comparability of research variables see: A. Przeworski and H. Teune, *op. cit.,* 'Equivalence in Cross-National Research', *Public Opinion Quaterly,* Vol. 30 (1966–67). See also: S. Verba, 'Cross-National Survey Research – The Problem of Credibility'; in: I. Vallier (ed.), *Comparative Method in Sociology,* University of California Press, 1970.

[3] S. Verba, *op. cit.*

[4] See: Chapter I.

[5] E. Jacobson, 'Methods Used for Producing Comparable Data in the OCSR Seven Nations Attitude Study', *Journal of Social Issues,* Vol. I (1956).

[6] For the discussion of distinction between absolute and relational properties see: P.F. Lazarsfeld and H. Menzel, 'On the Relations between Individual and Collective Characteristics', in P.F. Lazarsfeld, A.K. Pasanella and M. Rosenberg, (eds.), *Continuities in the Language of Social Research,* New York, 1972.

[7] Ph. Jacob et al., *Values and the Active Community,* New York, 1971, p. 242.

[8] Ph. Jacob et al., *op. cit.,* p. 243.

[9] See: Chapter I.

[10] See: S. Verba, *op. cit.*

[11] For an excellent analysis of different correlational techniques involved in the assessment of this kind of equivalence see: A. Przeworski and H. Teune, *The Logic of Comparative Social Enquiry.*

[12] Ch.E. Osgood, G.J. Suci and P.H. Tannenbaum, *The Measurement of Meaning,* Urbana, Ill., 1957.

[13] For a more detailed analysis of the content of Student's Attitudes see: S. Nowak, 'Social Attitudes of Warsaw Students', *The Polish Sociological Bulletin,* 1-2(3-4), 1962. See also: S. Nowak, 'Factors Determining Egalitarian Attitudes of Warsaw Students', *The American Sociological Review,* April 1960.

[14] We understand of course that formulations like 'less sensitive' and 'more in favor of' are used here in a rather metaphorical sense; they refer to the differences in frequencies of the given answer between diverse groups and not – at least directly – to differences in intensity of the given attitude at the individual level.

BIBLIOGRAPHY

Jacob, Ph., *et al., Values and the Active Community,* New York, 1971.

Jacobson, E., 'Methods Used for Producing Comparable Data in the OCSR Seven Nations Attitude Study', *Journal of Social Issues,* Vol. I, 1956.

Lazarsfeld P.F., and H. Menzel, 'On the Relations between Individual and Collective Properties', in P.F. Lazarsfeld, A. Pasanella, and M. Rosenberg (eds.), *Continuities in the Language of Social Research,* New York, 1972.

Merrit, R.L., and S. Rokkan (eds.(, *Comparing Nations; The Use of Quantitative Data in Cross-National Research,* New Haven, 1966.

Nowak, S., 'Correlational Control of Meaning of Attitudinal Variables in Cross-Cultural Survey', *The Polish Sociological Bulletin* 5–6, 1962.

Nowak, S., 'Factors Determining Egalitarian Attitudes of Warsaw Students', *The American Sociological Review,* April 1960.

Nowak, S., 'Social Attitudes of Warsaw Students', *The Polish Sociological Bulletin*, 1–2 (3–4), 1962.

Nowak, S., 'The Strategy or Cross-National Survey Research for the Development of Social Theory': paper presented at the International Conference on Cross-National Survey Research experienced in Budapest in 1973 by Centre Europeen de Coordination de Recherche et de Documentation en Sciences Sociales in Vienne.

Nowak, S., 'Concepts and Indicators in Humanistic Sociology', in this volume, p. 1.

Osgood, Ch. E., G.J. Suci, and F.H. Tannenbaum, *The Measurement of Meaning*, Urbana, Ill., 1957.

Przeworski, A. and H. Teune, 'Equivalence in Cross-National Research', *Public Opinion Quarterly*, Vol. 30 (1966–67).

Przeworski, A., and H. Teune, *The Logic of Comparative Social Enquiry*, New York, 1970.

Verba, S., 'Cross-National Survey Research, The Problem of Credibility', in J. Vallier (ed.), *Comparative Method in Sociology*, Univ. of California Press, 1970.

CHAPTER IV

COMPARATIVE SOCIAL RESEARCH AND
METHODOLOGICAL PROBLEMS OF
SOCIOLOGICAL INDUCTION*

1. DIFFERENT ASPECTS OF GENERALITY OF THEORETICAL SOCIAL PROPOSITIONS

In the last decades more and more social scientists seem to agree with the opinion expressed by R. Marsh in his book, *Comparative Sociology*, that:

Cross-social comparative analysis is fundamental to any general sociological or anthropological theory.[1]

Evidence of this understanding may be found in the increased number of researches of cross-national type, in the development of many data archives, of which the Human Relation Area Files at Yale University are the most important example, and finally in the moral and financial support for this type of study by international organizations like UNESCO. Even if we agree with those who say that so-called comparative research is only a special category of techniques of inductive verification of social hypotheses[2], the increased number of studies, testing the validity of theoretical propositions against broad cross-national or cross-cultural samples is in itself an important phenomenon in the development of social sciences.

Cross-social comparative studies may be undertaken for different purposes. Sometimes they are made because the scientist simply wants to know the degree of similarity or difference between the structures or cultures of two or more societies in which he is interested. Sometimes a large number of societies are being compared because the scientist is interested in establishing the range of possible variation and the number of possible sub-types of the given category of behavior or institutional pattern. But in most cases a more or less direct goal in such studies is the testing of validity and *evaluation of the degree of generality* of a theoretical proposition.

But it should be recalled here that the term 'degree of generality' of a proposition may have different meanings.

First, propositions may differ in the degree of their *theoretical generality*. This is the case when the propositions refer to roughly similar mechanisms (i.e. they have the same dependent variable) but some of their concepts (i.e. those referring to the independent variables) are so related that the concept of one proposition is found (either on the basis of some terminological conventions or due to empirical findings) to be more general than the other one. Thus a 'working team' may be defined as a special category of a 'task-oriented group'; 'racially prejudiced person' may be defined as a special category of 'ethnocentrically oriented person'; 'army' or 'corporation' may be characterized as a special subclass of 'organizations'; 'relative deprivation' may be found as a special category of 'frustrating situations'; etc.[3] Then the proposition referring to a more general class of phenomena or situations is theoretically more general than the other one.

Secondly, the propositions we formulate in any science (including social science) may be classified into two groups – namely *general* propositions and *statistical* propositions. The general propositions attribute a certain property or regularity of behavior to all objects or phenomena of a certain class – the statistical propositions describe them in terms of relative frequencies or relative frequencies 'in the long run', i.e. in terms of probabilities. This is of course a strictly dichotomic classification, but when looking at the statistical propositions, we can classify them in terms of the degree of their empirical generality, i.e. by the degree of their approximation to general propositions (to the probability equal to 1).

Finally, both general and statistical propositions may be classified into two categories – *historical propositions*, i.e. propositions which refer to relationships which are limited to some time-space areas only (e.g. they are valid for one society and not valid for others) – and universal propositions – called also the *universal laws of science*. Their validity is free from any time-space limitations. By their definition they apply to any area of space and to any moment of time.

When two historical propositions differ in the degree of their *historical generality* it means that they differ in the extension of time-space coordinates of their applicability which, on the other hand, is determined by the degree of historical universality of concepts appearing in these propositions. If one generalization describes regularities of political behavior within the 'historical area' called 'Contemporary Northwestern Europe' while the other refers to the same mechanisms within 'Scandinavian Countries' only, the first generalization is 'historically' more general than the second. The fact that the time-space coordinates have been replaced by proper names equivalent to the given time-space coordinates does not change anything here.[4]

When we say that comparative social research is undertaken usually with the aim to test the degree of generality of a proposition, the term 'generality' refers here usually to the 'historical' dimension, i.e. it refers to the problem whether a given regularity can be termed a universal law of science or a historical generalization only, and — if it is a historical proposition — how universally it should be formulated, given the assumption that in science we try to formulate the propositions of the most universal character possible.

The last formulation deserves some comments. The scientist may legitimately aim his research toward the verification of a historical hypothesis, or a set of such hypotheses. The main object of study presented in *The American Soldier* was the analysis of the functioning of one specific organization at one specific moment in time, i.e. the American Army in World War II. The main goal of Jacob Burckhardt's famous study *Renaissance Culture in Italy* was the study of this culture itself, while American racism was the object of interest in Gunnar Myrdal's *An American Dilemma*.

But sometimes a scientist is compelled to formulate a historical proposition (let us call it historical generalization, although we remember that it may also be a statistical one) even when he would like to have a theoretical proposition of unlimited validity — a universal law of science. The folk-urban continuum — interpreted as a set of propositions about the covariation of a set of community traits — was advanced as a universal theory, but as is now becoming obvious, its validity is historically limited. Migration patterns within the city and from the city to suburbia are obviously limited in their historical generality even though urban sociology would be very happy to have as general a theory as possible. Many regularities of economic behavior seem to end at the frontiers of contemporary industrial states. In all these cases, the scientist — even if he aimed toward a universal law — has had to present his hypothesis as a generalization of limited historical validity because he knows that outside certain time-space coordinates his generalization is obviously false.

As we know, most of the findings in social sciences, besides being historically limited, are also statistical in their nature, i.e. they refer to regularities which are neither general nor universal. And now the problem can be posed: how can this fact be interpreted in terms of causal principles — to what degree this can be explained as congruent with the notion of the

causal character of social phenomena — which by definition postulates
regularities of both general and universal kind? This is one of the questions
which I would like to discuss in my paper. The other question — of more
practical character — is the problem of fruitful strategy in comparative
social research which in my opinion is strongly related to the discussion and
proper understanding of the nature of regularities governing social phenomena and human behavior.

2. CONDITIONAL CAUSAL RELATIONS AND
THEIR OBSERVABLE CONSEQUENCES

By proper understanding of the notion of casuality I mean here the clear
distinction between *unconditional* and *conditional* causal relations. Let us
suppose that we are considering causal relations between two classes of
events or properties, S and B. For example, S might be understood as
'stimulus' and B as 'behavior', although these symbols may also mean any
type of cause and its effects. When we say that S is the cause of B, we
usually assume that S *precedes B in time*. Secondly, we usually assume that
their time sequence is necessary. The idea of the necessary character of
causal relations seems to imply such terms as 'always' of 'never'. Therefore
we would expect that if S is the cause of B, S is always followed by B,
which means that there is a *necessary relation of a positive kind* between S
and B, in other words, S *is a sufficient condition for B*. Or we could have in
mind a *necessary relation of a negative kind*, saying that 'non-S is never
followed by B' which means that S *is a necessary condition for B*. Relations
which imply the use of the terms 'always' or 'never', I call here *unconditional causal relations*. Their proper meaning is: "Whatever occurs
jointly with S and in whatever context or situation S occurs, it is always
followed by B". The same holds for unconditional causal relations in the
negative sense of this term: "Whatever would occur jointly with non-S, or in
whatever context non-S would occur, it will never be followed by B".

For reasons of brevity, let us discuss only causal relations in the positive
sense.[5] It is obvious that in empirical sciences we very seldom meet
propositions which would correspond to the notion of the unconditional
causal law. The propositions which we usually meet in sciences have usually
either explicitly the form: 'S is followed by B, *provided* that another event

D also occurs' or they should be understood as implying that 'S is followed by B only when some other events $D_1, D_2, ..., D_n$ also occur'. Therefore S is not a sufficient condition for B, but only an essential component of such sufficient condition. Supposing that the condition to which the causal relation between S and B is relativised is the property D (one could exemplify it by the term 'disposition' to the given response B to the stimulus S) – the joint occurrence of S and D is of course a sufficient condition for B.

Let us call them conditional causal relations. The conditionality of a relation between S and B might be stressed by the following formula:

$$\underset{x}{\wedge} D(x) \rightarrow (S(x) \rightarrow B(x))$$

which is of course equivalent to another formula, more suitable for stressing that S and D are sufficient jointly for

$$\underset{x}{\wedge} (S(x) \cap D(x)) \rightarrow B(x).$$

If S is only an essential component of a sufficient condition for B, S may be followed either by B or by non-B depending upon the occurrence or non-occurrence of D.[6]

Suppose now that in the empirical reality the relation between S and B is conditional and depends on the occurrence of D, but the scientist does not know the role of D for this relation. Will he then be able to formulate any valid scientific proposition about the relation between S and B?

This depends on the *way of occurrence or existence of D in empirical reality*. What do I mean by this?

Let us assume first that the condition D occurs in such a way, that we may say that it is randomly distributed throughout the entire population of x's, so that wherever a certain $S(x)$ occurs it has a definite probability p to occur jointly with $D(x)$.

Since the occurrence of B (provided that S has occurred) is conditional on the occurrence of D, we may say that S will produce B as often, as it occurs jointly with D. But it implies that the *frequency* of the occurrence of B in relation to the occurrence of S is equal to the frequency of the occurrence of D in relation to the occurrence of S. Since in this case we assumed that the relative frequency of D when S occurred is roughly constant in the long run, we may use here the term 'probability'. We can therefore say that in this pattern of conditional causal relations the probability of B given S is equal to the probability of D gives S or symbolically

$p(B/S) = p(D/S)$.

It means that if D is randomly distributed in the populations of x's with a constant probability p that S will occur jointly with D, even without discovering the importance of D the scientist can legitimately formulate a universal statistical law of science saying that there is a definite probability p of the occurrence of B, given the occurrence of S.

But this is not the only way in which our D can occur or exist. D may sometimes occur in such a manner that it is the characteristic of *all human beings but only within some definite time-space region H*, that is, outside these limits D does not occur at all. Sociologists could give many examples of such dispositions with definite 'historical' localizations. It might be, for instance, a disposition which is shaped by the cultural heritage of a given nation in an actual socio-political system, and therefore common to all (or: almost all) members of one nation H in the given time and nonexistent outside these 'historical boundaries' H. If we assume now that the true causal law which could be used to explain the occurrence of B has the form $(S \cap D) \rightarrow B$, this relation is of course *theoretically* both general and universal. It applies without any exception anywhere and to any time-moment. But its *factual operation* is limited to the time-space area H, because only there D occurs even if S could occur anywhere. Also B can occur only within the historical boundaries H.

Suppose now that we don't know the importance of the property D for the relation between S and B and try to find a generalization about the relation of S and B only. In this situation we will not be able to formulate a universal causal law describing the sequence of occurrences of S and B, because, S may be followed by both B and non-B. But even *without discovering the importance of D we can formulate a general historical proposition* which says that within the time-space boundaries H (in the group population, nation, culture etc. H) S is always followed by B.

The historical coordinates H play a substitutional role in this kind of historical generalizations: they are the substitutes for the unknown conditions D of a universal conditional causal relation, permitting us to formulate a true general proposition — although limited in its time-space validity. They also inform us *where* our unknown conditions hold, and where they do not hold. This plays an essential role in the strategy of verificational studies aimed toward discovery of these unknown conditions.

In the above analysis we assumed that 'our' D occurs as a characteristic of *all* members of a given historical population H. As a result we are able to formulate a true historical generalization $S \rightarrow B$, even if we don't know the importance of D for the sequence $S \rightarrow B$. But it may well happen that D exists or occurs in a way which corresponds to a statistical random distribution with probability p, but only in one population H.

If this is the case, our observation permits us to formulate a proposition about the sequence $S \rightarrow B$ which will be both *historical and statistical* (probabilistic) at the same time. This proposition may be again transformed into a universal law of science, if we identify D as a condition which co-determines this regularity.

I would risk the statement that most of the empirical findings in contemporary social research belong to this last category, i.e. we discover at the best definite probabilities or correlations between some independent and some dependent variables and the validity or strenght of these probabilities or correlations is limited to one society only. As we can see, even these propositions can be interpreted as 'first approximations' to some conditional causal relations of universal validity.

3. HISTORICAL AND UNIVERSAL CONCEPTS AND HYPOTHESES IN COMPARATIVE SOCIAL RESEARCH

Let us discuss now a special, but nevertheless important problem, namely the precise difference between historical generalizations and universal laws of science. According to our understanding a proposition is historical when it contains terms *the meanings of which* involve a definite localization of their referents. It is universal when it has been formulated only by using of concepts, the meaning of which is free from such time-space reference or its equivalent (by the use of some proper names). If the concepts of our propositions are defined in universal terms, but we know independently of their meaning that the phenomena or objects denoted by them have definite 'historical' localization, it still does not make our proposition a historical one.

This is not the only way of understanding the universality of a sociological concept, however. Another approach to this problem was proposed by Reinhard Bendix, when he wrote:

The concept of division of labor is universal because we know of no collectivity without such a specialization... Stratification is present in all societies, but stratification by class in only some... Again, the exercise of authority requires subordinate agents everywhere, but their organization in bureaucracy is a more specific phenomenon. Bureaucracy in the sense of Max Weber's concept of governmental organization under the rule of law applies principally to the conditions of Northwestern Europe from the nineteenth century onwards.[7]

One could say that instead of referring to the historical nature of such concepts, it is better to say that some of the universally defined phenomena may possess *limited time-space localization*. This may happen in both the social and natural sciences. 'Automobile' and 'atomic bomb' are universally defined objects, but we are able to demonstrate the historical limitations of their existence. There is no reason why the concepts of social science should be treated differently in this respect.

But when we look closer, the problem becomes more complex. It seems that in all sciences two categories of concepts should be distinguished. The first category of concepts is defined in such a way that the definition specifies a set of properties necessary and sufficient for the recognition of certain objects and phenomena, and nothing more. We define the 'face-to-face group' as a group in which everybody is able to communicate directly with 'all other members of the group'. We say that 'traditional action' means 'action based on habit rather than on perceived relation between goals and behavioral means', etc. Since the concepts of this category only enumerate the minimal set of characteristics sufficient for the recognition of their referents, and since their meaning does not include or imply any *propositions* about the objects defined, we may call them *enumeratory* concepts.

Concepts belonging to the second category to be distinguished here are defined in such a way that we not only specify the characteristics necessary and sufficient for the recognition of the extension of a given concept, but also include in their meaning many *additional characteristics* of their referents. These additional characteristics are not necessary for the recognition of the objects denoted by the concept. They create the 'surplus meaning' which is *equivalent to a set of theoretical propositions* about the objects or phenomena defined in a strictly enumeratory way. We might call them *syndromatic concepts*.

It seems to me that many social concepts are defined in such a syndromatic way. When Max Weber defined the ideal type of bureaucratic authority, he included into the meaning of this concept not only those properties which were necessary and sufficient for the recognition of bureaucratic authority among other types of power (obedience to some legal principles), but also many others which were only *correlates* of the given type of legitimacy. We could translate this concept into a set of propositions about correlates of this type of legitimacy. When Etzioni defines 'coercive organizations', he writes:

Coercive organizations are organizations in which coercion is the major means of control over participants and high alienation characterizes the orientation of most lower participants to the organization.[8]

It is obvious here that the first property is sufficient for recognition of the referents of the concept of coercive organization, whereas the second one, alienation, is there the correlate of the coercive type of organizational structure. Here, again, one could translate these into one definition and one theoretical proposition about the object thus defined. But such translations are not necessary, because such 'uneconomic' syndromatic definitions (possessing broader or narrower surplus meaning) are met in many sciences as *another method of systematization of theoretical knowledge about some natural classes of phenomena*. It is especially characteristic of the 'typology' of chemical elements when all new properties of the given element are simply added to the meaning of the syndromatic concept defining this element. Classification of species in biology is another example of this type of systematization of theoretical knowledge. Therefore there is no reason why such syndromatic concepts should be avoided in the social sciences.

But this has certain important consequences for the problems of concept formation. While concepts defined in an enumeratory manner are neither true nor false, they may be only more or less useful, the syndromatic definitions of theoretical models (the term 'model' seems to be the most adequate here) of certain complex phenomena or objects *may be true or false to the degree to which propositions about the relationships between the components of the given syndrome involved in the meaning of the given concept are true or false.*

Moreover, the proposition about the relationships between the components of the given syndrome may have – as all other general propositions – either universal or historical validity exactly to the same degree as it applies to the propositions formulated in a 'normal' theoretical manner. It means that such syndromatic concepts may be 'historical' also when the relations implied by their meaning are historical. Let me here quote R. Bendix again:

> If we are to refer to social structure, we must define a cluster of attributes which distinguish one structure from another. It is fiction to suppose that these attributes generally occur together... Hence comparative sociological studies are to delimit applicability of these concepts: here we are back to the space-time dimension of sociological concepts. Even more important, such studies would enable us to examine critically the implied – and to my opinion unjustified – generalization according to which several attributes of 'urbanism' tend to occur and vary together.[9]

To say that concepts defining the model of a certain class of phenomena in a 'syndromatic' way possess limited historical applicability does not imply that the elements of the given syndrome appear (empirically or by definition only) within a limited time-space area; it refers to the 'historical limitation' of validity of relationships between these elements involved in the meaning of the given syndromatic concept. As we know it means that the *relationships implied by the meaning of the given concept are conditional, depending upon some additional unknown factors.* As we remember from the analysis made above, these unknown factors are equivalent to the time-space limits of applicability of our generalization (here: relationships involved in the meaning of our syndromatic concept). In other terms, it means that our syndrome is theoretically incomplete, omitting some conditions necessary for universal relations between its other elements.

If we were able to define (or better, to describe) a theoretical model of a type of social fact in all its complexity so that we believed that we had spelled out all essential elements involved in the social mechanisms of the given type, our model would possess all the properties of a universal theory (or universal syndromatic theoretical concept), independent of where or how often the referents of the given model appear.

But in order to build such a complete model we probably will have to apply the theoretical propositions from a higher level of generality within the vertical structure of our knowledge, which shall be discussed later.

4. THE STRATEGY OF INDUCTIVE COMPARATIVE STUDIES

Let us now look at the problem how we can discover the nature of our unknown condition D which determines the relation $S \rightarrow B$, or in other terms how we can transform a historical generalization into a general law of science.

In general it means that either by experimental manipulation or by the observation of natural sequences of events we should try to discover these additional conditions D (or their complete sets D_1, D_2,... D_n) such that whenever D is present (or D_1,..., D_n are present) S is always followed by B, whereas the absence of D (or D_1,..., D_n) effects the lack of regular sequence of the type $S \rightarrow B$.

If we can manipulate experimentally the conditions D_t,..., D_n and at the same time produce the event S, the control of the hypothesis $SD \rightarrow B$ is fairly simple.[10] It is much more difficult when our experimental manipulations of properties of men and societies are restricted (either for moral or technical reasons) and we have to study these conditions as they exist naturally, by the method of cross-cultural or cross-historical comparisons, because then we often meet certain natural obstacles in finding instances necessary to test the hypotheses that some specific factor D is relevant to the relationship $S \rightarrow B$.

If in their natural occurrence or existence the properties of human beings and the characteristics of the societies in which people live were distributed *randomly* in time and space, the formulation and verification of universal and general social theories by the comparative social research would be an easy task. It would then be possible to take a sufficiently large random sample of men (or groups) in order to determine what differentiates the persons or the societies for which a certain regularity holds true from those in which it does not, and to ascertain what is common to all those situations for which the given generalization holds true. But for many reasons, the internal uniformities within societies and the external differences between them are usually *syndromatic* in their nature. It means that the societies are usually similar on many traits at once and many of their features vary simultaneously. When a sociologist wants to determine the additional unknown factor D involved in a historical generalization which describes regularities of events characteristic for one society and wants to discover the conditions D according to the method of 'only agreement' − to use J. St. Mill's terminology of inductive analysis − he is faced by too many agreements, i.e. by a great number of similar characteristics of all the members of the given society or social group − or too many similiarities of the social context in which they live: characteristic of culture, social structure, technology, level of consumption, position among and relations with other societies, etc. Each of these factors may be important or irrelevant for the given regularity $S \rightarrow B$. On the other hand, when he looks for the 'only difference' responsible for the fact that a certain regularity is limited only to the given time-space area, and does not occur outside this area, he usually finds many factors differentiating the compared societies, or – in other terms – he is met by too many differences instead of only one.

To find two societies which would *differ only on one given characteristic* which might be involved in limiting the validity of a certain generalization to a certain historical area is a difficult (and in many cases even impossible) task. It is nevertheless obvious that one such comparison is much more important for the discovery of unknown factors and for the transformation of a historical generalization into universal law than the comparison of several hundreds of societies described in the Human Relation Area Files. For example, it is much more important for the evaluation of the degree of universality of the Oedipus Complex Theory and for discovery of additional factors involved in it to study one society in Melanesia (where the father is not the central person in the family) than to study two hundred societies where he is the central person. For evaluation of the degree of universality of the 'functional theory of stratification' it is much more important to study a few really egalitarian communities than to study all non-egalitarian history of mankind.

Any type of sampling in terms of randomness, geographical representation, etc. is therefore not the best research strategy in such cases. Verificational studies based on this type of representation may be justified in two types of situations only: (1) when we try to test the degree to which a certain regularity known from previous study is a universal one, or (2) when – already knowing its historical limitations – we lack knowledge (or even tentative guesses) concerning the unknown factors involved. But in these situations (for reasons which I mentioned above) the chance of random discovery of conditions co-determining social regularities is not a good one.

In order to avoid the inefficient technique of random sampling in verificational studies and soundly base our *purposefully chosen samples* for comparative verificational studies of theoretical hypotheses, on conscious selection of properties for comparative inductive study, we must formulate all possible alternative hypotheses for such studies in such a way that they spell out all possible factors, both obvious and hypothetical, which may be relevant for the given regularity. It means that whenever we have formulated a hypothesis about the relation of two classes of events or properties, we should ask ourselves; Does this hypothesis seem to hold true for all human beings or groups or only for those who possess some special characteristics? If it seems to hold true for only some people or some kind of groups, then – what characteristics might also play a role in this regularity? Are they some psychological properties, characteristic of human behavior, or are they of a more situational character? Do some conditions external to the feeling, thinking and behaving individuals seem to be involved too? If yes, are they of a cultural, organizational or physical nature? What is their nature in most specific terms possible?

After these questions have been answered, the next problem arises: Where can we find or how we can produce instances of human, psychological, behavioral or situational characteristics or properties of groups or societies which, being similar in most of their aspects, will differ in just one, the importance of which we wish to investigate for our theory, in order to choose from all alternative hypotheses the correct one?

If we happen to find in a natural setting such configurations of properties, which are necessary for the test of different alternative hypotheses, we can say which of the alternative possible conditions is essential in the case of our hypothesis. If not – we must try to apply another approach, which will be discussed in the Section 5 of this chapter.

5. THE ROLE OF REDUCTIVE SYSTEMATIZATION OF THEORIES IN FORMULATION AND INDIRECT CONFIRMATION OF HYPOTHESES

Specification of all essential factors involved in a certain hypothesis (whether it has been formulated in the 'normal' way or is involved into the meaning of a syndromatic concept) may be somethimes rather difficult, especially when we want to choose from a great number of characteristics of a certain 'local' culture or to select those characteristics them from the properties of social structure of a certain society which might be relevant for the given regularity. One possible solution to this problem may be an attempt to derive our hypothesis from propositions of a more general nature.

This is sometimes possible because in the science of man's social behavior, as well as in most other sciences, theoretically more general, more abstract laws may be often discovered and tested easier than the laws referring to more complex phenomena of lower level of theoretical generality. Science may be roughly compared to a pyramid, in which the laws of the less general level are empirical foundations for more general theories, where the more general laws furnish – at least sometimes – explanatory principles for the less general mechanisms. But unlike the real pyramid, the structure of scientific knowledge may be sometimes easier to build from the top than from the foundations and the verification of more general theories might be easier than that of those which are located fairly low in the vertical structure of logically interconnected propositions.

In such a structure, if it is sufficiently consistent, each law of less general level may be reduced to one or more laws of more general level or explained by them. But by the same token, each factor of regularity of the less general level must have a corresponding phenomenon involved in the mechanism of a more general nature. Therefore, if we try to spell out all possible factors involved in a certain regularity of a fairly low level of theoretical generality, one way of testing the completeness of our list is to try to *explain* this regularity in terms of some more general theories. And, as I have said, such explanation requires that for each factor of more general theories we find its counterpart or less general level. Application of frustration-aggression theory to the explanation of intolerance points out the necessity of finding frustrating situations on the social level. Application of stimulus-response theory to the pattern of interpersonal interaction leads to the importance of finding proper 'reinforcers' of human relations, etc.; to give the simplest examples.

After specifying these factors which seem to be essential on the less general level in light of more general theories, we may undertake the next step described in the previous section of this paper, namely the choice of the proper sample and research design. If it happens that we are unable to find or to produce experimentally instances of persons or groups which differ on one given property only, we may feel a little more secure about the correctness of our hypothesis if this property seems to be important according to some more general theories than if we are unable to present any evidence for its importance. In other words, the explanation of a less general regularity by more general laws may sometimes furnish arguments about its validity – if, of course, more general theories seem to be sufficiently well verified.

In other terms, the reductive explanation of a hypothesis may sometimes – with all necessary reservations – be used as a substitute for its direct empirical test – as its indirect confirmation, including the identification of conditions of its validity.

Let me quote here an example of analysis in which a more general (in the case, psychological) theory was applied to identify among a great number of social characteristics of a certain group those characteristics which play an essential role in determining the pattern of ideological orientations of this group. The following are Martin Lipset's comments on 'working class authoritarianism':

A second and no less important factor predisposing the lower classes toward authoritarianism is a relative lack of economic and psychological security... Such insecurity will of course affect the individual's politics and attitudes. High states of tension require immediate alleviation, and this is frequently found in the venting of hostility against a scapegoat and the search for a short-term solution by support of extremist groups... The lower classes' insecurities and tensions which flow from economic instability are reinforced by their pattern of family life. There is a great deal of direct frustration and aggression in the day-to-day lives of members of the lower classes...[11]

Since the most fruitful use of such 'reductional approach' in social sciences consists of applying – as I have mentioned – 'psychological theories' describing individual human behavior for a better understanding of the functioning of groups and institutions, formulation and indirect confirmation of social laws may profit from application of psychological theories to the explanation of social phenomena. But in order to do this we must be able to identify social stimuli and social conditions determining individual social behavior as a special subclass of stimuli and conditions described in psychological theories.[12]

6. HISTORICAL DIMENSION OF SOCIAL PHENOMENA
AND THE PROBLEMS OF SOCIOLOGICAL INDUCTION

While contemporary social sciences more and more often surpass the frontiers of local cultures and national states in seeking broader evidence for their propositions, they relatively seldom cross the frontiers between historical areas, especially those which divide contemporary societies (accessible to a direct study and observation) from the past ones. The neglect by social theory of cross-historical comparisons even led to the formation of 'comparative historiography', a term which would have sounded like a contradictio in adiectu to Henrich Rickert.[13] Of course, no textbook of sociology denies that the laws of our discipline are meant to cover social phenomena of the past as well as those of the present and future. But attempts to test social theories against longer periods of history or sample of societies taken from different epochs seem to be relatively less frequent now than in the times of Comte, Marx, Weber or Toennies, although one could mention many important studies in which the sample of cases taken into account is not limited to the history of mankind after World War II, most efforts in comparative social research is oriented toward the study and theoretical analysis of *contemporary* social phenomena.

The importance of greater stress on historical aspects in comparative social research seems to be too obvious to require additional arguments in its favor.

The most important argument is that this kind of study is necessary for the confirmation of hypotheses claiming universal validity, or for the limitation of this validity to some periods of human history only, or some epochs in the history of the given society. Besides, in the past societies we may often find instances differing (or similar) on such characteristics which are essential for the test of certain hypotheses, even though we don't find them among contemporary societies. But if generalizing findings from one contemporary society to another and discovering additional conditions of validity of such generalizations creates many problems, the generalization of results across historical periods is usually even more difficult.

One of the reasons for lack of interest of contemporary, 'empirically minded' social scientists in historical data is the relatively poor quality of many historical data when compared with highly standardized, representative and controlled data of surveys and experiments conducted in contemporary societies. But if social science really is to consist of universally valid propositions, the theoreticans have to learn to make use of any existing knowledge even if it is not very satisfactory from the point of view of representativeness and standardization of data and their fitting into the elaborated conceptual scheme of the testing of sociological hypotheses.

This requires a much more flexible approach to the problem of verification which would make it possible to include among the empirical data favorable or unfavorable to a given social theory both very precise observations and measurements of contemporary sociology or social psychology, and the (usually much less precise and often more doubtful) data found in historical sources. There are two inherent dangers in the existence of unprecise and doubtful data about the social facts of past periods. One consists of attributing to these reports greater validity and reliability than they deserve. The other one consists of losing all potential informations which might be used in checking the limits of validity in social theories. It seems that both of these dangers should be avoided.[14]

The logic underlying such comparisons is in general the same as that used in surpassing space limitations of the generality of our findings. This means that we must look for those relevant factors which exist in contemporary society and which did not exist in those periods in which the regularity seems to be invalid or vice versa. After discovering by the method of a agreement and difference the conditions of our regularity we will be able to reformulate our historical proposition into a universally valid law of social sciences. But the historical extension of social phenomena and the *directional character* of social processes pose some additional methodological problems.

The first point is that the factor D. necessary for the 'operation' of the conditional causal sequence $S \to B$ may occur at the certain point of time and from this moment on it may be a general characteristic of all human beings, or at least a general characteristic of large civilizational areas. Its impact may be also of a negative character – it may create conditions which 'cancel' factual operation of some regularities of human behavior – 'valid' before its occurrence. It means that at least some regularities of social behavior must be so restricted that we are able to show a certain moment before which they did not operate or from which they cease to operate. The existence of a certain 'terminus ab quo' or 'terminus ad quem' would be therefore one of the effects of this type of historical localization of conditions co-determining the regularities of social behavior or man which was characterized above.[15]

There is no reason to believe that all possible conditions of regularities of human behavior have already occurred in the past or occur in contemporary social situations. We have to face the possibility that some such basically new conditions will occur in the future. If we remember that their effects may result either in co-determining some regularities of social behavior, or in 'cancelling' their operations, we must accept the possibility that:

(a) Some of the regularities which according to our theoretical and empirical knowledge would not occur until now, might be still possible in the future, if in the course of historical transformations of societies the necessary conditions of their operation will occur.

(b) On the other hand, we also have to admit the possibility that the historical change, by creating some basically new conditions, will 'eliminate' some regularities of social behavior believed until now to be 'iron laws' or social and behavioral sciences.

In terms of factual sequences of events it would mean the occurrence at a certain moment, e.g. in the future of some basically new regularities of human behavior, or elimination of some old ones. It also means that at least some of the regularities of social behavior which we could formulate on the basis of some more general theories (e.g. by deriving them from more general theories), may be basically unverifiable or basically unfalsifiable by the known – past and present – empirical data. They must wait for their confirmation or rejection until the course of historical transformations (or purposeful human action) brings to existence the conditons under which these generalizations are supposed to be valid.

To discover the unknown condition of our regularity besides all the problems discussed above may present a special difficulty if this condition is an element of a *complex set* historical event which occurred in the past but the influence of which may be observed (directly or indirectly) until now. The complex syndromatic sets of consequences of events like the appearance of Christianity, ideas formulated in the French Revolution or in the Communist Manifesto probably play essential roles in many of the regularities of human social behavior. But it is not easy to say, and even more difficult to prove, our guesses in empirical research which of the elements of these complex sets of consequences of these events (if any) are playing a part in the given regularity, and which of them are only unessential correlates of casually important conditions. Of course, the theoretical analysis of the reductionist type may lead us to the conclusion that of all the components of Calvinist ideology the only one which really counted in the genesis of contemporary capitalism was – as Max Weber postulated – *die innerweltliche Askese*, but then the empirical proof of such hypothesis is not an easy one.

This is especially difficult if all the observable sequences $S \to B$ are occurring only within the area of influence of the complex syndromatic set of potential conditions $D_1, D_2,..., D_n$, and do not occur outside of it. To give an example, it is not easy to say to what degree the relation between a certain stage of development of technology and the development of economy characteristic for the capitalist system was co-determined by the type of ideological orientation mentioned by Max Weber in his work on the relation between Protestant Ethics and the Rise of Capitalism, since outside the area of influence of the complex syndrome of more or less direct consequences of Reformation we do not find either the development of modern technology nor economic system characteristic for early capitalism. Thus we are unable to prove whether any of the elements of cultural inheritance of Reformation was essential here and if so, which of them. Every argumentation can here be based only upon indirect evidence of more or less 'reductionist' type. And since the cumulative character of historical processes is constantly adding new complec sets of potential modifiers of social regularities, this fact does not make easier the life of the theoretically-oriented sociologist who would like to extend his generalizations in time as much as possible, to make them really universally valid.

7. THE PROBLEM OF SPURIOUSNESS AND THE ROLE OF GENETIC EXPLANATIONS IN SOCIAL THEORY

Until now we have assumed that the relation between S and B, being either unconditional or conditional one, is nevertheless causal. But it may also happen that it is not a causal but a spurious relation. We call a certain relation (SB) spurious when S and B are not causally related to each other and their joint (or consecutive) occurrence may be explained as the result of certain other causes which produce both S and B and thus are responsible for the relation between them.

Let us distinguish two kinds of spurious relations. In one of them the event or property C is producing both S and B according to the schema:

$$C \to (S \wedge B)$$

In the other of them the syndrome $S_c B_c$ as a whole produces another syndrome SB, when each of the elements of the first syndrome is the cause of a corresponding element of the other syndrome. The relation between S and B is then also spurious. Graphically one could present this as follows:

The spurious or causal character of the relation between S and B can again be most efficiently proven experimentally in the way discussed above. But if we cannot do that, we can try to solve our problem by observation of naturally existing cases. If the variables S_c and B_c are not the elements of a perfect syndrome, i.e. they occur sometimes jointly, sometimes separately from each other, we can observe that:

$$S_c B_c \text{ is followed by } SB$$
$$\overline{S}_c B_c \text{ is followed by } \overline{S}B$$
$$S_c \overline{B}_c \text{ is followed by } S\overline{B}$$
$$\overline{S}_c \overline{B}_c \text{ is followed by } \overline{S}\overline{B}$$

and then we can almost certainly say that the relation between S and B is a spurious one.[18] If the syndrome $S_c B_c$ is very consistent, we have to prove the causal character of the relation between S and B in a more indirect way, e.g. by showing or assuming that $(S_c B_c)$ and (SB) are independent from each other.

This is not a new observation. At the meeting of the Royal Anthropological Association where E. Taylor presented his paper on application of correlational analysis to discovery of causal relation between different cultural traits. Galton, who was chairman of the meeting, made the following comments:

It was extremely desirable for the sake of those who may wish to study the evidence for Dr. Taylor's conclusions that full information should be given as to the degree in which customs of the tribes and races, which are compared together, are independent.

It might be that some of the tribes had derived them from a common source so that they were duplicated copies of the same original.[19]

The fact that the compared instances of relationships are not independent from each other has one important implication for their use in inductive analysis. It reduces the number of cases from a great number to a smaller number — sometimes to one case only. Suppose that someone would like to make a generalization about the relationship between some ideological values in the contemporary press by presenting as evidence for this relationship the results of content analysis of one hundred thousand copies of *the same issue* of *The New York Times*. In this case most of us would probably be inclined to treat such evidence as one instance rather than as one hundred thousand instances.

Let us assume that we observe a syndrome of four traits T_1, T_2, T_3, T_4 usually occurring jointly (or consecutively) in a constant pattern in different cultures or societies. Suppose now that the occurrence of this syndrome is based upon spurious relations between T_1, T_2, T_3, T_4. We can present graphically two different patterns of their interdependence. In one of them this syndrome, once shaped in history, in the culture 1 is consecutively transmitted in time from one culture or epoch to another:

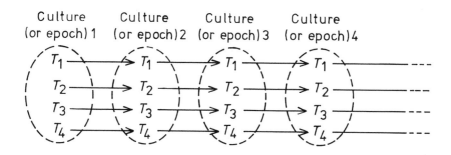

Culture (or epoch) 1   Culture (or epoch) 2   Culture (or epoch) 3   Culture (or epoch) 4

Persistence of a complex cultural or institutional pattern in one society through many generations or centuries corresponds roughly to this schema. Diffusion of some customs or institutions between different societies but along one diffusion path only also corresponds to it. Another schema corresponds to the pattern of diffusion going to different directions at once:

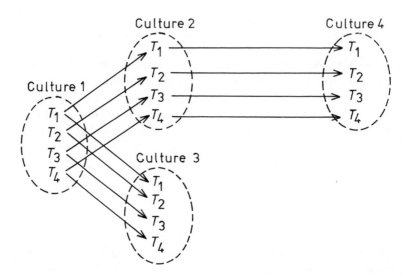

If we are unable to test experimentally whether relations between the elements $T_1,..., T_4$ are causal or spurious, we must rely upon more indirect evidence. If the syndromes $T_1 T_2 T_3 T_4$ are not perfect ones we may apply some statistical techniques of analysis which permit us to conclude to what degree the relation between different elements of our hypothetical 'causal chains' are spurious or causal by multivariate analysis or by some of its equivalents, e.g. by application of the coefficients of partial correlation or of other measures of partial dependence in so-called 'path analysis'.[20]

In some cases we are able to assess almost a priori that some of the relations we observe are spurious. It means that our knowledge or understanding of social phenomena tells us, that some of the complex configurations of elements occurring jointly and creating a certain syndrome can be a priori classified as spurious. This is e.g. when we know that a certain complex of thinking and behavior characteristic for the members of a certain society has been learned in the process of socialization, the joint occurrence of its elements thus being due to their joint occurrence in the cultural pattern 'sent' by the society to its members. The same holds when we see that a certain complex institutional pattern existing in one society has been borrowed in all its complexity from another society. In all these cases the non-causal nature of connections between the different components of the complex syndrome can be easily assessed.

Sometimes common sense or relatively simple scientific knowledge may play a similar role in suggesting quite the reverse i.e. the independence of the cases under comparison, and the causal nature of their relation. For example, it seems that for many regularities we study and describe in the theory of social behavior and for many regularities of small social systems to which many theories of sociology or social psychology refer, the assumption of independence of particular instances is justified. We do not think e.g. that frustrated persons are intolerant *because* such a pattern of reaction was created once in the past, and is now 'duplicated' by imitation in contemporary societies but we treat it as causal relations between frustrating situations and behavioral reaction to them and each case is treated as independent from all others. We do not think — to quote the famous study by Lippit and others — that the democratic groups have greater chance of persistence in the periods of crisis of leadership than authoritarian ones because it once happened so in the past, and is now inherited by other groups, but we believe that it is a natural regularity of events and all instances of this regularity are causally independent from each other.

Finally, sometimes we have to use the reductive approach in order to prove that the relation between T_1 and T_2 at a given time moment is spurious, when the causal relation connects these elements with their 'isomorphic correspondents' in the culture which is the source of the given configuration in many others. One essential argument in this type of reasoning is the demonstration of cultural contacts between the two cultures or demonstration that such contacts between the two cultures or demonstration that such contacts were impossible.[21] This requires very thorough descriptive knowledge of the history of the societies under comparison.

The procedure of explaining some social patterns by showing that they come (were taken) from a certain past culture or society is well known to historians of all times, who used to call it *genetic explanation*. Therefore it seems at first that the adequate explanation of social phenomena should consist of combination (in proper proportions) of two different approaches.

One of them should apply to causal relations. This approach would explain the phenomena in terms of their causal antecedents according to the well-known schema of *deductive-nomological explanation*. Another approach would apply to the category of relations which arise from the fact of persistence or diffusion of cultural or institutional patterns (and therefore are spurious ones), and could be termed *genetic approach*. These two approaches would of course be not contradictory but complementary to each other.

But this conclusion is justified only at the first step in explanatory procedures. Even when we find out that the relationship between a set of elements of a cultural pattern is spurious and we are successful in tracing by genetic analysis the historical sources of this spuriousness, we still face the problem of nomological explanation of the causal relation between this pattern and its historical source. Suppose that we denote the primary configuration of elements by M (let us call it a 'genetic matrix' for its eventual 'reprints') and the isomorphic configurations of similar elements produced by this 'matrix' M in other societies, or in later periods in the history of the same society, by I_M. To show by genetic-historical analysis that I_M was shaped by M is not equivalent with a theoretical explanation of I_M. For that purpose we need a proper general law or laws which say that whenever configuration M occurs; it will be 'duplicated' by isomorphic configurations of the type I_M. But such a law will almost always be a conditional one, i.e. it will be relativised to a set of conditions D which are necessary for the occurrence of the isomorphic relation between M and I_M. Symbolically we could write that.

$$(D.M) \rightarrow I_M.$$

The scientists studying the processes of cultural diffusion or historical heritage could mention many conditions which are necessary for the relation of isomorphism between a given 'matrix' and its 'reprints' in other societies or later stages of development of the same society, for instance the functional requirements which the persisting or borrowed custom or institution has to fulfil. It would be beyond the scope of this paper to analyse them more thoroughly.[22] What I would like to stress here is the fact that

(a) When we discover by the method of historical-genetic analysis that there is a causal relation between two (or more) complex syndromes of elements M and I_M, we treat the internal relations between the elements of the syndrome I_M as spurious ones.

(b) Such conclusion does not end our task but changes its character; from now on we have to formulate such a law (or laws) which specify under what conditions D the 'genetic matrix' M can shape its 'isomorphic reprints' of the kind I_M.

(c) The confirmation of such hypothesis should be carried out according to all rules of experimental, multivariate or comparative induction and will meet all the problems already mentioned in this paper.

Thus we see that spurious relations of this kind can be dealt with by the use of the same nomological-causal thinking as all the other relations between the social phenomena. But because of the importance of both actual context and of past social configurations in the shaping of subsequent social phenomena, and relationships between them, the final future structure of social theory may be more similar to biology with its combination of genetic [23] and 'normal nomological' approach than to the simpler nomological structure of Newtonian physics[24] – which seems to have been the ideal pattern of good theory to at least one or two generations of methodologists in the social sciences.

And in general we may say that if *history as a process* poses many problems and difficulties for sociological induction, then *history as science* may help us in solving them. Let us hope that sociologists will soon come to the conclusion that historical competence is equally as important for a theoretician as is competence in statistics or in knowledge about the logic of experimental design.

8. COMPARATIVE INDUCTION AND THE PROBLEM OF ONE-CASE MACRO-THEORIES

I have so far discussed in this paper primarily the problems of inductive confirmation of laws and theories applying to phenomena of the *sub-societal level:* to individuals or to smaller social units such as groups, institutions, etc. Only in such cases can we formulate a generalization on the basis of observation of a multiplicity of cases within one society and later on extend its validity in a cross-national or cross-historical study, eventually discovering additional conditions specific only for some societies and which codetermine a given regularity. But what happens when our generalization refers to the relationships between variables that describe the *society as a whole?*

In this case we need, of course, a multiplicity of societies in order to substantiate it empirically, and not only for the *extension* of its validity. Even here we still expect to find a sufficient number of societies that can be classified as identical with respect to the variables to which our macro-theory refers, so that we will be able to test our theory. Actually, there are numerous such studies in contemporary comparative social sciences. One category comprises those which try to relate the variables of nation-states, the other deals with relationships among the features of 'cultures' and here we may cite primarily the whole trend in comparative cultural anthropology started by P. Murdock using the data from the Human relations Area Files.[25] In spite of all the differences these two kinds of studies have one feature in common − namely, the use of correlational analyses of various kinds for testing their theories about whole societies or cultures.

But when we look at the results of such correlational cross-national or cross-cultural studies, we usually find that the correlations they establish aside from the fact that they may be suspected of being spurious only, as I pointed above, are rather low. This means that certain important variables which codetermine them have been disregarded in the analysis, and these theories are highly incomplete.

Even if we reexamine the data and finally obtain a fairly high probability of the occurrence of the phenomenon in question by increasing the number of independent variables, the trouble does not end here. For it may happen that the number of those cases which satisfy a definite configuration of the values of all essential independent variables and which imply the explained property with a fairly high probability turns out to be very small (if not just equal to one), so that while we obtain a very high probability (or even certainty) of the occurrence of a given property, it holds for few or only one single culture or social system. And them we start to worry about the statistical and theoretical significance of our findings.

This is a not unusual state of affairs, even if we take as the dependent viariable of our generalization a fairly simple trait and try to increase its probability by increasing the number of the independent variables. The situation is even more serious when we take a more complex dependent variable, constituting for example, *a configuration of the structural features of whole societies* or *a whole hierarchy of values characterizing their cultures*. Here we are likely to start at the very beginning with one single case. How can we deal with such situations in our theories, and how do we test such theories?

We face here two problems. One is how to distinguish the essential relationships from the non-essential ones, how to build a theory *of such case*. The other problem is of a more terminological nature. We usually treat as a theory a set of system of *general* propositions, whereas by the nature of social reality — by its syndromatic variation -- our theory seemingly *applies to one case only*.

There are some sciences which deal with complex configurations of structural features and with complex patterns of the functioning of certain 'wholes', biology being one of them. But the biologist is lucky in having an infinite number of cases of each species (at least theoretically), so that assuming a basic uniformity within the species he is able to observe as many examples as he wishes. Moreover, he may undertake certain experimental manipulations of a given number of cases so as to determine the consequences of such manipulations. But try to imagine the difficulties of biology as an inductive, theoretical science if it had to study the laws of the functioning of organisms in a situation wherein each of the species was represented *by one case only,* and had to base its generalizations on the observation of these cases throughout their whole life cycles? Which of the features are essential for the given 'species', i.e. are causually related to the whole configuration, and which are only accidental? What are the relationships among the various structural and functional characteristics of a particular case? Which of the changes constitute the 'natural internal dynamics' of the given species, and which are its reactions to external stimuli and changing conditions? These are the questions with which our hypothetical biologist would have to deal, and we agree that his situation would not be easy.

Unfortunately this is precisely the situation we face in sociology when we try to deal in our theories with macro-phenomena from the societal level. We are actually faced by a situation in which we are dealing with strongly internal interrelated systems of variables, with syndromes, each of which may be represented by one case only. And at the same time each of these 'species' has its own dynamics with the other, surrounding societies, as well as with the natural conditions in which it exists and develops.

As I noted above, when speaking of the *theoretical generality* of a proposition we do not refer to the number of cases to which it actually applies (i.e. cases which in reality), but to the number of cases to which it *intentionally* applies. If it is formulated in universal terms, then it is intended to apply to an infinite number of cases. If it is *theoretically valid,* i.e., *complete*, then it can explain any case of the given type and be used for prediction. The number of existing cases is from the standpoint of theoretical generality irrelevant, as long as our universally formulated theoretical model of the given, is adequate for the entire intentional class.

A scientifically valid design for the construction of a building or a car is from the methodological standpoint a theoretical ('structural-functional') model of a generally defined class of phenomena, a theoretical model that *satisfies the condition of adequately describing* the structure and functioning of the phenomena in question. From the theoretical standpoint it is irrelevant whether that theoretical model has only one referent or thousands of referents. It is irrelevant whether the designed building is unique, as the Parthenon is, or is one of serially-built family houses; and whether the designed car is unique, as is the car specially built for a United States President, or is one of the many thousands of serially-produced cars. Likewise, if we were to succeed in working out a model of a social system, using theoretical categories precise enough to make the model satisfy the condition of adequate explanation of the structure and functioning of society of a certain type, such a model would be called a theory. Should that model have one referent only, e.g. Polish society under the socialist system, then that model would have nothing to do with 'idiographism', provided it were theoretically adequate. If, for theoretical or practical social reassurance have to concentrate in our theories upon whole configurations of features of our societies, and these features are complex enough to make the studied cases 'unique', this does not imply that we have to resign from building a theory of such a system.

But such 'theories of single cases' raise serious methodological problems connected with their *verification*. At the observation level all the regularities found are by definition exception-free, but the theorist is interested in finding out which consequences and co-occurrences, out of those found in a single case, may be taken to be 'essential', non-incidental, reflecting the natural laws of sequence of events, etc., and which are merely coincidences, effects of chance, etc., and are thus spurious in nature. The basic difference between such a model and a purely idiographic description is that a theoretically valid model brings out only those elements which are essential for a certain property or regularity in which we are interested, whereas an idiographic description simply lists events and properties without making any selection among them. But how are we to verify a hypothesis of properties and/or event if this co-occurrence is observed in one case only?

First, it should be clear that even if a configuration of the concrete values of a multiplicity of variables is unique, this does not mean that these variables are not represented *by some other values* in other societies. So it may happen that by analysis of the 'monotonic' relations between the variables occurring in their different values in different societies, and assuming that they act together according to a simple additive pattern, we find relatively strong evidence for the theoretical validity of a single case model. Moreover, as I said above, the societies usually change more or less over time, and the observation of temporal relations among their different or aggregate characteristics may constitute additional evidence for the validity of our models. Finally, societies are *deliberately* changed, which may constitute a case of experiment under natural conditions, if the results of such planned changes are put into the context of the conclusions of the two above kinds.

But still the degree of confirmation of our theory may not be very high, even after all these 'tests'. And here again I would like to stress the importance of the reductive approach to theory construction, which should be regarded as a supplementary approach to inductive analysis. If it happens that we have at our disposal a whole set of theories applying to the regularities of phenomena from the subsocietal level, we can try to derive our macro-theory from these lower level theories. The theoretical validity of a macro-theory is greatly enhanced if it can additionally be demonstrated that it is in agreement with the laws of the lowel level, that it can be deduced more or less completely from them. As I said in another essay of this volume,[26] the reduction is profitable both for the explaining and the explained theories. In our case it would imply that we have to build a reductive model of the processes we observe on the macrolevel, and then try to estimate the degree to which what we actually observe corresponds to the model derived from the laws of the lower level. If the correspondence is considerable, we can have great confidence in our macrotheory and use it both for the explanation of our 'single case', and for the prediction of its future course and for planned change.

In order to have at our disposal a sufficient number of theories from the lower level, we have to apply for the test of their validity the rules of comparative induction, which has been the main topic of this paper.[27]

NOTES

* Reprinted with certain modifications from *Synthese* 24, (1972).

[1] R. Marsh, *Comparative Sociology: A Codification of Cross-Social Analysis,* Harcourt, Brace, World, Inc., 1967, p. 6.

[2] Let me quote here Oskar Lewis: 'It is part of this paper that there is no distinctive... comparative method... in anthropology and that the persistence of this expression leads to unnecessary confusion and artificial dichotomies in much of the theoretical writing on this subject.' O, Lewis, 'Comparisons in Cultural Anthropology', in *Readings in Cross-Cultural Methodology* (ed. by F.W. Moore), HRAF Press, New Haven, 1961.

[3] For a much more detailed discussion of relations between the propositions of different degree of theoretical generality see: Chapter XI.

[4] See: S. Nowak, 'General Laws and Historical Generalizations in the Social Sciences', *The Polish Sociological Bulletin* 1, 1961.

[5] For a detailed discussion of different patterns – both of positive and of negative type – of conditional causal relations see: Chapter V.

In this paper I shall analyse the observable consequences of one type of conditional causal relations, i.e. when S and D are jointly the sufficient condition of B, each of them being essential component of the cause SD of the event B.

[6] Under assumption that some other, alternative causes of B (e.g. event A being also the sufficient condition for B) do not occur at the same time. For the analysis of consequences of existence of alternative causes of the same effect see: Chapter V.

[7] R. Bendix, 'Concepts and Generalizations in Comparative Social Studies', *American Sociological Review*, August 1963.

[8] A. Etzioni, *Comparative Analysis of Complex Organizations*, The Free Press, New York, 1961, p. 27.

[9] R. Bendix, *op. cit.*

[10] It should be noted here, that for the full test of the causal hypothesis $SD \rightarrow B$ both its antecedents should be accessible to the experimental manipulation. If we are unable to control experimentally the occurrence of S, then the relation $S \rightarrow B$ may be a spurious one, what is discussed in the last section of this chapter.

[11] M.S. Lipset, *Political Man*, Anchor Books, Doubleday and Company, New York, 1963, p. 106.

[12] See application of this type of approach in G.C. Homans, *Social Behavior, Its Elementary Forms,* Harcourt, Brace, World, Inc., New York, 1961.

[13] See F. Redlich, 'Toward Comparative Historiography: Background and Problems', *Kyklos*, 1958, 3.

[14] I would like to mention here two books in which the authors had the courage to use 'weak' historical data as evidence for hypotheses which are usually tested today in the laboratory of social psychology. One of them is the book by D. McLelland, *Achieving Society,* where the content anlysis of tales in different cultures was used as an indicator of achievement motivation, in different historical societies. Another is the book by Swend Ranulf, *Moral Indignation and the Middle Class Psychology*, where socio-psychological theory of intolerance was developed more than thirty years ago and tested against more than 2000 years of human history on the basis of 'normal' historical sources and monographs.

[15] See: S. Ossowski, 'Two Conceptions of Historical Generalizations', *The Polish Sociological Bulletin* 1 (9), 1964.

[16] See P.F. Lazarsfeld, 'Interpretation of Statistical Relationships as a Research Operation', in *The Language of Social Research* (ed. by P.F. Lazarsfeld and M. Rosenberg), New York, 1955. See also S. Nowak, 'Some Problems of Causal Interpretation of Statistical Relationships',

[17] It is essential that we try to test experimentally the causal nature of $S \rightarrow B$ in the same social situations or populations in which it occurs spontaneously, because if it is causal it means, that other conditions for the sequence $S \rightarrow B$ also do occur in this population. If we tested $S \rightarrow B$ in basically different conditions, we wouldn't know (if the test fails) whether the relation $S \rightarrow B$ is spurious or it is causal, but the conditions necessary for it do not exist in this situation.

[18] For simplicity I assume here that $Sc \equiv S$ when $Bc \equiv B$. If we assume that the causal relations between Sc and S and Bc and B are conditional in any sense of this term, the observable relations are more complicated. It is needless to say that the same kind of spuriousness may be discovered by the technique of multivariate statistical analysis, but in this case we have to keep constant two test variables at once.

[19] Quoted from F.W. Moore, *Readings in Cross-Cultural Methodology*, p. 22. See also a much more detailed analysis of Galton's dilemma in a paper by R. Naroll, 'Two Solutions of Galton's Problem', *ibid*.

[20] See: H. Blalock, *Causal Inference in Non-Experimental Research*, The University of North Caroline Press, Chapel Hill, 1961.

[21] See: R. Naroll, 'Two Solutions of Galton's Problem'.

[22] For a much more extended discussion of this problem see: Chapter IX.

[23] R. Marsh writes in his book that 'the future of comparative sociology is intimately related to the development of neo-evolutionary theory — development already set in motion by Stewart, Service, Parsons, Eisenstadt and others'. R. Marsh, *op. cit.*, p. 322. I don't think that the evolutionary kind of theorizing is here as essential as genetic — nomological explanations with the application of laws of isomorphic relations between the 'genetic matrices' and their 'reprints', but I agree that historical approach of a special kind is essential for the future social theory.

[24] It seems that this model of theoretical physics as an ideal type for theory construction in social sciences was introduced most clearly by G. Lundberg in his *Foundations of Sociology*, New York, 1939.

[25] See L. Merrit and S. Rokkan, (eds.), *Comparing Nations: The Use of Quantitative Data in Cross-National Research*, New Haven, 1966. See also P. Murdock, *Social Structure*, New York, 1960.

[26] See Chapter XI.

[27] For the analysis of some more general logical problems of induction, not necessarily relate to comparative studies in sociology, see Chapter VII.

BIBLIOGRAPHY

Bendix, R., 'Concepts and Generalizations in Comparative Social Studies', *The American Sociological Review*, August 1963.

Blalock, H., *Causal Inference in Non-Experimental Research,* The University of North Caroline Press, Chapel Hill, 1961.

Etzioni, A., *Comparative Analysis of Complex Organizations*, The Free Press, New York, 1961.

Homans, G., *Social Behavior, Its Elementary Forms*, Harcourt, Brace, World, Inc., New York, 1961.

Lazarsfeld, P.F., 'Interpretation of Statistical Relationships as a Research Operation', in *The Language of Social Research*, P.F. Lazarsfeld, M. Rosenberg (eds.), New York, 1955.

Lewis, O., 'Comparisons in Cultural Anthropology', in F.W. Moore (ed.) *Readings in Cross-Cultural Methodology*, HRAF Press, New Haven.

Lipset, M.S., *Political Man*, Anchor Books, Doubleday and Company, New York, 1963.

Lundberg, G., *Foundations of Sociology*, New York, 1935.

Marsh, R., *Comparative Sociology: A Codification of Cross-Social Analysis,* Harcourt, Brace, World, Inc., 1967.

McLelland, D., *Achieving Society*,

Merrit, L. and S. Rokkan (eds.), *Comparing Nations, The Use of Quantitative Data in cross-National Research,* New Haven, 1966.

Murdock, P., *Social Structure*, New York, 1960.

Naroll, R., 'Two Solutions of Galton's Problem', in F.W. Moore (ed.), *Readings in Cross-Cultural Methodology*,

Nowak, S., 'Causal Interpretation of Statistical Relationships in Social Research', in this volume, p. 165.

Nowak, S., 'Cultural Norms as Explanatory Constructs in the Theories of Social Behavior', in this volume, p. 319.

Nowak, S., 'General Laws and Historical Generalizations in the Social Sciences', *The Polish Sociological Bulletin* 1, 1961.

Nowak, S., 'Logical and Empirical Assumptions of Validity of Inductions', in this volume, p. 256.

Nowak, S., 'The Logic of Reductive Systematizations of Social and Behavioral Theories', in this volume, p. 376.

Ossowski, S., 'Two Conceptions of Historical Generalizations' *The Polish Sociological Bulletin*, 1 (9), 1964.

Ranulf, S., *Moral Indignation and the Middle Class Psychology,* New York.

Redlich, F., 'Toward Comparative Historiography: Background and Problems', *Kyklos* 3, 1958.

CHAPTER V

CAUSAL INTERPRETATION OF STATISTICAL RELATIONSHIPS IN SOCIAL RESEARCH*

1. THE PROBLEM OF CAUSALITY

The scientist, who has observed in his research that the event S was followed by the event B or that the class of events S was (or usually is) followed by the events belonging to the class B (when both classes may be finite or infinite in their number), may formulate on the basis of such findings two different types of conclusions:

(a) He may simply assert that in his population or on a much larger scale (if he makes an inductive generalization from his findings) *S is followed by B;* or

(b) He may also say that in his population or on a much broader scale (if not on a universal scale) *S is the cause of B*, or — in other terms — *B is the effect of S.*

As we know, philosophers differ strongly in their opinions as to whether the propositions of the last category are acceptable at all in the body of science. Moreover, if they accept the notion of causal relations as a special category of connections between the observable events or variables, they have different ideas of the notion of causality.[1] It is not the aim of my chapter to analyze these differences of definition of causal relations. I would rather like to stress here one very important similarity between all these notions. Both philosophers and research workers seem to agree on the 'operational definition' of causal relation, namely that the causal connections between the events are those connections which under conditions of a controlled experiment lead (or would lead) to the observable sequences of events S and B prescribed by the given design of the experiment.[2]

The words in parentheses: 'would lead' stress that the above definition is a conditional operational definition of a causal connection. It distinguishes in· a literary sense as causal connections only those which have been submitted to experimental control and 'behaved' in those conditions in the 'prescribed' way, but it permits us to characterize as causal all other connections for which we have other (i.e., non-experimental) premises permitting us to predict that, if taken under experimental control, these relationships would also lead to the sequence of events conforming to the rules of the experiment.

As we know, the experimental manipulations of variables preceding in time a given effect B permit us to obtain such combinations of variables characterizing the initial conditions for the given effect B, which are hardly obtainable and observable in natural situations. Thus if the experiment delivers us the instances of 'only agreement' or of 'only difference' prescribed by the rules of inductive analysis, we are able to say whether after the occurrence of 'S only' B always occurred or after the disappearance of 'S only' B also disappeared. Then we are able to say whether the occurrence of B was or was not dependent on the occurrence of S.

But the experimental manipulation of antecedent variables of a certain effect B has another very important implication for our conclusions, usually underestimated in methodological analyses of experimental research. By producing (or changing) the necessary values of the independent variables of a given effect B in different time and space coordinates, in different situations, we also prove (usually unconsciously) that all other variables which were not taken under experimental control, but which have different values for different experimental situations, are irrelevant for the occurrence of B.

This is true both for the variables which are occurring at the same time as S occurs and for those which are prior to S. Anticipating what will be said below, we may say that the other (uncontrolled and different in different instances) events occurring together with S cannot be the supplementary factors for S. It also proves that the relation between S and B cannot be a spurious one, i.e. caused by a prior factor.[3]

The last conclusion is true only when we accept one assumption, namely that the time-space coordinates for the given experimental control of the relation between S and B were not (knowingly or unknowingly) correlated with some events essential for the occurrence of B and prior to S or occurring at the same time of S.

This assumption seems to be valid for most rigorously conducted experiment and its fulfillment is even included in some experimental schemes (e.g., in the form of the random selection of the experimental sample) but it is not necessarily true for natural situations. Let us look a little closer at the uncontrolled conditions antecedent to S. It is often true that a certain S occurs in nature mostly or only as the effect of a certain A. When we then observe that S is followed by B we do not know and we are not able to say whether S is only an intermediary link in the causal chain of events $A \rightarrow S \rightarrow B$, or the relation BS is spurious, caused by the common cause A. In other words we are not able to say whether S would also be followed by B in the case where S would follow after non-A.[5]

The notion of a causal connection is sometimes defined or better presented in the form: 'S produces B'[6], because some authors seem to assume that the term 'to produce' has a more understandable intuitive meaning for the reader. The equivalence of these two terms seems to be based on the same definition of causality which was presented above and which I would like to state now in a little different form:

To say that S is the cause of B means that: wherever and whenever within the limits of validity of the given causal generalization S occurs (or would occur) it is (or it would be) followed by B, independently of whether S occurs (or would occur) 'spontaneously' or was (or would be) brought into existence by some 'voluntary' action of any 'actor' or 'producer'.

2. TYPOLOGY OF CAUSAL RELATIONS

In the first section of this chapter we have analysed the situation in which a certain S, exemplified for the purpose of convenience as 'stimulus', was causally connected with a certain B, 'behavior', in a special way — S was a sufficient condition for B. But the causal relationship between S and B may be of a different, conditional type — S may produce B only when some other conditions are fulfilled: e.g. when a certain D[7] occurs. In other terms: S and D jointly are the sufficient condition for B, being at the same time necessary components of this condition.

We see that the term 'causal connection', may have different meanings. Let us analyze here different possible meanings of the statement: 'S is the cause of B' under assumption that both S and B are defined in such a way that they are 'dichotomous attributes', i.e. they may either occur or not occur.[8]

When we say 'S is cause of B', we may think either of an *unconditional* or of a *conditional* type of causal relation. By unconditional type of causal relation, I mean the type of relation between variables which is usually characterized in general terms by 'always' or 'never'. By a conditional type of causal relation, I mean the type of relation between attributes which is characterized by statements like 'unless', 'provided that', 'when some additional conditions are held', etc.

When we say that between S and D there is an unconditional causal relation, it may mean:

(a) that S is a *sufficient condition* for B. It means that B always occurs when S occurs; we may say here that the relation between S and B is unconditional in its positive sense.

(b) that S is a *necessary condition* for B. It means that B never occurs when S does not occur; we may say here that the relation between S and B is unconditional in its negative sense.

(c) that S is both *necessary and sufficient* condition for B It means that B never occurs if S does not occur and that B always occurs when S occurs. This relation is unconditional both in the positive and in the negative sense.

When we say that A is necessary but not sufficient condition for B, it implies:

(d) that there exists at least one *supplementary factor* (D) which is such that D together with S is the sufficient condition for B.

When we say that S is a sufficient but not necessary condition for B it implies:

(e) that there exists at least one *alternative sufficent condition* (A) which is such that B always occurs when A occurs.

Let us come to the case when S is *a necessary but not sufficient condition*. The statements of this type are unconditional in their negative form (because non-S is never followed by B) but they are conditional in their positive form. Only in the conditions when D occurs is S always followed by B. We could write in the following implicational form:

$$D \rightarrow (S \rightarrow B) \text{ and } (\overline{S} \rightarrow \overline{B}).$$

In the case when S is a *sufficient but not a necessary condition* we have a general positive causal relation, but a conditional negative one. Suppose that A and S are the only possible alternative sufficient conditions; we may say that S is always followed by B, but non-S will be always followed by non-B

only in the conditions, when A does not occur. Or symbolically:

$$\overline{A} \rightarrow (\overline{S} \rightarrow \overline{B}) \text{ and } (S \rightarrow B).$$

It is often so that there is no unconditional causal relation between S and B either in the positive or in the negative sense, but we may still say that in some specific conditions S is causally related to B, both in positive and in negative sense of this term. It means, then, that S *is an essential component of one sufficient condition (DS) for B*, but in the reality (or in the population of events being analyzed) there exists at least one alternative sufficient condition of B (A). This relation is conditional both in negative and in positive sense: S will be always followed by B only if D also occurs or another alternative cause, A, occurs. And non-S will be followed by non-B only in the conditions when another alternative cause does not occur. Or symbolically:

$$D \rightarrow (S \rightarrow B) \text{ and } \overline{A} \rightarrow (\overline{S} \rightarrow \overline{B}).$$

So we see that the causal relations between two variables may be:

(a) unconditional both in positive and in negative form (S is necessary and sufficient condition for B);

(b) unconditional in positive but conditional in negative form (S is sufficient but not necessary condition for B);

(c) unconditional in negative but conditional in positive form (S is a necessary but not sufficient condition for B);

(d) conditional both in positive and in negative form (S is an essential component of one alternative sufficient condition for B, when B also may be caused (produced) by some other alternative sufficient conditions.

These four types of causal connections may be defined in a more formal way with the use of the language of the set algebra. Let us use

$S \overset{\doteq}{\rightarrow} B$ to denote the situation in which B is equivalent to S and follows S in time;

$S \overset{\supset}{\rightarrow} B$ to denote the situations when S includes B and is followed in time by it;

$S \overset{\subset}{\rightarrow} B$ to denote the situation when B includes S and follows it in time.

Let us also use

$S \cap B$ to denote the situations when both S and D occur;
$S \cup A$ to denote the situations when either S or A occur.

Now we may say that:

S is necessary and sufficient condition for B, which means

$$S \overset{\cdot}{\Rightarrow} B$$

or graphically

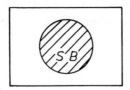

S is sufficient but not necessary condition for B when B may be also caused by another alternative sufficient condition A

$$(S \cup A) \overset{\cdot}{\Rightarrow} B$$

or graphically

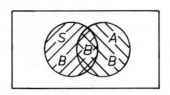

S is necessary but not sufficient condition for B, when the other supplementary factor of the sufficient conditions is D

$$(S \cap D) \overset{\cdot}{\Rightarrow} B$$

or graphically

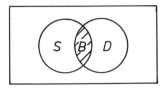

S is necessary component of one alternative sufficient conditions $S \cap D$ when B may be also produced by another cause A

$$(S \cap D) \cup A \overset{.}{\to} B$$

or graphically (assuming that $(S \cap D)$ and A are mutually exclusive)

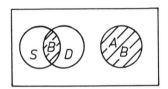

3. STATISTICAL LAWS AND HISTORICAL GENERALIZATIONS

The existence of conditional causal relations between classes of events has some interesting implications for the nature of the theoretical conclusions which we may draw from the analysis of empirical data. Let us assume that we are analyzing the situation in which in reality the necessary and sufficient condition for a certain B is composed of two classes of events (or traits) – S and D – but we only know about the importance of S for B, and the existence of D or its role in producing B is unknown to us. The analysis of the empirical relationships between the variables S and B may lead us to the conclusion that S and B are somehow interrelated, but the nature of our generalizations is here strongly dependent on the way in which the unknown factor D exists in empirical reality.

Let us analyze some possible ways in which the factor D may exist and their implications for the theoretical conclusions based on the empirical analysis of the relationship between S and B:

(A) D may exist 'practically everywhere' so that every S will act under the conditions of a certain D. If this is the case we will be able, without discovering the importance of D, to formulate a universal law saying that 'S is a sufficient condition for B' and this law will not be falsified by empirical research even though for an 'adequate' law of science there should be two components of a sufficient condition for B (S and D).[9]

(B) D may exist practically everywhere but in such a way that its occurrence may be characterized as a relatively even random distribution for S. This means that only some S's will act under condition D, but for every sufficiently large sample of S's the proportion of those occurring together with D to all S's will be relatively constant and approximately equal to p.[10] If this is the case, we will be able, without discovering the role of D in affecting B, to formulate a *universal probabilistic law* stating that there is a constant probability (p) that S will be followed by B.

(C) D may exist in such a way that it characterizes all S's in a certain population defined in terms of some time-space coordinates, but outside of these coordinates D either occurs irregularly or not all.[11] If this is the case then we will be able to formulate, without discovering the importance of D for B a *'historical generalization'* saying that within the time-space limits h, S is always followed by B.[12]

Thus we see that there is an interesting analogy between the two categories of statements which till now usually have been classified in quite different methodological categories — namely between universal statistical laws and historical generalizations valid only for one given population. *Both types of propositions refer to conditional causal connections in which some essential factor is unknown to the scientist.*[13] *They differ only in the way this unknown factor exists.* In the case of historical generalizations this factor characterizes all events within the limits of one 'historical population', whereas in the case of universal statistical laws it occurs regularly with a constant probability 'practically everywhere'.

(D) D may exist in such a way that it is relatively evenly distributed (and its probability of occurrence together with S is equal to p) but only within some time and space limits h. If this is the case we will be able to formulate, without discovering the importance of D for S, a *probabilistic historical propositions* saying that 'within the time-space limits h the probability that S will be followed by B is equal to p'.[14]

(E) But it may happen that the sufficient condition for B is composed of, let us say, the three factors S, D_1, and D_2, and at the same time:

D_1 is characteristic for all events in one historical population h;

D_2 is characterized by a relatively even random distribution and occurs practically everywhere (with probability p).

If this is the case, then we will be able to formulate, as in case (4), a historical probabilistic statement, but the situation will be quite different if we later discover the importance of one of the two supplementary factors for B, namely:

(a) if we discover the importance of D_1, *we will be able to formulate a* universal satistical law saying that the probability that SD_1 will be followed by B p;

(b) if we discover the importance of D_2, we will be able to formulate a historical generalization saying that in the population h all SD_2 are followed by B.

Only if we discover the importance of both previously unknown factors (D_1 and D_2) will we be able to transform our generalization into a universal law of science of the type: SD_1D_2 is always followed by B.

4. STATISTICAL RELATIONSHIPS IN UNCONDITIONAL AND CONDITIONAL CAUSAL PATTERNS

In Section 2 we distinguished unconditional and conditional causal relations between 'dichotomous' variables. It should be noted here that such relations may have a universal or a historical character. When I mean that a relation between S and B is unconditional in its positive form, it may mean:

(a) either that S is always followed by B practically everywhere because it is a sufficient condition for B (a universal law); or (b) that S is always followed by B in one population h, because in this population all events (persons) are under the influence of all other supplementary factors, which implies that the occurrence of S is a practically sufficient (in this population) condition for B (a historical generalization).

When we say that the causal relation between S and B is an unconditional one in a negative form, it may mean:

(a) that S is always and everywhere a necessary condition for B (a universal law); or (b) that in the population h, S is a necessary condition for B, e.g., for the reason that all other theoretically possible alternative causes of B do not occur in the population h (a historical generalization).

All of what will be said below on the statistical relationships between different classes of events treated as the functions of the causal connection between these events refers to both types of situations, i.e., both to the situation where a connection of the given type has an historical and where it has a universal character.[15]

Let us look now at the consequences of 'experimental manipulations' of our independent (i.e. antecedent) dychotomic variables or their 'natural' occurrences or non-occurrences, for different patterns of causal relations distinguished above.

(A) If S is a necessary and sufficient condition for B, then of course S is always followed by B, and non-S is always followed by non-B, which can be written:

$$S \overset{C}{\rightarrow} B, \overline{S} \overset{C}{\rightarrow} \overline{B}. \text{ or more simply } S \overset{\cdot}{\rightarrow} B.$$

Let us now introduce the symbol p to designate the *relative frequency* of B when S has occurred. Instead of relative frequency we shall speak also about *probability* of B/S (although this notion is here a little questionable as we shall see later).[16] The relations $S \overset{C}{\rightarrow} B$ and $\overline{S} \overset{C}{\rightarrow} \overline{B}$ can be now written as follows:

If S is a *necessary and sufficient* condition for B, the probability (understood here as relative frequency) of B when S has occurred is equal to 1, and the probability of non-B when non-S has occurred is also equal to 1 – or symbolically.

(1) $p(B \mid S) = 1, p(\overline{B} \mid \overline{S}) = 1$

or graphically:

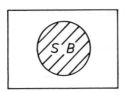

This is a very simple situation, and the only thing we can say about it is, that although logically clear, it seems to be an empty class in the area of empirical social research.[17]

(B) If S is a *necessary but not sufficient* condition for B, it means that there is at least one supplementary factor which forms jointly with S sufficient condition for B. Suppose that there is only one such factor D. Then we may say for sure, that non-S is always followed by non-B, but S may be followed either by B or by non-B depending upon occurrence or non-occurrence of D:

$$\overline{S} \overset{C}{\rightarrow} \overline{B}, S \overset{C}{\rightarrow} (B \cup \overline{B}).$$

Graphically we may present it as follows:

$$\begin{array}{l} S \\ \vdash \!\!\! \longrightarrow B. \\ D \end{array}$$

If we assume e.g. that the relation between the extensions of S and D corresponds to the following scheme

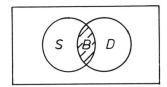

we may say that

(2) $p(\overline{B} \mid \overline{S}) = 1, 0 < p (B \mid S) < 1.$

Since the occurrence of B (provided that S has occurred) is conditional on the occurrence of D, we may say that S will produce B 'as often', as it occurs jointly with D. But it implies, that the relative frequency of the occurrence of B when S occurs is equal to the relative frequency of the occurrence of D when S occurs.

Or symbolically

(3) $p(B \mid S) = p(D \mid S).$

Between the two supplementary factors: D and S of the above scheme may be, in different populations, quite different statistical associations. They may be associated positively or negatively; they also may be statistically independent.

We should consider here especially the last category of the above situations – in which S and D are statistically independent, i.e., in which $p(D \mid S) = p(D \mid \overline{S})^{18}$ because then

(4) $p(B \mid S) = p(D)$

or in a more descriptive language:

In the case when two supplementary factors S and D are statistically independent, the probability of the occurrence of effect B when one factor (S) occurs is equal to the relative frequency of another factor (D) in the population.

(C) If S is *sufficient but not necessary condition* for B, it means that there is at least one alternative sufficient condition for B. Suppose there is only one such alternative cause of B (denoted here by A). Graphically we might present it as follows:

In this situation S will be always followed by B, but non-S will be followed by non-B only if at the same time A does not occur:

$$S \overset{\subset}{\to} B \text{ and } \overline{S} \overset{\subset}{\to} (B \cup \overline{B}).$$

We may characterize more specifically the statistical relations between S and B in this situation, namely:

(5) $p(B \mid S) = 1 \quad p(\overline{B} \mid \overline{S}) = p(\overline{A} \mid \overline{S})$.

If we assume that A is not included in S, e.g., according to the following scheme:

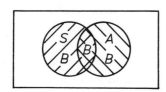

we will have

(6) $0 < p(\bar{B} \mid \bar{S}) = p(\bar{A} \mid \bar{S})$.

Here again, if we assume that S and A are statistically independent, then

(7) $p(\bar{B} \mid \bar{S}) = p(\bar{A})$.

Therefore: if two alternative sufficient conditions are statistically independent, probability of non-occurrence of the effect, when one of the causes did not occur is equal to the relative frequency of non-occurrence of the other alternative cause.

(d) Finally if *S is neither necessary nor sufficient condition for B*, it means, as we remember, that S is an essential component of one sufficient condition for B (namely $S \cup D$), but there is at least one alternative sufficient condition of B (assume here again that there is only one such alternative sufficient condition denoted by A).

Graphically it will look as follows:

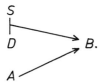

In this situation there will be no general relation of the 'always-never' kind between S and B. S may be followed by B or non-B depending upon the joint occurrence of D and non-S. Non-S also may be followed by B or non-B depending upon the occurrence or non-occurrence of A, or symbolically:

$$S \overset{\subseteq}{\to} (B \cup \bar{B}) \text{ and } \bar{S} \overset{\subseteq}{\to} (B \cup \bar{B}).$$

Here we will have (assuming for simplicity sake that S and D on one side and A on the other are multually exclusive)[19] the following pattern of graphical and statistical relations:

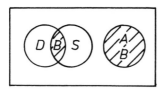

$$0 < p(B \mid S) < 1, 0 < p(\bar{B} \mid \bar{S}) < 1 \quad \text{and also}$$

(8) $p(B \mid S) = p(D \mid S), p(\bar{B} \mid \bar{S}) = p(\bar{A} \mid \bar{S}).$

It means that the situation where a certain S has the probability greater than 0 and lower than 1 to be followed by a certain B, and the probability of occurrence of non-B, when non-S has occurred is also higher than 0 and lower than $1 - S$ may be at the best[20] the essential component of one alternative sufficient condition of B. It should be noted here, that this is the kind of relation which we usually discover in empirical social research. It means that in social research we usually discover causal relations which are conditional both in their positive and in their negative form. *The statistical value of our relations is here the substitute for the other unknown causes,* i.e. both for the other components of the same sufficient conditions and for the other alternative sufficient conditions of our effect. If we discover them, then again we will have a general unconditional causal law with its antecedents composed of many elements jointly, e.g.

$$(S \cap D) \overset{C}{\to} B, [(\bar{S} \cup \bar{D}) \cap \bar{A}] \overset{C}{\to} \bar{B}.$$

Thus we see the *notion of conditional causal relations permits us to apply the idea of causality to the situations in which we do not observe general, necessary relations between S and B* (neither in positive nor in negative sense), *but where we discover statistical relations only.*

Let us assume now that S and A are not mutually exclusive – but D and A are still mutually exclusive – which is represented by the following graphical schema.

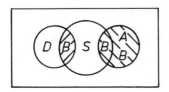

The probabilistic relations between S and B are here partly different than in (8) because we have:

(9) $p(B \mid S) = p(D \mid S) + p(A \mid S), p(\bar{B} \mid \bar{S}) = p(\bar{A} \mid \bar{S}).$

The existence of alternative causes of B (A) occurring at least sometimes jointly with S, introduces here a certain element of 'spuriousness' into probabilistic relations because the overall probability $p(B \mid S)$ is here the sum of the probabilities $p(D \mid S)$ and $p(A \mid S)$. In this situation $p(A \mid S)$ may be called the non-causal or *spurious component* of the overall observed probability $p(B \mid S)$, when $p(D \mid S)$ may be called its *causal component*. One could say that from all B's which occurred after the occurrence of A only a certain fraction of them was produced by S and this fraction is represented by $p(D \mid S)$, constituting the causal component of $p(B \mid S)$ when all the other B's were not produced by S but were produced by A when A was occurred jointly with S. This spurious component of $p(B \mid S)$ is represented by $p(A \mid S)$. The spuriousness does not refer here, of course, to the correlation between S and B (to be discussed later) but to the *probabilistic relation* between S and B. The probabilistic relation between S and B is in such case 'contaminated' by the additional occurrence of the alternative cause A which, occurring jointly with S, increases the probability of B when S occurs. As we remember from (3) the 'genuine' probabilistic relation between S and B, i.e. such relation which expresses only the causal connection between S and B (and the association between S and D) should be equal to $p(D \mid S)$. We may generalize this by saying that in this configuration of two alternative causal relations the probability of the occurrence of the effect when one component of one cause occurs, is the sum of a 'genuine probabilistic relation' between this component and the effect (when 'genuine probabilistic relation' means the probability of the occurrence of other components of this cause) and of spurious probabilistic relation which is equal to the probability of occurrence of other alternative causes – when this component occurs.[21]

Since the existence of alternative sufficient conditions of the same effect is a rather general phenomenon – at least in the social sciences – we may say *that in most probabilistic relations between social phenomena which we interpret in causal terms, we should try to distinguish and to evaluate the relative strength of their spurious and causal components and to systematize them in different theoretical propositions, relating different sufficient conditions to the same effect.* This can be done by means of 'multivariate analysis' of non-experimental data (that shall be discussed later) and especially by multivariate experimental design – the analysis of which would go beyond the scope of this chapter.

Let us assume now that none of the causes of B excludes the others, which may be graphically presented as follows:

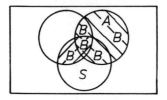

In this situation we shall have

(10) $p\,B \mid S = p(D \mid S) + p(A \mid S) - p(AD \mid S),$

$p(\overline{B} \mid \overline{S}) = p(\overline{A} \mid \overline{S}).$

Finally, let us assume a population in which all three causes of B, i.e., S, D, and A are statistically independent, and the conjunction of any two of them is also independent from the third variable. Then we will have

(11) $p(B \mid S) = p(D) + p(A) - p(D) \cdot p(A),$

$p(\overline{B} \mid \overline{S}) = p(\overline{A}).$

(E) Suppose now that S is a necessary condition for B but that it may form two different sufficient conditions with two different supplementary factors, D_1 and D_2. Graphically we may present it as follows

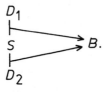

Or, under assumption that D_1 and D_2 are mutually exclusive:

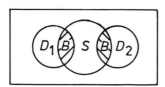

In this case the probabilistic relations look as follows:

(12) $p(B \mid S) = p(D_1 \mid S) + p(D_2 \mid S),$

$p(\overline{B} \mid \overline{S}) = 1.$

The first equation is identical as in (9), but the second is different because here we are certain that if S does not occur, B does not occur either, while in (9) we had $p(\overline{B} \mid \overline{S}) = p(\overline{A} \mid \overline{S}) < 1.$ We can summarize it by saying that the effect of occurrence of another alternative supplementary factor (D_2) and alternative sufficient condition (A) upon the probabilistic relation $p(B \mid S)$ may be the same. The difference between these two kinds of situations can be discovered by the observation of the relations between non-S and B.

If D_1 and D_2 are not mutually exclusive, their relation to S and B are presented by the following graphical scheme:

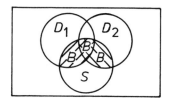

Here again we have:

(13) $p(B \mid S) = p(D_1 \mid S) + p(D_2 \mid S) - p(D_1 D_2 \mid S).$

And if these events are all mutually statistically independent, we have

(14) $p(B \mid S) = p(D_1) + p(D_2) - p(D_1) \cdot p(D_2)$

In both these situations, of course, $p(\overline{B} \mid \overline{S}) = 1.$

(F) Finally, let us take the case where S may have two alternative supplementary factors D_1 and D_2, and at the same time B may have another alternative sufficient condition A. Graphically we may present it as follows:

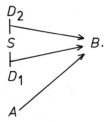

Assuming that D_1, D_2 and A are mutually exclusive we can present it as follows:

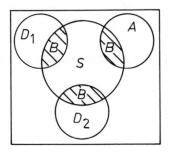

Here the probabilistic relations look as follows:

$$(15) \quad p(B \mid S) = p(D_1 \mid S) + p(D_2 \mid S) + p(A \mid S)$$

$$p(\overline{B} \mid \overline{S} = p(\overline{A} \mid \overline{S}).$$

One could present similar equations for any combination of relations *between* A, D_1, and D_2 other than their mutual exclusions, but it seems to be useless in this short presentation. We can do it only for the situation, where all four antecedent variables S, D_1, D_2, and A are completely statistically independent.[22] They we can write:

$$(16) \quad p(B \mid S) = p(D_1) + p(D_2) + p(A)$$

$$- p(D_1) \cdot p(D_2) - p(D_1) \cdot p(A)$$

$$- p(D_2) \cdot p(A) + p(D_1) \cdot p(D_2) \cdot p(A)$$

and

$$p(\overline{B} \mid \overline{S}) = p(\overline{A})$$

5. STATISTICAL RELATIONSHIPS IN MULTISTAGE CAUSAL CHAINS

On the previous pages we analyzed the case in which two different variables that were causally related to effect B were either different components of the same sufficient condition (different supplementary factors) or components of two different alternative causes (or two alternative causes themselves). But two different variables may both be causally connected with effect B and causally connected with each other because they are elements of *different links in the same causal chain*. Suppose that we analyze variables C, S and B, and that t_1, t_2, *and* t_3 indicate three consecutive cross-cuts of time. The most simple case will be:

$$t_1 \qquad\qquad t_2 \qquad\qquad t_3$$

$$C \longrightarrow S \longrightarrow B.$$

This means that C is a sufficient condition for S, and S is a sufficient condition for B. Let us call the causal relation SB the *direct causal relation*, and the relation CB the *indirect causal relation*.[23] But the case where the indirect cause is the sufficient condition of the direct cause is only one of many possible patterns of causal relations.[24] Let us analyze some other patterns.

Suppose that both C and S are essential components of two successive necessary and sufficient conditions of B. Graphically this may be presented as follows:

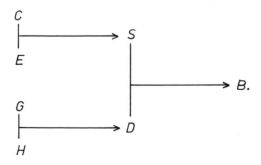

If this is the case, then the following equation is true:

(17) $p(B \mid C) = p(S \mid C) \cdot p(D \mid S)$.   or

         $p(B \mid C) = p(EGH \mid C)$.

Let us now assume that the pattern of causal relations is both multistage and alternative, according to the following graphical scheme:

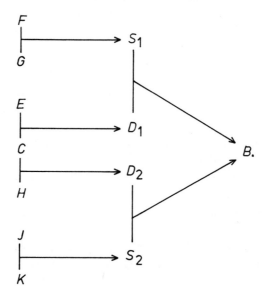

If this is the case, the following equation is true:

(18) $p(B \mid C) = p(D_1 \mid C) \cdot p(S_1 \mid D_1) + p(D_2 \mid C)$

         $\cdot p(S_2 \mid D_2)$ or

         $p(B \mid C) = p(FGE \mid C) + p(HJK \mid C)$.[25]

Suppose now that two different supplementary factors, D and S, have one common cause C when their other causes are different. Or graphically:

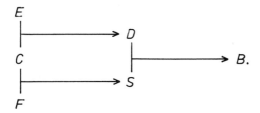

The following equation must be true:

(19) $p(B \mid C) = p(DS \mid C) = p(EF \mid C)$ or

$p(B \mid C) = p(D \mid C) \cdot p(S \mid D)$.

Lastly let us analyze the case where factor C (being equivalent either to C_1 or to C_2 for different cross-cuts of time) is the component of some consecutive sufficient conditions which form the necessary elements of several links in the causal chain. For example:[26]

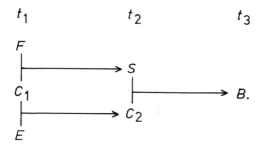

If this is the case then the following equations are true:

(20) $p(B \mid C_1) = p(S \mid C_1) \cdot p(C_2 \mid S)$ or

$p(B \mid C_1) = p(FE\ C_1)$.

6. RELATIVE FREQUENCIES AND RANDOM PROBABILISTIC RELATIONS

Until now we were using the term: 'probability' as equivalent to the term: 'relative frequency'. It is of course not always justified. Suppose that relative frequency of D when S occurs is equal to p in the whole population of S's (under assumption that we have a finite population of S's) but the time-space area occupied by the events of the type S may be divided into two sub-areas with clear time-space boundaries. In one of these subareas all S's always occur jointly with D, while in the other sub-area D *never* occurs jointly with S. Suppose now that the proportion of these S's which occur jointly with D among all S's is p. In this case we would still say that the relative frequency of D in relation to S is p in the whole population of S. But we wouldn't say that there is a certain *probability* of occurrence of D when S occurs (equal to p), because the notion of probability usually refers to such situations in which the distribution of D is 'relatively even' throughout the whole population of S's, in all time-space area occupied by the events S.

But this is not enough. We wouldn't say either that there is a certain probability of D when S occurs even if it is in the case when there is a certain 'even distribution' of D's throughout the all time-space area of occurrence of S's, but where we can discover a definite 'systematic pattern' of relations between S and D. Suppose that each second S (in space or time order) is accompanied by D. In this case the relative frequency of $B \mid S$ is exactly 0.5, but, knowing exactly which of the S's produce B and which of them do not produce B, we would not use the term 'probability' here because this term refers to the situations in which we are *basically uncertain* of the occurrence of B in relation to singular cases of S. This uncertainty is usually denoted by the term *randomness*. Therefore we would say, that in order to make justified the use of the term 'probability' for the pattern of joint occurrence of S and D, the occurrence of D in relation to S must be both relatively evenly distributed and random in its character. Only then we will observe such sequence of events S and B which may be characterized as *strictly probabilistic in terms of frequency theory of probability*[27], i.e. B will occur randomly in relation to S with p(D $\mid$ S) *as the limiting value of frequency of occurrence of B when S has occurred*, and with relative frequencies of B 'oscillating' around p in finite series of S's.

The same applies to all other causal situations characterized above. If the relations between our 'predictor' S of every B and its supplementary factors D_1, D_2 ... D_4 and (or) other alternative sufficient conditions of $B(A_1, A_2,...$ $A_4)$ can be characterized as random with the given $p(D \mid S)$ or $p(A \mid S)$ as limiting value of relative frequencies of their occurrence – then the above models can be transformed into the strictly probabilistic ones with $p(B \mid S)$ understood in strictly probabilistic terms. The meaning of the above models are not limited to such probabilistic situations; they can be used for any prediction if we dispose only relative frequencies of the conditional occurrence of the proper variables. But in such case in which the relations between the variables in our population are not random the formula $p(B \mid$ $S)$ should be read: 'The relative frequency of B when S occurs...' because the term 'probability' does not apply to them. We could nevertheless use the term 'probability' for any random sample drawn from such 'non-random population'. If in our population every second S occurs jointly with D and produces B, it would be useless to make a 'probabilistic' prediction of B with respect to S, when we are able to make a certain one. But the things look different for a random sample of S's from such whole non-random population. Here we may say justifiably, that in our random sample 0.5 is the limiting value of relative frequency of B when S occurs, or that *probability* of B when S occurs equals 0.5 using the term probability in its proper sense.

7. CAUSALITY, CORRELATION AND SPURIOUS INDEPENDENCE

In the last two sections we have analyzed the probabilistic relations between the variables viewed as functions of causal connections between those variables. But, as we know, statements about relative frequencies or probabilities of some events, when other events have occurred, are not the only category of statistical descriptions of relationships between the classes of events. We are often interested not only in what the probability of B is when S occurs, but also in their interdependencies, or in other words whether the occurrence of S increases the probability of the occurrence of B. Put still another way, we may ask whether the knowledge of the fact that S has occurred increases the accuracy of our predictions concerning the occurrence of S. In brief, we are often interested in whether S and B are

statistically independent or – if they are correlated – whether their correlation is positive or negative.[28]

As we know, two classes of traits or events, S and B, are statistically
independent when the following equation is true:

(21)　　　$p(SB) = p(S) \cdot p(B)$.

But the situation of statistical independence may also be described in
another way. The two classes of events, S and B are statistically independent
if the following equation is true.

(22)　　　$p(B \mid S) = p(B \mid \bar{S})$.

To say it in other words: two classes of events are statistically independent when the occurrence of one event does not increase (or decrease)
the probability of occurrence of the other event.

If the occurrence of one event increases the probability of occurrence of
the other one – then they are positively correlated according to the
formula:[29]

(23)　　　$p(B \mid S) > p(B \mid \bar{S})$.

Finally, S and B are negatively correlated when the occurrence of S
decreases the probability of occurrence of B according to the formula:[30]

(24)　　　$p(B \mid S) < p(B \mid \bar{S})$.

Thus we see that the correlation between S and B is here a function of
the probabilities $p(B \mid S)$ and $p(B \mid \bar{S})$. Let us consider on what these two
probabilities depend.

As we remember from Section 4, the probability of B when S does occur
depends on two factors:

(a) the relative probability of occurrence of the supplementary factor D,
when S has occurred i.e., $p(D \mid S)$;

(b) on the relative probability of occurrence of the alternative cause A,
when S has occurred i.e., $p(A \mid S)$ constituting its 'spurious component'.

But on the other hand both $p(D \mid S)$ and $p(A \mid S)$ are functions of the relative frequency of D and A in our population ($p(D)$ and $p(A)$) and their correlation with D in our population;

To put it briefly:

the more frequent the supplementary factors D in our population; the more frequent the alternative causes A in our population; and the more highly correlated both A and D are with S, the greater the probability that S will be followed by B.

Let us now look at the other side of the above equations and analyze which factors contribute to the probability that non-S will be followed by B. Here of course the supplementary factor D is irrelevant (because it may produce B only when it occurs with S, and the relative frequency of B when non-S occurs is a simple function of the relative frequency of the alternative cause A, when non-S occurs. But $p(A \mid S)$ is a function of the relative frequency of A in our population and of the correlation between S and A, To put it briefly:

the more frequent A is in our population; and the stronger its negative correlation with S, the greater will be the probability that non-S will be followed by B.

Thus we see that the frequency of the alternative cause A [31] and the sign and strength of its association with our 'predictor' S has an important role in determining what will be the predictive value of the variable S for the variable B. If A is positively associated with S it increases its predictive value. If S and A are statistically independent the existence of A in the population does not eliminate the prognostic value of S for B. (See below.) If they are negatively correlated it descreases, more or less, the prognostic value of S for B. Let me stress two special cases for which the existence of A diminishes the prognostic value of A for B.

The situation may be graphically presented as follows:

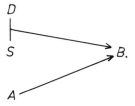

Let us suppose for simplicity's sake that there is a perfect negative correlation between S and A (i.e., that they are mutually exclusive).

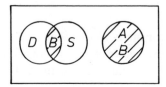

If in this situation

(25) $p(D \mid S) = p(A \mid \overline{S})$ we then have

$p(B \mid S) = p(B \mid \overline{S})$.

This is the situation which may be called *spurious independence*, because the two variables S and B are statistically independent in a situation in which they are causally connected.

If it happens that $p(A \mid \overline{S})$ is greater than $p(D \mid S)$ we would have:

(26) $p(B \mid S) < p(B \mid \overline{S})$.

This means that B and S are negatively correlated in a situation where they are connected causally. This situation may be called *spurious negative correlation*.

Thus we see that the relation between the fact that the variables are causally connected and the prognostic value of one variable for the other is highly conditional. The prognostic value of a cause for its effect is diminished by the existence of alternative causes of our effect in the studied population if they are negatively correlated with our predictor. In some cases — as we have seen — it diminished the prognostic value of S for B to 0 or even changes its sign to a negative one.

In all these cases we again obtain the positive correlation between S and B when we take the alternative cause A as a constant factor in multivariate analysis.[32]

Let us now look more systematically at this problem, studying the statistical relationships between S and B for the four kinds of causal relations between S and B distinguished above.

(a) If S is a necessary and sufficient condition for B, then according to what was said above, $p(B \mid S) = 1$ and $p(B \mid \bar{S}) = 0$. In such a situation the correlation between S and B will be positive and equal to $+1.0$.

(b) If S is a necessary but not a sufficient condition for B, then $0 < p(B \mid S) < 1$, but $p(B \mid \bar{S}) = 0$. In this case there will be a positive correlation between S and B.

(c) If S is a sufficient but not a necessary condition for B then $p(B \mid S) = 1$, whereas $0 < p(B \mid \bar{S}) < 1$. Here again there will be a positive correlation between S and B.[33]

Thus we may say that if the *causal relation between S and B is unconditional, either in its positive aspect or in its negative aspect or in both, there must be a positive correlation between S and B.*

(d) If S is a necessary component of only one of the alternative sufficient conditions, then S may be either positively or negatively correlated with B or independent of B.

Therefore *if the causal relation between S and B is conditional both in the positive and in the negative senses such a situation does not necessarily imply a positive correlation between S and B.*

Nevertheless there is a class of situations in which there will be always a positive association between S and B, even if S is not a sufficient condition for B and if there are some alternative causes A of our B, whatever the relative frequency of A in our population and whatever the probability of $(D \mid S)$. This will happen always when these alternative causes A are statistically independent of S. Let us consider the situation where D and A are mutually exclusive, and S and A are statistically independent.

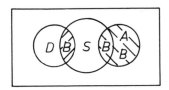

Here we will have:

(27) $p(B \mid S) = p(D \mid S) + p(A \mid S)$

but, because S and A are statistically independent, we may write

$$p(B \mid S) = p(D \mid S) + p(A).$$

On the other side

$$p(B \mid \overline{S}) = p(A \mid \overline{S}) = p(A).$$

As a result of this we have

$$p(B \mid S) > p(B \mid \overline{S}).$$

The same is true when all three antecedent variables are completely statistically independent. Let us look at the following scheme

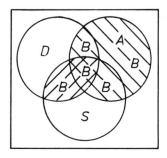

Here we have:

$$(28) \qquad p(B \mid S) = p(D) + p(A) - p(D) \cdot p(A)$$

when on the other side we still have

$$p(B \mid \overline{S}) = p(A).$$

Since $p(D) > p(D) . p(A)$ for any relative frequency of A which is lower than 1 we have

$$p(D) + p(A) - p(D) \cdot p(A) > p(A)$$

which implies

$$p(B \mid S) > p(B \mid \bar{S}).$$

We can summarize by saying that *if all components of all alternative sufficient conditions of our effect are mutually statistically independent, this will result in a positive association between any of the essential components of any of the sufficient conditions and their effect.* This seems to be a necessary assumption in everyday research practice, which takes for granted that the causal relations between variables can be assessed by the observation of positive associations and correlations between them. But this assumption should be clearly formulated in any inference from statistical to causal relations, especially because it is by no means universally valid, as we saw above.

Sometimes we can make a weaker assumption, namely that one of the causes − the variable S − that is a necessary component of one of several sufficient conditions of B is statistically independent both from all its supplementary factors $D_1, D_2, ... D_n$ and from all alternative sufficient conditions $A_1, A_2, ... A_n$, whatever the mutual statistical relationships between these variables $A_1, A_2, ..., A_n, D_1, D_2, ..., D_n$. It is obvious that in this situation there will always be a positive relationship between S and B, although not necessarily between B and any of its other causes.

It should be noted that there is a fairly common research situation in which we may assume that the variable S is statistically independent of all other variables involved in producing B, i.e. both from all supplementary factors $D_1, D_2, ..., D_n$ and also from all alternative sufficient conditions $A_1, A_2, ..., A_n$. Therefore S will always be positively associated with B. *This is the situation of random experiments,* in which we randomly assign the subject to either the 'experimental' or the 'control' group. Randomization makes the distributions of all relevant variables approximately equal in both groups. If we then apply the 'stimulus' S to the experimental group only, S will be statistically independent of all its supplementary factors as well as all other alternative causes of B. Then, according to the above reasoning the causal connection between S and B must be revealed by comparison of the frequencies of B in the two groups with $p(B \mid S) > p(B \mid \bar{S})$. It seems that the basic logical foundations for the validity of random experiments in revealing causal connections consist of the fact that its procedures guarantee us that the assumption of independence discussed above is fulfilled.

8. SPURIOUS CORRELATIONS

In the previous section of this chapter we mentioned that a positive association of the alternative cause of B with our 'predictor' S increases the prognostic value of S for B, contributing, more of less to the positive correlation between S and B. This correlation is in general a function of the two factors:

(a) the relative frequency of the occurrence of supplementary factor D, when S occurs, and

(b) the relative frequency of the other cause of B (let us designate it here by C) and the association between C and B, which constitutes the spurious component of this relation.

The relative size of the 'contribution' of these two factors to the total strength of the correlation between S and B may vary; the one extreme of this variation is the situation where the correlation between our predictor S is completely determined by the factor C and its association with S. This means that in our population (or on a universal scale) B is not the effect of S. This may happen either when:

(a) S is not causally related to B in any possible conditions or

(b) S may be a conditional cause of B but does not 'meet' in this population its supplementary factor D together with which it forms a sufficient condition for B.

This situation where S is positively correlated with B but is not connected causally with S for any of these reasons is called 'spurious correlation'.

Let us analyze different types of spurious correlations.

By a spurious correlation we often mean the correlation which is produced by the fact that two variables that are causally unrelated have one or more causes in common. If this is not a 'perfect' (i.e., 1.0) correlation this means that the common cause of S and B (or in our terminology, common essential component of two different sufficient conditions of S and B), in connection with different supplementary factors, forms two sufficient conditions for the two original variables. Or, graphically, the correlation SB is spurious when, for example:

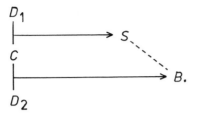

Here we have of course:

(29) $p(S \mid C) = p(D_1 \mid C)$

$p(B \mid C) = p(D_2 \mid C).$

It should be stressed that the fact that two variables S and B have one cause (C) in common does not yet imply that they have to be positively associated. The association (correlation) between S and B is here primarily a function of the correlation between D_1 and D_2, and we can well imagine a situation where there will be a perfect negative correlation between S and B even though they have one common cause C. This happens when D_1 and D_2 are mutually exclusive. This is illustrated by the following graphical scheme:

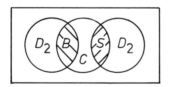

Then, of course, S and B are also mutually exclusive.

Even when D_1 and D_2 are not completely mutually exclusive, but are strongly negatively associated, it may result in a negative correlation between S and B, or (if their negative association is not too strong) it may result in statistical independence between S and B.

Let us now analyze our favorite situation in which C, D_1, and D_2 are statistically independent. In this case we will have:

(30) $p(B \mid S) = p(D_2 C) / p(D_1 C).$

But, since all variables are independent and S and B have as one of their causes the same variable C in common, it turns out that

$$p(B \mid S) = p(D_2).$$

On the other hand

$$p(B \mid \overline{S}) = p(CD_2 \mid \overline{CD}_1)$$

But now we have another situation, namely, that due to the independence of (CD_2) from $(\overline{CD}_1)$

$$p(B \mid \overline{S}) = p(CD_2) = p(C) \cdot p(D_2)$$

and since

$$p(D_2) > p(C) \cdot p(D_2),$$

Therefore

$$p(B \mid S) > p(B \mid \overline{S}).$$

Let us now see what will happen if we apply Lazarsfeld's multivariate analysis[34] , keeping C constant and looking for partial relationships between S and B both for C and for non-C.

For C: $p(B \mid S) = p(D_2)$

(31) $p(B \mid \overline{S}) = p(D_2)$

For non-C: $p(B \mid S)$ cannot be computed because
              S is absent for non-C

$$p(B \mid \overline{S}) = 0.$$

We see here that in the case *where C, D_1 and D_2 are statistically independent, the relationship between S and B must be positive in the whole population, and it must disappear when C is taken constant as the 'test variable' in multivariate analysis.*

The same will happen when C is not a necessary condition for B, but when B has in this population another alternative sufficient condition, namely A, if we assume that A is statistically independent of CD_1 and CD_2 (which are also mutually independent, as above). Graphically we may present this as follows:

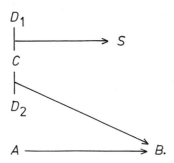

Here we will have:

(32) $p(B \mid S) = p(D_2) + p(A) - p(D_2) \cdot p(A)$

on the other side:

$$p(B \mid \bar{S}) = p(D_2) \cdot p(C) + p(A) - p(D_2) \cdot p(C) \cdot p(A)$$

and since

$$p(D_2) - p(D_2 \cdot p(A) > p(D_2) \cdot p(C) - p(D) \cdot p(C) \cdot p(A)$$

therefore

$$p(B \mid S > p(B \mid \bar{S}).$$

We see that in this case also there will be a positive association between S and B. When we take C as constant the partial relationships will look as follows:

For C: $p(B \mid S) = p(D_2) + p(A) - p(D_2) . p(A)$

(33) $p(B \mid \bar{S}) = p(D_2) + p(A) - p(D_2) \cdot p(A)$

For non-C: $p(B \mid S)$ cannot be computed because S is absent in this
        situation

$$p(B \mid \bar{S}) = p(A).$$

Finally if we assume that S and B may in addition have some alternative
causes A_1 and A_2 to which C does not belong, e.g.:

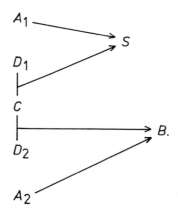

and all five variables (A_1, A_2, C, D_1, D_2) are statistically independent in all
possible combinations, we will obtain the typical results as described by
Lazarsfeld, in which the relationship between S and B is positive in the total
population and disappears both for C and for non-C. But $p(B \mid S) > 0$ in all
groups of the test variable, namely:

For C: $p(B \mid S) = p(D_2) + p(A_2) - p(D_2) \cdot p(A_2)$

(34) $p(B \mid \bar{S}) = p(D_2) + p(A_2) - p(D_2) \cdot p(A_2)$

For non-C: $p(B \mid S) = p(A_2)$

                $p(B \mid \bar{S}) = p(A_2).$

But now suppose that we have our previous most simple situation in
which:

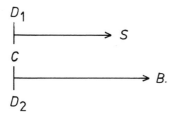

But D_1 and D_2 are positively associated, so that $p(D_1 \mid D_2) > p(D_1 \mid \bar{D}_2)$. If they are in addition either independent of or related to C in the same way, it is obvious that the correlation between S and B will not disappear for the group C. It will still be positive and will 'disappear' only for the group non-C where both S and B are absent. The same conclusion can be extended to all of the more complicated situations analyzed above.

Let us again consider the situation described above in formula (34), but under the assumption that A_2 is positively associated with CD_1, with all the other variables being mutually independent. Without further analysis we can see that the relationship between S and B will not disappear in the group C if we keep C constant; it will only decrease, due to the 'contribution' of the variable A_2 to the relationship between S and B in group C, since $p(B \mid S) = p(B \mid \bar{S})$ is equal to $p(A_2)$ for this group.

Thus we may say that *if S and B have one cause C in common, their relationship will disappear after making C constant only under the condition that all other variables responsible for the occurrence of S and B in this population are statistically independent in all possible combinations.*

And lastly, let us analyze the scheme where the spuriousness is caused by the fact that S and B have more than one cause in common. Graphically this may be presented as follows:

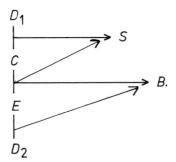

If this is the case, it is quite obvious that under all assumptions mentioned above the spurious relationship will never disappear in the partial
correlation when we take as the test variable either C or E. It can disappear
only in a more complicated four-variable scheme of analysis in which we
keep both C and E constant.

This means that *the definition of the 'causal' statistical correlation as a
correlation which never disappears after introducing a third test variable*[35]
*is too broad because it also denotes all types of spurious correlation
analyzed above, and also many others omitted here.*

In all cases analyzed above the spurious correlation between S and B was
produced by the fact that both S and B have a common cause C. But it may
happen that there is no causal connection between C and S (and C is
causally connected only with B), but there is nevertheless a positive correlation between S and C resulting in a spurious correlation between S and B.
We can imagine, for example, the situation in which we have a pattern of
causal connections between the variables as follows:

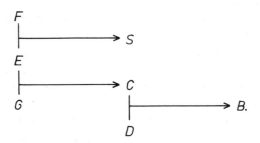

(a) S is not causally connected with B;
(b) S is not causally connected with C.

In this situation the spurious correlation between S and B will disappear
when we take C as a test factor in multivariate analysis.[36] But C does not
necessarily precede S in time.[37] It may precede it, it may occur at the same
time, or, in some cases, it may even occur after S and nevertheless, the
correlation between S and B will be spurious.

The situation where S and C occur at the time seems to be especially
worth mentioning. This is the situation where both C and S are elements of
a composed syndrome of variables, e.g., a syndrome of dispositional traits,

or of stimuli. It is well known that in many psychological researches the effects (B) first attributed to variables of the S type were later interpreted as the results of some other elements (C) of the same syndrome to which S belonged. If the correlation between S and C is not a perfect one, the 'spuriousness' of the connection between S and C might easily be discovered if we took C as a test factor in a multivariate scheme of analysis.

9. THE TEST VARIABLE AS A SUPPLEMENTARY FACTOR AND AS AN ALTERNATIVE CAUSE

In the previous section of this chapter we have analyzed hidden assumptions behind usually applied methods of discovering whether a certain correlation is a spurious or causal one. But it should be remembered that proof of causality or of spuriousness is not the only possible result of multivariate analyses. The other possible results may constitute the evidence that the 'test variable' introduced in order to interpret the initial relationship SB is either a supplementary factor (D) or an alternative cause of $B(A)$, or a necessary component of such an alternative cause A_1. Let us designate the alternative sufficient condition for B as A_1 and A_2

Let us assume now that:

(a) SD is a sufficient condition for B, and

(b) There is no alternative cause of B in this situation.

Then we have the following pattern of relationships between S and B, when we take D as constant:

for D $p(B \mid S) = 1$

$p(B \mid \overline{S}) = 0$

for non-D $p(B \mid S) = p(B \mid \overline{S}) = 0.$

In other words, we have a positive (1.0) correlation between S and B for subgroup D and no correlation for non-D.

In the situation where the relation between S and B is conditional upon the occurrence of D (or the relation between D and B is conditional upon S), when S and D are necessary components of the same sufficient condition, we may say that we have the case of *interaction* between S and D in producing the effect B.

In the situation where S and A are either two alternative sufficient conditions of B or are necessary components of two alternative sufficient conditions, we may say that S and A (or D and A) are *additive causes* of the effect B.

In the above case we assumed that there are no additive causes of B in our population, that all B's are the effects of the interaction between S and D only.

Let us now assume that there is another alternative cause (A) of B in our population, all factors: S, D and A, are statistically independent and the relative frequency of A is equal to p_1.

Here we will have:

for D $p(B \mid S) = 1$

              $p(B \mid \overline{S}) = p_1$

for non-D $p(B \mid S) = p_1$

              $p(B \mid S) = p_1$.

If the sufficient condition for B is more complex, namely SD_1D_2, but we control only D_1 (assuming that the alternative cause is here A with relative frequency p_1 and all variables are statistically independent), we will have:

for D_1 $p(B \mid S) = p(D_2) + p_1 - p(D_2) \cdot p_1$

              $p(B \mid \overline{S}) = p_1$

for non-D_1 $p(B \mid S) = p_1$

              $p(B \mid \overline{S}) = p_1$.

In general it may be said that:

(a) if D_1 and S are components of the same sufficient condition;

(b) if the probability of the joint of occurrence of all unknown components (D_n) of the same sufficient condition to which S and D_1 belong is equal to p_1;

(c) if all the probability of the occurrence of all alternative causes (A_n) of B is equal to p_2; and

(d) if all the variables involved in producing B in this population are statistically independent; then

(e) the pattern of probabilities for B may be presented in the following fourfold table with two independent variables:

	D_1	$\bar{D}_1$
S	$p_1 + p_2 - p_1 \cdot p_2$	p_2
$\bar{S}$	p_2	p_2

If S, D_1, D_2 and A are not independent, the pattern of relationships may be a little different, but we will usually still have the situation in which we have a visible positive correlation between S and B for one value of the test variable (for D_1) and a lack of relationship between S and B for non-D.[38]

The situation will be different if the test variable A is an alternative *additive sufficient condition of B* or, what is more interesting, when it is a necessary component of another alternative sufficient condition of B.

Now suppose that we have a situation in which B may be produced in our population by two alternative sufficient conditions (S and D_1), or (A and D_2). Suppose additionally that all four variables are statistically independent and the relative frequencies of the unknown supplementary factors D_1 and D_2 is:

$$p(D_1) = p_1$$
$$p(D_2) = p_2.$$

If we then control for S and A in a multivariate analysis, we obtain following frequencies for the effect of B.

	S	$\bar{S}$
A	$p_1 + p_2 - p_1 \cdot p_2$	p_2
$\bar{A}$	p_1	0

If we denote $p(B)$ in the four cells of our table as follows:

	S	$\bar{S}$
A	p_{I}	p_{II}
$\bar{A}$	p_{III}	p_{IV}

we will of course have

$$p_{\text{I}} > p_{\text{II}} \quad \text{and} \quad p_{\text{III}} > p_{\text{IV}}$$

but also

$$p_{\text{I}} > p_{\text{III}} \quad \text{and} \quad p_{\text{II}} > p_{\text{IV}}.$$

Now suppose that in addition to the causes SD_1 and AD_2, B has another alternative cause C in our population, completely unknown to us, independent of the partially controlled causes S and A and having a relative frequency equal to p_3. If we then control for S and A as above, the multivariate frequencies of B will be as follows:

	S	$\bar{S}$
A	$p_1 + p_2 + p_3 - p_1 \cdot p_2$ $-p_1 \cdot p_3 - p_2 \cdot p_3$ $+ p_1 \cdot p_2 \cdot p_3$	$p_2 + p_3 - p_2 \cdot p_3$
$\bar{A}$	$p_1 + p_3 - p_1 \cdot p_3$	p_3

Also if we denote the frequencies of B in our table by:

	S	$\bar{S}$
A	p_{I}	p_{II}
$\bar{A}$	p_{III}	p_{IV}

we will have: $p_I > p_{II}$ and $p_{III} > p_{IV}$

and also $p_I > p_{III}$ and $p_{II} > p_{IV}$

I think that this represents a case which is *typical of that we obtain from multivariate analysis in which we control for two additive causes.* It should be noted that *this situation is only partly additive.* In each case when we are ad-ding the probabilistic values with which any or two or three different causes produces the given effect in the absence of all other causes, this sum has to be diminished by the combinatorial product of the probabilities, relating to all causes involved in the given situation.

The fact that the additive causes are only 'partly additive' in their probabilistic effects deserves special attention in statistical analyses of social data.

10. SOME OTHER FUNCTIONS OF THE TEST VARIABLE

P.F. Lazarsfeld in his study mentioned several times above describes a certain situation in which

(a) there is a positive correlation between the test variable and the independent variable;

(b) the independent variable precedes in time the test variable; and

(c) the initial correlation between independent and dependent variables desappears

(c) the initial correlation between independent and dependent variables disappears

In this situation the initial correlation is by no means a spurious one; it is only *less direct* than the causal relation between the best variable and the dependent variable. If less directs than the causal relation between the test variable and the dependent variable. If situation as follows:

$$S \longrightarrow M \longrightarrow B.$$

If we assume that between the variables belonging to different links of a causal chain there exist probabilistic relationships, we might present them as follows:

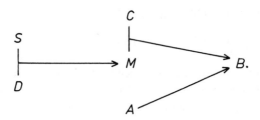

The statistical associations between S and B are here quite obviously functions of the causal connections between S and B and of the statistical association between M and the alternative cause A of the variable B. The reader may easily formulate the assumptions on which the patterns of simple and partial relationships described by Lazarsfeld for this case will hold true. Accepting for simplicity's sake these assumptions, I would like to present some other more complicated patterns of causal connections between S and B, on the one hand, and a test variable designated as T (the test variable), on the other.

The first pattern, in a simplified form, may be presented as follows:

In this situation S is both an indirect cause of B (by producing its other cause T) and a direct one, because S and T are interacting causes of B, belonging to the same sufficient condition of B.[39] Here the relationship between T and B is partly spurious and partly causal. In this situation we have the following patterns of relationships between the three variables:

(a) S precedes T in time;

(b) there is a positive correlation between S and T;

(c) after holding T constant the relationship between S and B increases for T and disappears for non-T.

A different case is the situation in which the relationship between S and B does not disappear for both subgroups of the variable T (all other relationships being the same as in the previous case).[40] If this occurs we may say that S produces B both directly and indirectly by the mediating factor T, and T and S are (or belong to) two alternative additive causes of B. Graphically we may present this as follows:[41]

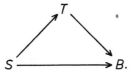

In both these cases the relationship between S and B is by no means a spurious one; moreover it is indirect for only one group of B, whereas for the other group of B it is as direct as the relationship between T and B.

Another type of situation is the case where the correlation SB is partially spurious and partially causal. In this situation the pattern of relationships between the three variables is as follows:

(a) T precedes S in time and
(b) T is correlated with S.

There are three possible categories of situations in which S produces B both directly and indirectly by producing T.

If after holding T constant the relationships SB descrease in both subgroups of T but are still positive and approximately equal, this means that T and S belong to alternative causes of B according to the following scheme:

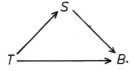

If the partial relationship SB is much stronger for T and disappears for non-T, this means that T and S are necessary components of the same cause according to the scheme:

Let me give two examples illustrating the two situations presented above. In a study of social attitudes of Warsaw students I found that a student's equalitarianism (B) is related both to his parents' income (S) and to his parents' occupational group (T). T and S were strongly correlated. When we held either of these variables constant there was a viable correlation between egalitarianism and the other variable in all subgroups of the test variable. This means that we had the following pattern of causal relations between these three variables:

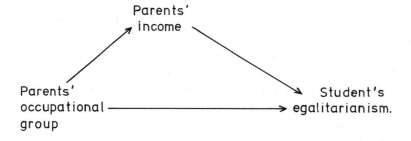

In other words, occupational group and income (although causally connected to each other) were in this population components of alternative additive causes of students' egalitarian attitudes.

Another example may be the relationship between egalitarianism, parents' occupational group and the place of the student's habitation. The last two variables were also strongly correlated; the dormitories were inhabited mostly by students coming from lower social strata. When we held the place of living constant we found that the correlation between family background and a student's egalitarianism increased visibly for students living with families and disappeared for those who lived in dormitories.[42] The pattern of causal connections may be presented as follows:

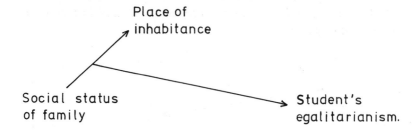

11. A TYPOLOGY OF THREE-VARIABLES ANALYSES

In an earlier section of this chapter I analyzed the probabilistic and cor-relational relationships between an independent variable x and a dependent variable y under different patterns of causal relationships between x and y, and in different patterns of statistical associations between x and a set of alternative causes of y in the given population.[43] Then we analyzed the results of the introduction of test variables (t) into these patterns of relationships, taking into account the time order of the variables x and t, their statistical association and the partial relations between x and y when t was held constant. The results of the above analyses may be briefly summarized in a table in which we have the following criteria of classification:

(1)   the time order of t and x:

   (a) t precedes x;
   (b) x precedes t;
   (c)   time order of x and t irrelevant;

(2)   statistical association between x and t:

   (a) t and x statistically independent;
   (b) t and x positively associated;

(3)   the partical relationship between x and y when test variable t is con-stant:

   (a)   disappears in both groups of the test variable t;
   (b) xy positively associated in both groups of the test variable t;
   (c) xy disappears for non-t and increases for t.

The results of the above analyses may be presented as in Table I.

TABLE I

Typology of three-variable analyses

Statistical relationships between the independent and the test variables (x,t)	time order between independent variable x and test variable t
x and t statistically independent	time order of x and t irrelevant
x and t positively associated	t precedes x
	x precedes t

Partial relationship between x and y when t is held constant

xy disappears in both groups of the variable t	xy increases for t and disappears for non-t	xy is positive and approximately equal for t and for non-t
introduction of another test variable (e.g., z) explaining this inconsistent pattern necessary	t and x belong to the same sufficient condition for y x $\vdash\!\longrightarrow y$ y both xy and ty causal relations	t and x belong to different alternative causes of y $x \longrightarrow\!\!\!\!\searrow$ $\longrightarrow\!\!\!\!\nearrow y$ $t \longrightarrow$ both xy and ty are causal relations
original relationship xy spurious $\nearrow x$ $t \longrightarrow y$ relationships tx and ty causal	t produces y both directly and indirectly by producing x $\nearrow x$ $t \longrightarrow\!\!\!\!\searrow y$ x and t belong to the same sufficient condition of y. Relationship xy partially spurious, partially causal. Relationship ty causal	t produces y both directly and indirectly by producing x $\nearrow x \searrow$ $t \Longrightarrow y$ x and t belong different alternative causes of y xy partially spurious, tx causal
t is a mediating variable between x and y $x \longrightarrow t \longrightarrow y$ x is an indirect cause of y when t is a direct cause of y. All relationships, xt, ty, xy are causal	t is a mediating variable between x and y and a supplementary factor for x $\nearrow t$ $x \longrightarrow y$ Relationship xy is causal, when ty is partially spurious and partially causal	t is a mediating variable between x and y belong to alternative cause of y $\nearrow t \searrow$ $x \Longrightarrow y$ Relationship xy is causal, when ty is partially spurious and partially causal

12. ADDITIVITY AND INTERACTION BETWEEN THE
QUANTITATIVE VARIABLES

In the earlier sections of this chapter I analyzed the statistical relationships between dichotomous attributes as functions of the causal relations among them. The typology of possible causal connections was based on the idea that we are looking for either conditional or unconditional causal relations (in both the positive and the negative senses of this term) of a certain class of dichotomously defined events B. We also wish to discover all other (dichotomously defined) supplementary factors of the given conditional cause and also all alternative causes of B existing in the given population.

This type of reasoning seems to be useful both for discovering the existing causal connections between a set of variables in a given population and for analyzing hidden assumptions and limitations of methods of statistical analysis usually encountered in empirical research. The fact that the models presented above are only valid for dichotomous events or attributes is nevertheless a serious limitation with respect to their applicability. It may also be, as some authors have already indicated, a source of more basic methodological difficulties because a dichotomous classification is usually based on more or less artificial cutpoints in some quantitative variable.[44]

It seems that a way of reasoning similar to that we applied for dichotomous attributes may also be applied to quantitative variables, where we can interpret the statistical correlation as a function of the given type of causal connection between the quantitative variables. Let us present some premises for this type of analysis.

Let us assume that we have four quantitative variables S, D, A, B, which for simplicity's sake have been given some measurable values:

the variable S may have as its values S_0, S_1, S_2;
the variable D may have as its values D_0, D_1, D_2;
the variable A may have as its values A_0, A_1, A_2;
the variable B may have different values depending on the
functional relationship between D, S, A and B.

It seems that, looking for different patterns of causal relations among these quantitative variables and taking, for simplicity's sake, only monotonic 'rectilinear' functional relations between them, we might distinguish quite a few different causal patterns. These patterns can be classified into two large groups depending on whether S, A, and D act *additively* or *interact* with each other in determining the value of B.

(1) Before we do that we should distinguish the simplest pattern of causal connection between S and B: the situation in which the relation $B = f(S)$ is an *unconditional* or *completely functional causal relation*. This means that in all possible conditions (including all possible values of D) the value of B is completely determined by the value of S and the values of D are irrelevant. This may be presented as in Table II.

TABLE II

	S_0	S_1	S_2
D_0	B_0	B_1	B_2
D_1	B_0	B_1	B_2
D_2	B_0	B_1	B_2

Looking for an analogy to this type among the types of causal connections between dichotomous attributes, this situation corresponds to the case where S is a necessary and sufficient condition for B because we have a complete correspondence of values of the two variables S and B according to the formula:

$$S_0 = B_0$$
$$S_1 = B_1$$
$$S_2 = B_2$$

For a sociologist a much more realistic situation is when the given value of D does not change the strength of the statistical relationship between S and B (i.e., R_{SB}) according to the formula.[45]

$$R_{SB}(D_0) = R_{SB}(D_1) = R_{SB}(D_2) = R_{SB}.$$

It should be noted, moreover, that if D and S are statistically independent, we shall have $R_{DB}= 0$ in the total population. If they are correlated positively (or negatively) there will be a positive (or negative) correlation between D and B in the total population.

(2) Then we have a class of *additive situations* in which

$$B = f_1(A) \pm f_2 (S)$$

(Remember that additivity means that A and S are either two alternative causes of B due to two functional relations: $B = f(A)$ and $B = f(S)$, or they are the essential components of two such alternative causes of B.)

(2a) A deterministic pattern of additive relations between SA and B might be presented in Table III when the sign of the above formula is positive.

TABLE III

	S_0	S_1	S_2
A_0	B_0	B_1	B_2
A_1	B_1	B_2	B_3
A_2	B_2	B_3	B_4

(2b) It may also happen that $B = f_1(A) - f_2(S)$ when there are two factors which *counteract* each other in determining the value of B. If we also introduce for B some negative values, the result of two (negatively additive) counteracting factors could be presented for deterministic situations as in Table IV.

TABLE IV

	S_0	S_1	S_2
A_0	B_0	B_1	B_2
A_1	B_{-1}	B_0	B_1
A_2	B_{-2}	B_{-1}	B_0

It should be remembered that for two (positively or negatively) additive factors the following formulas are true:

$$A_{const} \longrightarrow B = f(S)$$

$$S_{const} \longrightarrow B = f(A).$$

This means that when the values of one independent variable are kept constant, we obtain a functional relationship between the other variable and the effect, and vice versa. But in both cases the relationship between one independent variable and the dependent variable will be slightly different for different values of the test variable. We know that a general formula for a linear functional relation is as follows:

$$y = ax + b.$$

In our case (for the situation 2a) when holding constant the variable A, the relation between S and B will be as follows:

for A_0; $B = S$

for A_1; $B = S + 1$

for A_2; $B = S + 2$.

If we take S as the test variable, we will have

for S_0; $B = A$

for S_1; $B = A + 1$

for S_2; $B = A + 2$.

We can observe this when we control both additive variables at the same time. But now suppose that we observe in total population only the overall relation between S and B. In this case it is very likely that for any given value of S we will observe a certain range of values of B, but not lower (if S and A are positively additive) than the pure effect of the given value of S upon B, and not higher than the sum of the impact of this value of S upon B and of the impact of the highest possible value of A. Thus, the range of values of B

for S_0 will be B_0, B_1, B_2

for S_1 will be B_1, B_2, B_3

for S_2 will be B_1, B_3, B_4

as can be seen from Table III.

We therefore see that the range of variation of the values of the dependent variable for the given value of the independent variable is a significant indicator of the range of variation of the dependent variable due to alternative, additive causes of this dependent variable in the given population.

It is interesting to study what the observed mean value of B will be for given values of S. This of course depends upon the correlation between S and A in our population. We can distinguish some different situations:

(a) S and A are positively correlated and their correlation is equal to 1.0. In this case we will have

$$\text{for } S_0 - B_0$$

$$\text{for } S_1 - B_2$$

$$\text{for } S_2 - B_4$$

or in other words the observed functional relation between S and B corresponds to a theoretical relation of the type $B = 2S$.

(b) If the correlation between S and A is positive but not perfect in the total population we will observe a certain relation between S and the mean values of B, the regression line of which has a steeper slope than this determined by the genuine functional relation between them only. The correlation between S and B is here partly spurious.

(c) We can also assume that there is a perfect negative correlation between S and A. In this situation there will be no correlation between S and B in the total population because there will be a constant value of B for all values of A. This will be a clear case of spurious independence for quantitatieve variables.

(d) We can also assume that S and A are correlated negatively but $R_{SA} > - 1.0$. In this case the mean values of B for S will indicate a regression line with a slope which is less than their 'genuine' functional relation. In this case their observed relations should be regarded as having a component of a spurious independence.

(e) Finally, we can assume that S and A are statistically independent. In such a case for a given value of S we obtain the mean value of B which is the sum of two components:

(1) the value of B which is produced by the given value of S due to their functional relation, and

(2) the mean value of B which is due to the action of the other alternative cause A in our population.

Suppose, for example, that in our population A has such an impact upon the dependent variable B – due to the given functional relation $B = f(A)$ and the given distribution of values of (A) – that as a result of the action of A along B (i.e., for S) has in our population the mean value = 1.8. Then the relationship between S and B in this population corresponds to the formula (if S and A are independent):

$$B = S + 1.8$$

or for specific values of S we will have:

$$\text{for } S_0 \qquad B = 1.8$$
$$\text{for } S_1 \qquad B = 2.8$$
$$\text{for } S_2 \qquad B = 3.8, \text{ etc.}$$

All this refers to situations in which we observe the total relation between S and B in our population, which is here a function of its 'genuine' and 'spurious' component. This is similar to the case of dychotomous attributes. When we observe the impact of the two variables S and A in a multivariate design, the nature of the 'disguised' functional relation behind the statistical relations among any of them and their effect can be clearly established – under assumption, of course, that no other causes of B are acting in this population.

If there are some other additive causes involved in determining the values of B in this population we will observe in our multivariate design similar effects in terms of variations of mean values of B for two independent variables jointly, to those we discussed above for the situation where we observed only one independent variable and B was a function of two additive causes. Needless to say, this is what we usually have in sociological data, even if we can believe that all variables involved have additive effects.

(3) When we say that S and D are *interacting* in producing B (here changing the symbol for the other variable) this means that the existence, shape, direction, or strength of the causal (functional or statistical) relation between S and B depends upon the value of the variable D.[46] It may be the case that:

(3a) The functional relations between S and B exist only for some specific values of D (e.g., when D_1 occurs) according to the formula $D_1 \longrightarrow [B = f(S)]$ whereas

$$\bar{D}_1 \longrightarrow [B | \neq f(S)].$$

Here (see Table) D_1 is the *determiner of the existence of the relation* between S and B.

TABLE V

	S_0	S_1	S_2
D_0	B_0	B_0	B_0
D_1	B_0	B_1	B_2
D_2	B_0	B_0	B_0

This seems to be only one of an infinite number of interaction situations which correspond to the case analyzed above for dichotomous attributes, where the two variables were supplementary factors for each other. We may say that D_1, together with different values of the variable S, creates sufficient conditions of different values of the variable B according to the formula:

$$D_1 S_1 = B_1$$
$$D_1 S_2 = B_2$$
$$D_1 S_3 = B_3 \text{ etc.,}$$

when for the other values of the variable D there is no relation between S and B.

This causal pattern has very clear probabilistic implications for situations in which we observe only one independent variable, namely S. What we must take into account is the relative frequency of D_1 for different values of S. Suppose that the mean frequency of D_1 for all values of D is equal to 0.5 and its occurrence is relatively evenly distributed for all values of S (i.e., that S and D_1 are independent). In this case the mean value of B for S will be:

$$\text{for } S_0 - B = 0$$

$$\text{for } S_1 - B = 0.5$$

$$\text{for } S_2 - B = 1.$$

But D_1 does not necessarily have to be independent of S, and different values of S may have different probabilities of joint occurrences with D_1. This may have an impact upon the shape or strength of the relationship between the values of S and the mean values of B. If $p(D_1)$ increases regularly with increases in S, then our linear relation between S and the mean values of B will be much 'steeper' than in a conditional scheme of analysis with D_1 held constant. If $p(D_1)$ decreases with the values of S, it may eliminate any linear relation at all. We can also imagine such a pattern for the distribution of $p(D_1)$ for different values of S that will result in a U-shape curvilinear relation between S and mean values of B. As in the case of multivariate determination for dichotomous attributes, everything is possible for the situation where the variables involved are not independent.

(3b) In some cases the values of the variable D may determine the strength of the functional relation between S and B, according to the formula:

$$D_1 \longrightarrow [B = f_1(S)] \quad \text{e.g.,} \quad D_1 \longrightarrow (B = S)$$

$$D_2 \longrightarrow [B = f_2(S)] \quad\quad\quad\quad D_2 \longrightarrow (B = 2S)$$

$$D_3 \longrightarrow [B = f_3(S)] \quad\quad\quad\quad D_3 \longrightarrow (B = 3S)$$

Let us call the variable D the *modifier of the strength* of the relation between S and B. Its role can be observed in Table VI.

TABLE VI

	S_1	S_2	S_3
D_1	B_1	B_2	B_3
D_2	B_2	B_4	B_6
D_3	B_3	B_6	B_9

Here again the observable statistical relation between S and B is additionally a function of the correlation between S and the variable D, which is the modifier of the strength of $B = f(S)$. If S and D are perfectly correlated we have a case of the function $B = S^2$. If they are correlated negatively, with $R = -1.0$, the observation will show in this example no relationship between S and B. Finally, if S and D are independent we will observe the effect of the mean impact of the modifier D upon the strength of the correlation between S and B in the given population for specific values of S.

(3c) The values of the variable D may also *determine the direction* of the functional relation between S and B, according to the scheme of Table VII.

TABLE VII

	S_0	S_1	S_2
D_0	B_0	B_1	B_2
D_1	B_2	B_1	B_0
D_2	B_2	B_1	B_0

Here D will be a *modifier of the direction* of the relation between S and B.

(3d) B may be functionally related to S only when the values of the other variable D covary in a certain way. S and D are here *necessary covariants for B* according to the following scheme: $(D = S) \longrightarrow [B = f(S)]$ or $(D = S) \longrightarrow [B = f(D)]$ (Table VIII).

TABLE VIII

	S_0	S_1	S_2
D_0	B_0	B_0	B_0
D_1 B	B_0	B_1	B_0
D_2	B_0	B_0	B_2

We have here:

$$D_0 S_0 = S_0$$

$$D_1 S_1 = B_1$$

$$D_2 S_2 = B_2 .$$

(3e) It may also happen that the value of B is a function of whichever independent variable, in any given case, takes on the greater (or smaller) value. We can say that in this case B is a function of the more (less) intensive variable, according to the schemes of Tables IX and X respectively.

TABLE IX

	S_0	S_1	S_2
D_0	B_0	B_1	B_2
D_1	B_1	B_1	B_2
D_2	B_2	B_2	B_2

Here again, for all these situations one could present the relation between the mean values of B and the values of S as a function of the distribution of the values of the interacting variable D and of the correlation between the interacting variable D and our predictor S. For reasons of brevity we will not discuss all these details, assuming that the main ideas of this type of reasoning have been sufficiently illustrated above.

These are mere examples of different categories of non-additive, inter-active causal relations. It is obvious that an infinite number of other situations could be defined if we were to depart from the linear character of our functional relations between S and B and also to take into account other patterns of dertermination of this relation by different values of the variable D. It is equally clear that with an increase in the number of interacting variables the patterns of their possible interaction become more and more complicated.

There is one important thing about all these situations involving inter-action between S and D that distinguishes them from the additive patterns, namely that *in these cases we cannot expect that the relationship between S and B will hold if we keep D constant.* It is equally important to know the level at which the value of D has been established. Suppose, for example, that the pattern of interaction between S and D corresponds to type (3a) according to which:

$$D_1 \longrightarrow [B = f(S)]$$
$$\bar{D}_1 \longrightarrow [B \neq f(S)].$$

In this case it is not enough to know that the value of D is constant (e.g., by controlling it in experimental conditions) in order to expect that there will be in our situation, characterized by the given constant value of D, a functional (or probabilistic) relation between S and B. If it happens that we 'stabilize' the value of D on the level D_1, then we will observe the relation between S and D which is codetermined by this value of D. But if we stablize the value of D on the level D_2 (or D_o), There will be no relationship between S and B in those conditions.

It may also happen that the variable D has only one constant value D_1 in natural conditions (or in our population). Then we may come to the conclusion that there is an unconditional causal relation of the form $B = f(S)$. Attempts to verify this relation in other populations where the (constant) value of D is different will show us that this relation is conditional. If on the other hand the constant value of D has been naturally stabilized in our population on the leve! D_2 or D_o, then we may come to the conclusion that S and B are not causally related to each other.

The same is true for all other patterns of interaction mentioned above. Suppose that we have a situation of type (3b) in which

$$D_1 \longrightarrow (B = S)$$
$$D_2 \longrightarrow (B = 2S)$$
$$D_3 \longrightarrow (B = 3S).$$

Suppose that it happens that the value of D is kept constant on the level D_2. Then we observe a functional relation $B = 2S$. But it would be incorrect to say here, 'other things equal, $B = 2S$', because this is not true if the interacting D has the value D_1 or D_3.

In the formulation of hypotheses (especially in the behavioral sciences) we often encounter the following type of statement: 'Other things equal, $B = f(S)$'. It is clear from the above analysis that this formula can be applied only in such situations where these 'other things' refer to additive (alternative) causes $A_1, A_2, \dots A_n$ of B, because on whatever levels $A_1, A_2 \dots A_n$ have been stabilized (held constant), there will always be the same definite pattern of observed relationships between S and B plus (minus) the total effect of all additive causes upon the mean value of B in our population.

The situation is quite different for interacting causes D_1, D_2, ..., D_n, because here it is not enough to assume that their values are held constant. We have to assume that they are held constant on just that level which determines the given direction or strength of the relation between S and B, which is necessary for the truth of the formula $B = f(S)$.

We can omit all additive causes in the formulation of a law describing the given relation between S and B and substitute for them the formula, 'Other additive factors being equal, B = f(S)'. But we cannot do this with nonadditive causes. They have to be specified in the formulation of our law as the necessary components of its antecedent.

And a final comment: until now we have been discussing some statistical effects of functional relations between one dependent and several independent variables, separately for additive and for interacting independent variables. It is obvious that one could also present the statistical relationships between the quantitative variables for situations in which some of the independent variables are additive, whereas when others are interacting with them according to the different possible patterns of interaction, just as we did it for the dichotomous attributes.

NOTES

* The following paper is the continuation of ideas presented in S. Nowak, 'Some Problems of Causal Interpretation of Statistical Relationships', *Philosophy of Science* **XXVII**, 1960: 'Causal Interpretation of Statistical Relationships in Social Research', *Quality and Quantity* I-1967, and in the modified version of this paper published in H. Blalock *et al.* (eds.) *Quantitative Sociology*, Academic Press, New York, 1975. See also: S. Nowak, 'Conditional Causal Relations and their Approximations in the Social Sciences', in: Patrick Suppes *et al.* (eds.), *Logic Methodology and Philosophy of Science*, 1973.

[1] See H. Blalock, *Causal Inference in Nonexperimental Research*, Chapel Hill, 1969.

[2] For convenience we may take as an example of the class S a 'stimulus' and as an example of the class B a 'behavior', but we should keep in mind that the following analysis does not refer only to these two classes of variables. S means 'any cause' and B 'any effect'.

[3] I am using the traditional meaning of the term spurious correlation, as introduced by P.F. Lazarsfeld, 'Interpretation of Statistical Relations as a Research Operation', in: P.F. Lazarsfeld and M. Rosenberg (eds), *The Language of Social Research*, Glencoe, 1955. A correlation is spurious when the sequence of the two classes of events S and B is produced by the fact that they have a common cause A, which must be prior in time to S. But we should stress that the correlation of S and B is also spurious in the

situation where A occurs jointly with S (being its constant correlate) and where A is causally related to B and S is not. Thus, e.g., in the famous Hawthorne experiment the presence of the experimenter (A) would be located in the same time period as all other experimental stimuli (S), but it was the only factor producing all changes in the workers' behavior (B) and also producung a spurious correlation SB.

[4] A man who would like to 'produce' the flute of a river at the time when he knows that it will occur anyway would be in the simplest example of this category.

[5] This is the reason why it is so difficult to make causal inferences concerning, e.g., natural developmental sequences of biological phenomena 'in vivo' until they are controlled 'in vitro',

[6] See: Blalock, *Causal Inference in Nonexperimental Research,* Chappel Hill, 1967.

[7] For the purpose of convenience D may be exemplified here as 'disposition' to 'react' in the way B when 'stimulus' S occurs.

[8] Some problems connected with typology of causal relations for 'continuous' variables will be analyzed in the last section of this chapter.

[9] In some cases we may be able to discover the importance of D for B by means of theoretical analysis, e.g., trying to deduce the relationship $S \rightarrow B$ from a more general theory.

[10] We may take as an example of D some personality trait for which we assumed this relatively even random distribution (it should be rather strongly connected with the genetic determination of human dispositions) and take an example of behavior for which this D is an essential factor.

[11] An 'internalized' pattern of behavior characteristic of the culture of a given historically defined population may be a good example of this category of 'dispositional traits'.

[12] These space-time coordinates in this case play a special role: they are substitutes for the unknown factor D and their coverage should be equivalent to those areas of reality in which the factor D exists practically everywhere. See Chapter IV.

[13] At least some of them refer to the conditional causal connections, because some historical or statistical relations between S and B may involve so-called 'spurious relationships between variables' which will be analyzed below.

[14] All insurance companies base their predictions on this type of relationship, i.e., probabilistic propositions with validity limited for one population only.

[15] Although in practical analysis of empirical social data we shall usually meet the cases in which the given relation between S and B is usually limited to some 'historical coordinates' e.g., for one given society.

[16] See Section 6, this chapter.

[17] This is true, if we have in mind universal laws of science, of course. We could give a great number of narrow empirical regularities corresponding to the finite classes of objects or events, but here we are interested in the relationships of universal or at least extended historical character.

[18] Which is of course equivalent to another definition of statistical independence-more frequently used: $p(SD) = p(S) \cdot p(D)$.

[19] The consequences of the situations when they are not mutually exclusive will be discussed below.

[20] I.e. when the relation between S and B is a causal and not a spurious relation, which shall be discussed later.

[21] To give a trivial example, we may say that the probability of death of a tubercular patient is usually higher than is determined by the pure mechanisms of tuberculosis, being increased e.g., by the probability of a tubercular patient's dying in a traffic accident.

[22] I.e., both pairwise and in all their possible combinations.

[23] It would be more correct to say that the relation SB is more direct CB instead of introducing a dichotomous classification. It is obvious that for any causal relation we may find one which is more direct than it.

[24] This is a rather rare case which might be called 'an isolated chain of causal transformations' which is the fulfillment of Laplace's deterministic ideas (even when limited in time and space).

[25] The equation applies only to situations in which S_1D_1 and S_2D_2, or FGE and HJK, are mutually exclusive. If they are not mutually exclusive, $p(B \mid C) = p(FEG \mid C) + p(HJK \mid C) - p(FEGHJK \mid C)$.

[26] Let us give an example to clarify this scheme. Suppose that in order to fulfill a difficult combat task (B) at time t_3 it is necessary for a soldier at time t_2 to have both courage (C_2) and a favorable orientation to the situation (D). But in order to be courageous at time t_2 it was necessary to be both courageous (C_1) and not shocked by the enemy's artillery (E) at time t_1. And in order to be favorably oriented to the situation (D) it was necessary to be both courageous enough (C_1) to look for information and to be in a favorable place where this information might be found (F) at time t_1.

[27] For a more detailed discussion of V. Mieses's theory and of the notion of randomness in general see: Chapter VI.

[28] It. would be more proper to use the term 'association' than 'correlation', because we are dealing here with 'dichotomous attributes' and not with 'continuous variables'. But the term 'correlation' often has a broader meaning which also denotes the case of association between dichotomous attributes, and in this broader meaning it is used here.

[29] This is of course, equivalent to the well-known formula of positive correlation: $p(SB) > p(S) \cdot p(B)$.

[30] This is equivalent to $p(SB) < p(S) \cdot p(B)$.

[31] Or, as often happens, the sum of frequencies of all different alternative causes of B, e.g., A_1, A_2, A_3 and the pattern of their associations with S.

[32] The problem of interpretation of partial relationships with the application of a 'test variable' was introduced into the methodology of social research by P.F. Lazarsfeld, *Interpretation of Statistical Relationships as a Research Operation.*

[33] One should remember the additional assumptions made here. In situation (2) we assume that the supplementary factor D occurs in our population at least with a certain proportion of S's; otherwise $p(B \mid S) = 0$. In situation (3) we assume that the alternative sufficient condition A does not always occur when non-S occurs; otherwise $p(B \mid S)$ would be equal to one. If these assumptions are not valid (as they usually are) there is no correlation between S and B in situations (2) or (3).

[34] P.F. Lazarsfeld, *Interpretation of Statistical Relationships as a Research Operation.*

[35] See Lazarsfeld (*op cit.*). According to the author, the correlation is spurious when, after introducing a 'test variable' (C) antecedent in time to the independent variable (S), we obtain independence in the partial relationship between SB when C is held constant.

[36] This assumes we make some additional assumptions which were analyzed above for the 'classical' type of spurious correlation.

[37] In the definition of spurious correlation by Lazarsfeld (*op. cit.*) it was formulated that the real cause of B precedes the 'spurious independent variable' S in time.

[38] Unless A is negatively associated with S and D_1.

[39] A more adequate pattern of relations would be as follows:

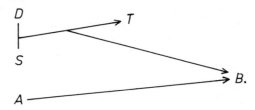

[40] This relationship usually decreases in both subgroups of T, and the amount of decrease is a function of the strength of the correlation between S and T.

[41] Or more adequately:

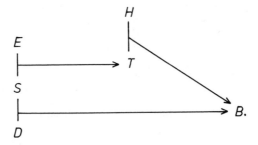

[42] See S. Nowak, 'Egalitarian Attitudes of Warsaw Students', *American Sociological Review*, April 1960, where the figures for these two illustrations are presented.

[43] The reader should note that I am changing the symbols for the variables in order to apply the symbols which were proposed for multivariate analysis by Lazarsfeld (1955). The typology of elaborations used here is different from that proposed by Lazarsfeld.

[44] For an analysis of this type of problem connected with artificial cuts in continuous variables, see H. Blalock, *Causal Inference in Non-Experimental Research.*

[45] Let us read $R_{SB}(D_o)$ as 'The correlation between S and B for the subgroup of the population characterized by the value D_o of the variable D', etc.

[46] The role of interaction as an obstacle in causal interpretations of multivariate statistical relationships between quantitative variables was strongly stressed by H. Blalock, *Causal Inference in Non-Experimental Research*, Chapel Hill, 1969.

BIBLIOGRAPHY

Blalock, H., *Causal Inference in Non-Experimental Research*, Chapel Hill, 1969.

Lazarsfeld P.F., Interpretation of Statistical Relationships as a Research Operation', in P.F. Lazarsfeld and M. Rosenberg (eds.), *The Language of Social Research*, Glencoe, 1955.

Nowak, S., 'Causal Interpretation of Statistical Relationships in Social Research', *Quality and Quantity* 1, 1967.

Nowak, S., 'Inductive Inconsistencies and the Problems of Probabilistic Predictions', in this volume, p. 228.

Nowak, S., 'Some Problems of Causal Interpretation of Statistical Relationships', *Philosophy of Science*, February 1960.

Nowak, S., Egalitarian Attitudes of Warsaw Students', *American Sociological Review*, 1960.

CHAPTER VI

INDUCTIVE INCONSISTENCIES AND THE PROBLEMS OF PROBABILISTIC PREDICTIONS*

1. 'INCONSISTENCIES' GENERATED BY STATISTICAL SYLLOGISM

It is rather generally believed that the application of statistical propositions for the indirect assessment of unobserved properties of certain individual objects or for the purpose of prediction of future events may create certain logical and practical problems discussed among others by Hempel in his paper on 'Inductive Inconsistencies'.[1] The nature of this problem can be illustrated by an example taken from Hempel's paper. Suppose that we meet a certain Mr. Petersen, and we would like to know whether or not he is a Roman Catholic. For some peculiar reason we decide not to use the method of direct (i.e. observational) assessment of Mr. Petersen's religion (e.g. by asking him about his religion, or looking into his documents) but prefer to approach the problem indirectly, by inferring the answer from other information. Such information will be of two kinds: propositions which describe some characteristics of Mr. Petersen other than his religion, and propositions describing in general terms the relations between these characteristics we observe as properties of Mr. Petersen and the property of being a Roman Catholic.

We are not surprised to learn that in this case these relations are of a *statistical* nature; therefore our assessment of Mr. Petersen's religion will not be *certain*, but will be made in terms of a certain *probability*. It happens that Mr. Petersen is a Swede. Knowing there is a very low proportion of Roman Catholics among Swedes (e.g. 2%) we say that the probability that Mr. Petersen belongs to this denomination is also very low, namely 2%. In other words, we cannot say that it is *almost certain that he is not* a Roman Catholic. On the other hand we know that Mr. Petersen has made a pilgrimage to Lourdes. But the proportion of Roman Catholics among those who have made a pilgrimage to Lourdes is very high — let us say 98%, which makes the probability that Mr. Petersen is a Roman Catholic very high. On the basis of the second piece of information we are inclined to believe it is *almost certain that he is* a Roman Catholic. Two different *facts* about Mr.

Petersen in connection with two valid statistical generalizations seem to lead to two *obviously inconsistent conclusions about the individual case* to which they have been applied.

The case of Mr. Petersen is of course not an exception but an example of a general problem. Every concrete individual object or event is, due to the infinite number of its 'absolute' and 'relative' properties, a member of an infinite number of classes, an element of an infinite number of sets. If we are interested in one of its properties only, this property is very likely to occur *with different frequencies* in the different classes of which our individual is a member. As a result of this our individual object will have simultaneously quite *different probabilities of occurrence* of this property in relation to its different characteristics.

Does this mean that our individual (and the whole world around us) is somehow 'internally inconsistent', or should we rather use the term 'probability' in such a way that it does not lead to inconsistencies and contradictions?

This chapter is an attempt to show that if the notion of 'probability' is properly understood, probabilistic propositions cannot lead to any inconsistencies of the nature exemplified above.

2. 'CONTRADICTIONS' GENERATED BY GENERAL CONDITIONAL LAWS OF SCIENCE

In contrast to statistical syllogisms, the syllogisms in which the relation between properties or events is described as general (deterministic) seem to lead to no inconsistencies, when applied to indirect assessment or prediction on the individual level.[2] If all F's are G, and our individual object a is F, then necessarily it has also to be G.

This is of course logically correct. If we apply to our predictions propositions which are *unconditionally general* in their character, they cannot lead to any contradiction. The problem is that among the propositions *which are actually used* for the purpose of indirect assessment and prediction in various sciences and which are usually called 'general laws of science' only very few (if any) may be called *unconditional general laws of science*.[3] A proposition may be called an unconditional law of science if it has the form: 'All F's are (followed by) G whatever might be their other characteristics and in whatever context or situation they may occur'. When we look at propositions which seem to have the character of general laws of science, it usually turns out that their truth is — explicitly or implicitly — relativized to the occurrence or non-occurrence of a great number of additional conditions C_1, C_2 ... C_n. These conditions are omitted in the formulation of the law for reasons of brevity, but every scientist knows that their non-occurrence (or their occurrence if they are 'disturbing factors') could falsify a given law. This means that these propositions are *conditional laws of science*. The complete formula for such laws is 'All F's are G *provided that conditions* C_1, C_2 ... C_n also occur (do not occur)'.

For most of the 'general laws of science', the number of potential conditions which could modify the occurrence of G when F has occurred is very great, and it would be practically impossible to include them all in the formulation of the law or theory. Therefore these conditions are omitted from it. As a result of this, general laws of science often look as if they were unconditional general laws of science.

If we forget about their conditionality and apply them as unconditional generalization to cases which are characterized by the presence of the conditions to which these laws in fact do not apply, *we will have logical contradictions of exactly the same type as in the case of statistical sullogisms*. We have, for example, in our theory a proposition which says that all bodies fall to the surface of the earth. We also have another proposition which says that all rockets, when fired with the second cosmic velocity toward outer space, continue in that direction. If we were to apply these two 'general propositions' in a naive way (i.e. disregarding their conditional character) to a certain object called 'Apollo space-ship' — which is both a 'body' and a 'rocket fired with second cosmic velocity' — we would obtain for that object two obviously contradictory predictions: that it will fall and will not fall toward the earth.

No one would make this mistake in such a case, because the conditionality of these two (and of most other) general laws of science is too obvious for us. We know at least the most typical limits of applicability of the first general proposition, knowing at the same time that it does not apply in this form to 'rockets fired with second cosmic velocity'. Therefore conditional general laws of science do not create such 'philosophical anxieties' as do the statistical propositions. But we often make the *mistake of disregarding the conditional character of our laws* when we try to predict future events (or to assess indirectly the existence of present ones) on the basis of the *statistical regularities* we observe in empirical reality, or when we try to apply many regularities at the same time to the prediction of a single case. This seems to be the main source of the 'problem of inductive inconsistencies'.

3. PROBABILITY AND RANDOMNESS

The predictions which are based upon unconditional general laws of science are *certain* because they assume a *necessary* relation between F and G. They are equally certain if they are based upon general conditional laws of science, assuming at the same time the presence of all conditions necessary for the occurrence of G when F has occurred, or the non-existence of any factor which might 'neutralize' the relation between F and G, because in this way they again refer to a necessary relation with a much more complex antecedent. If we omit from our application of a conditional generalization explicit or implicit reference to its modifying conditions, the relation ceases to be *necessary* and our prediction becomes *uncertain*.

In some situations we are able to express in numerical terms the 'strength of non-necessity' of the relation between F and G by saying what *proportion* of F's are accompanied (followed) by G's, or what is the *relative frequency* of the occurrence of G when F occurs. As a result of this we are also inclined to attribute a *numerical value to our uncertainty* that in a given case F will be followed by G, or in other words, a numerical value of *probability* of occurrence of G for our individual case in the class F.

If 'probable' here means 'uncertain to the degree p' — where p is the relative frequency of G's among all F's — it should be recalled that such non-necessary connections for which we know the relative frequency do not always imply that the prediction in which they appear *will be undertain in the probabilistic meaning of this term*. Suppose that the relation between F and G is a conditional one, and we know under which conditions it 'works'. Suppose also that we know that only 50% of F's occur in these conditions. We know then that the relative frequency of G when F occurs is 0.5, but we still are able to predict with certainty the occurrence of G by stating whether or not the given F occurs under the 'proper conditions'. These conditions may be defined in general terms — as when we say that a certain property or event C plays such-and-such a role and the proportion of F's occurring jointly with C is equal to 0.5. They may be defined in terms of time-space characteristics or 'coordinates'[4], as when we say that only those F's which occur withing given time-space limits are followed by G and at the same time we know that 50% of F's occupy this area. Finally, they may refer to some local characteristics of the given F in relation to other F's, as when we say that every second F in natural sequences of F's or in their 'natural existing clusters' (like e.g. the black squares in a chess-board) is followed by a G.

In all these situations, after including the information about the occurrence or non-occurrence of conditions whose relative frequency in relation to F is known, we are able to make a *prediction in terms of certainty* about the occurrence of G in the given case. Therefore it would not be reasonable to ignore this possibility and to make in such situations much less valuable and uncertain probabilistic predictions. The term 'probability' is reserved, as we know only for predictions of individual cases in which we are *really uncertain* whether G will occur in the given case, i.e., to situations in which we are unable to specify the conditions sufficient for the occurrence of G.

But when we use the term 'probability' we imply something more than the contingency of the given relation and the basically uncertain character of our prediction. The meaning of this term implies that there is a *relatively stable tendency* of the occurrence of G in relation to F. This tendency is such that the occurrence of G, although uncertain for any specific F, has approximately the same relative frequency in sufficiently numerous natural 'clusters' of 'sequences' of F's. To put it more precisely: We say that the probability of G relative to F is equal to p, if p is the limiting value of the relative frequency of G/F when the number of F's increases. The proportion of G's in different sequences or clusters of F's will of course be only *approximately equal*. In such situations we say that the relation between F and G is a *probabilistic random relation*. This notion refers both to the 'indeterminacy' of the occurrence of G on the individual level and to the relatively stable tendency of its frequency on the mass level. It should be remembered that this meaning of the term 'probability' corresponds to the meaning of 'necessity' mentioned above. It refers to the area of reality and characterizes the objective relations in this reality. The term 'probability', when applied to prediction of individual cases, belongs to the area of *knowledge* about this reality, and it characterizes the justified strength of our convictions. Sometimes it is said that the term 'probability', when applied to the description of 'objective regularities', is used in its *statistical meaning*, but when applied to the characterization of 'numerical value of the strength of our beliefs' is used in its *logical meaning*.[5]

The existence of empirical relations in the statistical meaning of the term probability is a necessary condition for its use in its logical meaning. Only if we assume that the phenomena are related in the way described above, with p as the limiting value of their relative frequency, will our expectation that G occurs in the given case with probability p (this is the adaptation of the strength of our belief to the level p) be a *rational* way of behavior. Such a conclusion has no certain implications for the next predicted case, but it has an almost certain implication for a sufficiently long sequence of individual predictions. If in each case we predict the occurrence of G, the proportion of correct predictions to all our predictions will be roughly p. The consequence of this is that the higher our probabilistic expectation, the smaller the proportion of incorrect predictions in the long run. Assuming that *this is all we can guess on the basis of really random relations*, the adaptation of the strength of our belief to the relative frequency of predicted events seems to be the *optimal solution* of the problem of prediction, and of decisions based upon such prediction.

4. THE UNCONDITIONAL AND CONDITIONAL PROBABILISTIC RELATIONS

Let us consider two different examples of probabilistic relations, in which a certain effect (or 'result') appears randomly with a given relative frequency in relation to some its 'predictors' (or in relation to the number of 'trails'). The first example will be a 'perfect coin', well known from textbooks of elementary statistics. It is known that the probability of tossing a 'head' in the given trail is equal to 0.5, which means that in the long run we will have approximately 50% heads among all trials. But the theory of probability says something more about such cases as that of a 'perfect coin toss'. It implies that *there is no way of changing this relative frequency because the results of these trials are basically independent of any other possible conditions*. This means that whatever should be the conditions in which a sequence of trials is undertaken, the limit of relative frequency of heads will always be 0.5. We may take a silver coin or a coin made from gold, it may be a Polish zloty or an American silver dollar, the trials can be made on a rainy day or in the sun, in the 15th or the 20th century, and *the result will be always the same probabilistic relations* with 0.5 as the probability of obtaining a head in an individual trail and the limit of relative frequency for a sequence of such trials.

Such probabilistic relations might be called *unconditional probabilistic relations*.

This notion of unconditional probabilistic relations corresponds to the notion of probability as proposed by von Mises in his frequency theory of probability. Von Mises calls a class of events in which a certain result is distributed randomly in the way characterized above a 'collective'. The basic requirement for a probabilistic relation in his theory is the principle of 'insensitivity to place selection' also called 'the principle of impossibility of a gambling system'[6]. Let us call such relations *collectives in the strong sense of the term*, and the occurrence of *G* in relation to *F 'random in the strong sense of this term'*. The reference to gambling is not accidental here, since most of the empirical sequences which correspond fairly closely to this definition of unconditional probabilistic relation are the results of action of various *artificial random devices*, especially those constructed in such a way that both prediction of individual results in terms of their certain occurrence and the alteration of the relative frequencies in the long run is impossible. There are some natural laws of science which at this stage of development of science *seem* to correspond to the notion of unconditional probabilistic relation — certain laws of microphysical phenomena are usually mentioned here. But scientists and philosophers differ in their opinions as to whether in the future the discovery of some 'latent parameters' will make possible the positive prediction of these phenomena on the individual level, or at least will permit us to modify the values of relative frequency of their occurrence depending on the conditions in which they occur.

If we agree that there are few examples of such laws in empirical sciences, we must also agree that the majority of *so-called probabilistic relations we discover in our scientific investigations and use in predictions are of a different kind.*[7]

If we take for example the probability of death for persons of a certain age in a certain population, we know that this probability can be modified by an almost infinite number of conditions. Besides age it will also depend on sex, occupation, place of inhabitance, state of health, and the hereditary factors of the person for whom we would like to predict life expectancy. This fact is well known to insurance companies, which use as great as possible a number of 'predictors' in such situations, knowing that for each combination of values of independent variables the values of probabilistic expectations will be different. Those probabilistic relations between phenomena which could be *modified* by including certain conditions into our predictive premises should be called *conditional probabilistic relations*.

In calling such relations 'probabilistic' we change the notion of probability in an essential way. Their conditionality means that for them the 'gambling system' is not prohibited as in von Mises' notion of probability. These 'systems' could be of various kinds. It is possible that introduction of a new, generally defined phenomenon or event C modifies the probability of occurrence of G when F occurs so that $p\,(G/F \neq p\,(G/FC)$, or that for some time-space sub-areas of the location of F's the probabilities $p\,(G/F)$ will be different. Finally, it is possible that a certain pattern of relative location of F's in relation to each other will be the modifying condition of their probabilistic relation to G such that this probability will be different, for example, for every second F.

5. TWO KINDS OF CONDITIONALITY OF PROBABILISTIC RELATIONS

One should distinguish two kinds of conditional probabilistic relations. Suppose, for example, that we have a conditional 'probabilistic'[8] relation modified by a certain C such that $p\,(G/F) \neq p\,(G/FC)$. Suppose now that this modifier C occurs only in a specified 'sub-area' of the time-space localization of F's and does not occur in other sub-areas of F's. In such a situation we will have two different probabilistic relations between F and G. One will be observed in the sub-area where C *does* occur, and will be equal to $p\,(G/FC)$. For the areas where C *does not* occur, we will have another probabilistic relation, namely $p\,(G/F\overline{C})$. But *there will be no constant overall probabilistic relation between F and G valid throughout the whole population of F's*. The notion of probability of occurrence of G when F has occurred is in such situations *meaningless*. We can use here the notion of probability only in respect to two variables at once: F, C. The sequences of events F and G do not form by themselves a 'collective' in von Mises' sence of this term.

Let me recall an old anecdote which illustrates the misuse of the term probability for such relations: A passenger sits in a bus and counts other passengers who enter the bus: 'One, two, three, four'. Then when passenger number four takes a seat opposite him he says, obviously surprised: 'Excuse me, sir, but you really don't like one'. 'What do you mean?' 'Well, I have just read that each fourth member of mankind is a Chinese'.

Let us suppose now that C is a modifier of the probabilistic relation between F and G (so that $p\ (G/F) \neq p\ (G/FC)$), but C is distributed in the timespace area of localization of F so that it *occurs at random in relation to the locational time-space parameters of individual F's*. In this situation if we observe and compare probabilities $p\ (G/FC)$ and $p\ (G/F\overline{C})$ – we will still find that they are different, because they depend on the occurrence of C. Suppose now that we do not know that C is the modifier of this relation, or are unable to observe or control it, or purposefully disregard C in our observations. In such a situation we will observe a *relatively stable probabilistic relation between F and G*. For any natural, relatively long sequence of F's the relative frequency of those followed by G will be approximately constant and equal to the 'weighted mean' of $p\ (G/FC)$ and $p\ (G/F\overline{C})$. We see then that a conditional probabilistic relation may also sometimes create a 'collective', if we disregard these modifying conditions and if these conditions occur at random. We could call them *conditional collectives* or *'collectives in the weak sense of the term'*. The condition necessary for their 'collective character' is here our disregard for the randomly distributed modifiers of the overall probability $p\ (G/F)$. *Conditionality of a probabilistic relation is therefore not necessarily contradictory to its random character* (in the weak sense of the term 'randomness') *if these conditions themselves are* distributed randomly in relation to our 'predictor', and are excluded from our prediction either on purpose or because of lack of knowledge of their role.

To summarize the foregoing: Collectives in the strong sense of the terms are characterized by their probabilities being independent of the time-space *locational* parameters of individual events which are the antecedents of given probabilistic relations, and of all other *generally defined* events or characteristics which might eventually accompany these antecedents. In collectives in the weak sense the probabilistic relations are still independent of the locational parameters of individuals, but they do *depend upon some of their generally defined properties* or other events which might accompany them. The fact that these modifiers occur at random in relation to time-space characteristics of our antecedents produces a random probabilistic pattern of occurrence of their consequents with a relatively stable probability – as long as we disregard the existence of a possible gambling system. This makes meaningful the application of the term probability to natural time sequences or special clusters of such antecedents with all its implications for decision making in uncertain situations.

If we take into account the fact that the probability value of most of the relations we observe in natural phenomena is modified by a great number of conditions, and that these conditions are often distributed in time and space in such a way that *their total impact* upon our relation may be more or less close to the notion of perfect locational randomness, we should not be surprised that we still observe an empirical reality relations between two variables which are fairly close to collectives in the weak sense. This means that *in empirical reality we meet fairly often relations which are random in the weak sense.* For such relations the idea of probabilistic expectation equal to the limit of the relative frequency of G/F is of course meaningful *to the degree to which they correspond to the notion of randomness as described above.* Even if we know or suppose that when we introduce additional predictors we will obtain a different probability for a given result, we can expect that the proportion of correct predictions of G in the long run on the basis of F only will be equal to the relative frequency of G/F for any sequence of such predictions.

Therefore one cannot fully accept the criticism formulated by von Wright (in connection with similar criticism made by Broad)[9] when he wrote:

that von Mises' theory which is sometimes called the *statistical* theory of probability, seems to be particularly illsuited to deal with probabilities in statistics... as a consequence of this instability [of limits of relative frequency – S.N.] in the conditions determining the frequencies in such classes, the notion of frequency limit and therewith also von Mises' notion of a collective threatens to become inapplicable to them.[10]

As I tried to show above, this criticism applies only to those conditional relations the modifiers of which are not randomly distributed and which therefore cannot be called random even in the weak sense of the term, such as the distribution of Chinese on the Earth. But it does not apply to the second category if we admit the use of the notion of randomness in the weak sense of the term.

For collectives in the weak sense of the notion of probability is a useful tool for dealing with empirical reality, including social reality. When the number of potential modifiers is sufficiently large, it may well happen that their total impact upon a certain regularity p (G/F) will be sufficiently close to a 'weak collective', i.e. to stabilize p (G/F) *at least within some historical and geographical populations*. Otherwise the insurance companies would be in danger of going bankrupt in some periods or in some geographical regions or of reaping an unexpectedly large income in the others. Such collectives – *or at least sufficiently near approximations may exist in social or 'natural' reality*. But since they are in no way a universal phenomenon, we must clearly formulate the assumption about the random locational occurrence of the 'sum' of all potential modifiers of the probability of the expected event in relation to the locational parameters of our 'predictor' whenever we use the term 'probability'. Only this category of non-necessary conditional relations, i.e. those which are collectives in the weak sense, deserve to be called *conditional probabilistic* relations. Only for them does the logical notion of probability have a clear meaning – i.e. can we justifiably expect that our predictions will be in the long run confirmed in a proportion equal to the relative frequency G/F, because for any sufficiently large natural sequence of F's the proportion of those which will be followed by G will be roughly equal to that frequency.

For the category of relations which lack any property of a 'collective' at all we are not entitled to say that there is a definite probabilistic relation between F and G in general. We can try to introduce for them the notion of probability of G for FC jointly, if for this combination of conditions we discover an approximately stable relative frequency of G. If we don't find it for these two variables either, we may say that we observe *non-necessary* conditional relation between FC and G, but *we cannot call it a probabilistic relation*. Neither can we attribute any nummerical (probabilistic) value to the degree of our uncertainty (logical probability) that G will occur when either F or FC has occurred.

6. PATTERNS OF PROBABILISTIC PREDICTIONS AND THE PROBLEM OF INDUCTIVE INCONSISTENCIES

Let us turn now to the problem of predictions on the basis of probabilistic relations of a different kind mentioned earlier.

If we have a proposition describing an *unconditional probabilistic relation*, i.e. a relation which corresponds to the notion of an 'ideal gambling device', we are entitled to make probabilistic predictions in which our degree of uncertainty of the occurrence of G in each individual case will be equal to the frequency-limit p (G/F), *regardless of any additional conditions which might occur*. For any possible configuration of other conditions, the relative frequency of G will in the long run always be p, and our probability of expectation in any individual case will also be p; *therefore there is no possibility here of facing the problem of inductive inconsistencies*, in exactly the same way as they could not occur as a result of application of unconditional general laws of sciences.

The second category of relations discussed above are the conditional probabilistic relations sensu stricto or 'collectives' in the weak sense of the term, i.e. relations for which there are some modifying factors of their frequency limit, but for which − due to the random distribution of these factors − we still have an *overall stable frequency limit p (G/F)*. It is obvious that if we treat them as unconditional relations, and try to apply them to situations which modify their probabilities, we are in the position of having 'inconsistent' conclusions. This is due to the fact that we *disregard their conditional character*. The same would occur, as we recall, if we applied the conditional general laws of science to situations which were outside the limits of their applicability. Therefore *if C is a condition which modifies the frequency limit p (G/F) we are not entitled to extrapolate*[11] *the overall probability p (G/F) to a situation of the type FC, under the threat of 'contradictions' or 'inconsistencies'*, because would be the application of a proposition describing an empirical regularity to situations to which it (explicitly or implicitly) does not apply. For these situations we have to apply another *probabilistic relationship*, namely p (G/FC).

On the other hand, we can 'extrapolate' the 'overall' probability p $(G/G/F)$ to all conditions I about which we know that they do not modify the result p (G/F) − i.e. which, *within the population of F's, are statistically independent* (or 'roughly independent') *of G*. Therefore if we know that I is statistically independent of G (within the population of F's), we are fully entitled to apply the probability p (G/F) also to the situations FI without any danger of inconsistent predictions. The frequency limit of G will remain the same.

This seems to be a rather banal conclusion, but it is the foundation of the applicability of *artificial randomness* to empirical research. If in a certain population of *F*'s *p* (*G*) is modified by a certain number of non-randomly occurring conditions, this, as we recall, implies that there will be no constant value of *p* (*G/F*) in 'natural time sequences' or 'spatially located clusters' of *F*'s. In such situations we are not entitled to expect *T* with probability *p* when we observe *F*, even if *p* is the relative frequency of *G* in the total population of *F*'s. But we can expect that *p* will be the frequency-limit of *G* for a *random sample* of *F*'s drawn from this population. Because the 'property of being a member of a random sample' tends to be statistically independent of any characteristics on which the members of the population may differ, the artificial randomness produces a sequence of observed events of a probabilistic character. Even if in the population from which the sample is taken, the frequency of their occurrence is non-randomly modified by different conditions and distributed non-randomly in the space and time area occupied by the members of that population, we can apply the notion of 'probability' to the observation of cases of a random sample from it. We can expect that the next member of the sample will be *G* with probability *p*, even if we cannot expect it natural sequences of events in this population.

Even if in the natural sequences of events the conditionality of the relation between *F* and *G* is of a *strictly deterministic character*[12] so that by taking proper conditions into account we can predict the occurrence of *G* with certainty, the sequence of *F*'s produced by the application of a mechanism of artificial randomness to this population will result in a probabilistic sequence in the random sample, with the relative frequency of *G* in the population as the frequency limit of their proportion in the sample.

This is very important from the point of view of the problem discussed in this chapter. As I mentioned above, we are not entitled to apply the notion of probability (either in its statistical or its logical sense) to situations which do not correspond to the notion of a strong or a least weak 'collective' with a stabilized frequency limit. This is true of the 'natural' sequences or clusters of events and the sequences of our observations and predictions of them. But we may again introduce the notion of probability in its statistical and logical sense for the sequences or clusters of cases produced from these populations by the mechanism of artificial randomness, and for predictions of properties of individual consecutive cases drawn at random into our sample.

Otherwise, when none of these patterns of randomness can be assumed, we can speak only of non-necessary (*contingent*) *connections* and *uncertain predictions* without any attempt to say that we can attribute to this contingency or uncertainty a stable value and, of course, without specifying the value of probability of the future occurrence and correct prediction of *G*.

7. MR. PETERSEN REVISITED

Let us now return to the case mentioned at the beginning of this paper, i.e. to our Mr. Petersen who has different 'strongly inconsistent' probabilities of being a Roman Catholic, due to his Swedish nationality and due to the fact that he has made a pilgrimage to Lourdes.

As the result of the above discussion we can say that Mr. Petersen is by no means 'internally inconsistent' because both these 'probabilities' we applied to him were obviously conditional ones, and the 'probability' that a Swede is a Roman Catholic is most certainly not equal to the probability that a Swedish visitor to Lourdes is a Roman Catholic. We do not think that among Swedes there is no statistical dependence between religion and visiting Lourdes, therefore the 'overall' probability of a Swede being a Roman Catholic does not apply to these modifying conditions, i.e. to Swedish visitors to Lourdes.

This conclusion differs of course from saying that there is a different probability for Mr. Petersen to be a Roman Catholic *in relation* to his nationality than in relation to his visit to Lourdes. Some writers try to solve the problem of inductive inconsistencies by treating 'probability' as a 'relational property'[13] and in this way save Mr. Petersen from the danger of being 'internally inconsistent'.

In the present understanding the overall probability of being a Roman Catholic known for all Swedes *does not apply to those Swedes possessing any property which modifies this overall probability in the situations where we take such modifying property into account*. It can be meaningfully applied only to such sub-categories of Swedes as are distinguished by a property statistically unrelated to and independent of their religion. Obviously, making a pilgrimage to Lourdes does not constitute such a property. Therefore, since the conditions of applicability of the overall probability are not satisfied in the group of which Mr. Petersen is a member, it does not make any sense to apply this probability in the estimation of the probability of his being a Roman Catholic. Therefore, Mr. Petersen is not 'inductively inconsistent'.

Unfortunately, the same is the case with the application of the second overall probability, i.e. the probability that a pilgrim to Lourdes is a Roman Catholic. It is not likely that this probability will be equal for Swedish and for, e.g., Italian visitors to Lourdes. Therefore, we cannot say that this probability is not modified by the visitor's nationality, and on the basis of the same reasoning as above we cannot apply it to Mr. Petersen. We are in a rather unpleasant situation in which we 'freed' our Mr. Petersen from the suspicion that he is internally logically inconsistent, and therefore 'does not exist', but at the cost of being unable now to say anything about his religion. Can we do anything more in this situation?

The first and most simple solution is to try to discover the probability (relative frequency) of Roman Catholics among Swedish visitors to Lourdes. Suppose that this probability is 0.65. Then we forget about the two overall probabilities, that for all Swedes and that for all visitors to Lourdes, and we say that we expect our Mr. Petersen to be a Roman Catholic with the probability = 0.65.

In doing this we are following the recommendation of Reichenbach, and many others, who have said that in such a case we should take into account the *smallest possible reference class for which the reliable statistics are known* and forget about all the other information. To put it in terms of our symbolism here: If pG/FC is not equal to $p\ G/F$ *and to* $p\ G/C$, we take into account only the first probability and forget all the others.

But even here I would like to add: provided that the term 'probability' can be applied to the smallest reference group.

Even if we knew the relative frequency of Roman Catholics among the Swedish visitors to Lourdes, this wouldn't automatically mean that the notion of probability is applicable to this group, and to Mr. Petersen as its member. It would be applicable on one of the three conditions only:

(a) There is a strictly random distribution of Roman Catholics (with a definite frequency limit) among the Swedish visitors to Lourdes, so that there obtains a probabilistic relation *independent of all* other possible modifying conditions, i.e. if they were a collective in the strong sense of the term; or

(b) the probability of a Swedish visitor to Lourdes being a Roman

Catholic is a conditional one, given that these conditions are randomly distributed throughout this sub-population so that the observation of *natural sequences or natural clusters* of Swedish visitors to Lourdes produces random probabilistic processes in the weak sense of this term.

In either of these two situations we could attribute to Mr. Petersen, after meeting him on our trip to Sweden and discovering about his visit to Lourdes, the probability that he is a Roman Catholic, on the basis of knowledge of the relative frequency of Roman Catholics in Mr. Petersen's *reference category*.

(c) A third possibility, disregarding the random character of the distribution of Roman Catholics among Swedish visitors to Lourdes, would be that we draw an artificial random sample from this population and learn that Mr. Petersen is a member of our random sample. In this case we would guess that he is a Roman Catholic with the probability of our guess being equal to the frequency of Roman Catholics in the population from which the sample was drawn, i.e. among the Swedish visitors to Lourdes.

The first assumption is obviously false for all relations in the social sciences. The second assumption, referring to natural random relations (in the weak sense) between all the phenomena which are Mr. Petersen's characteristics and religion within the category of Swedish visitors to Lourdes, seems very doubtful. Finally, the question of whether our Mr. Petersen can be treated as an element of a 'random sample' of this category is also rather questionable unless we specify the sequence of other 'random' observations for which we expect our guesses to be confirmed with the given relative frequency, and 'prove' that Mr. Petersen is an element of that sequence.

Therefore, instead of trying to estimate Mr. Petersen's religion with a definite probability, we should say rather that we are *uncertain* whether or not he is a Roman Catholic. Due to the nature of the relations between the variables which are used for our uncertain assessment and the potential modifiers of these relations (and due to the non-random character of the conditions under which we met Mr. Petersen), *we are unable to give a more definite, i.e. a probabilistic, value to the degree of our uncertainty*.

The last question is whether the expectation in terms of probability might be perceived as the best way of solving the problem of prediction, when we have *no knowledge about the pattern of time-space distribution* of a certain property in a given population and the only thing we know is its relative frequency in this population. We may then try to use the relative frequency of this property as the measure of our probabilistic expectation but then we implicite have to *assume* that this characteristic is distributed randomly at least in relation to time and space. The correctness of our probabilistic prediction in the long run will depend on the degree to which this assumption is true – to remind again the anecdote about the 'Chinese' in the bus, quoted above.

And in general it seems that *we should use much less frequently the term 'probability'* in our references to the relations between the phenomena and specify our assumptions about their 'collective' character in each case, because if relations which are not related to random collectives at least in the weak sense are called 'probabilistic', we are also in danger to formulate 'probabilistic' expectations or predictions in a way which does not correspond to the postulate of optimalization of predictions in uncertainty situations.

8. DERIVING THE PROBABILITIES FOR INTERSECTIONS OF ADDITIVE AND INTERACTING CAUSAL COLLECTIVES

Along with all the reservations made in the preceding section with respect to the possibilities of predicting (in probabilistic terms) Mr. Petersen's religion, we assumed that we are able to obtain information as to the probability (better: relative frequency) of Roman Catholics among Swedish visitors to Lourdes. In other words: that we have data which inform us as to the probability of occurrence of the predicted property G within a new collective CF,G formed by the intersection of the two previous ones: the collective of all Swedes and the collective of all visitors to Lourdes. But *in many situations this is just what we are looking for* and trying to predict, since we do not have any data on which we could base our prediction for the new collective. The question then arises: Can we, in any kind of situation, *derive this probability from the two previous one?* In other words: Knowing the overall probabilities for two initial collectives and knowing that they intersect, can we derive the value of our probabilistic expectation for cases which belong to the intersection of the two previous collectives?

I think that we can – at least for one special category of collectives which I propose to call *causal collectives*. By a causal collective I mean empirical series of events G and C in which:

(a) C is causally related to G, being a necessary component of a sufficient condition for G,

(b) the complete sufficient condition for G is composed from at least two variables (assume it is composed from the events C and D),

(c) the role of the variable D in causing G is unknown to us and *in our observations we are not taking D into account*,

(d) the events D are *randomly distributed* (whatever this might mean) with respect to C throughout the whole population of C's whether it is a finite or a non-finite population.

It is obvious that in such a situation the series of events C and G will constitute a collective in the 'weak' sense of the term characterized above. Now the question may be posed: Supposing that we have two such causal collectives with the same dependent variable (CG and FG) and the events C and F intersect, can we estimate the probability of G for the collective C, F, G being the intersection of the two previous ones?

Before answering this question, let me mention that the case of Mr. Petersen is not a good illustration for this type of case because at least one of the variables, 'going to Lourdes', is obviously the effect of the predicted (or better, postdicted) variable: being a Roman Catholic. We might imagine another situation that would constitute a good example of two causal collectives with the same dependent variable. Suppose that in a certain country we know the overall probability for a worker to have radical political views. Suppose also we know the probability that a member of a certain ethnic minority will have radical views. Now the question is: What is the probability for someone who is both a worker and belongs to this ethnic minority having radical views?

Our general problem can be formulated as follows: There are two collectives of the causal kind, CG and FG, in which we assume that both dichotomous variables, C and F, are the causes of G. We also know two probabilities, namely $p(G/C)$ and $p(G/F)$, and from these we want to derive the probability of the joint occurrence of these two causes: $p(G/FC)$.

It is obvious that such information is not sufficient for any derivations. In order to predict something, we must know (or be able to assume) *what is the nature of the causal relations* between G on the one side and C and G on the other.

Without going into the details of the assumptions on which the following reasoning is being made[14] let it be said:

(a) *C* and *F* may be *interacting causes* of *G*, by which we mean that they belong to the same sufficient condition for *G*, being the *necessary components of one sufficient condition*. Here we can distinguish two sub-categories of such situations: (a_1) *C* and *F*, when occurring together, make a sufficient condition for *G*; (a_2) *C* and *F* do not make a sufficient condition for *G* although they belong to one sufficient condition, in which case we might assume for simplicity's sake that a sufficient condition for *G* is composed of three events occurring jointly, namely *C, F, D*. Let us call the two sequences *CG* and *FG* in these situations two *interacting collectives*.

(b) *C* and *F* may be *additive causes* of *G*, by which I mean that they belong to two alternative, 'parallel', causally unrelated[15] sufficient conditions for *G*. Here we can distinguish three situations: (b_1) *C* and *F* constitute by themselves the alternative sufficient conditions for *G*. This situation is in our case impossible because it would mean that the relation between *G* and any of its causes is a general and not a probabilistic one. (b_2) *C* and *F* are necessary components of two alternative sufficient conditions for *G*; under the assumption that the other events supplementing *C* and *F* to make the corresponding sufficient conditions of *G* occur at random with respect to *C* and *F*, we will have the situation of two additive *causal collectives*. (b_3) One of the causes (let us say *C*) is a necessary component of one sufficient condition for *G* (e.g. of *CD*) while the other cause *F* constitutes an alternative sufficient condition for *G*.

Let us look now at the *interacting causal collectives*. If *C* and *F* constitute by themselves a sufficient condition for *G*, according to the formula *CF* = *G*, then we will have

$$p(G/C) = p(F/C) \quad \text{and} \quad p(G/F) = p(C/F) \quad \text{and at the same time}$$
$$p(G/CF) = 1.0,$$

which is a rather obvious conclusion for the situation of a sufficient condition. But it should be mentioned here that is it theoretically possible that two causal collectives, however small their probability of producing a certain effect might be, will result in a deterministic causal series if they happen to constitute a sufficient condition for this effect. (For instance, certain non-neutral medicines may result in the death of patients in a very small proportion of cases when each is taken alone, but will lead to certain death when taken together.)

A more realistic situation is that when two interacting variables (events) C, F, do not exhaust the whole sufficient condition for the given effect G. Suppose we have a population in which C occurs with the probability equal to p_1, F has the probability equal to p_2, and D, being the factor supplementing CF to the sufficient condition of G, has the probability equal to p_3. Here D may be either a single 'naturally distinguishable' event, or a conjunction of all the events necessary for the supplementation of CF to the sufficient condition of G. (If, for example, the sufficient condition for G is C, F, D_1, D_2, D_3... D_n, we may define $D = D_1$ and D_2 and D_3... D_n – i.e. D as the joint occurrence of all the other events D_2... D_n – and this does not change our conclusions here, if the assumptions made below are valid.) Here again, it is rather obvious that

$$p((G/FC) = p(D/FC).$$

But this does not help us very much in predicting G because we assume that neither the importance of a specific D nor $p\,(D/FC)$ is known to us. All we have at our disposal, besides the known relative frequencies of our antecedent variables (with $p\,C = p_1$ and $p\,F = p_2$), are the values of their predictive power for G. Let us assume that

$$p(G/C) = p_4$$
$$p(G/F) = p_5$$

Assume now for the sake of simplicity that all three variables, (events) C, F, D are statistically independent of each other both pairwise and in all possible combinations. If we now additionally assume that there are no other causes of G in our population (i.e. that CFD is both a necessary and a sufficient condition for G), then of course the probability of D and by the same token our prediction for the collective CF can be found quite easily. The whole picture can be presented in Table I.

We may, under the above assumptions, find the unknown p_3 either from $p(G/F)$ or from $p(G/C)$. It is obvious that $p_3 = \dfrac{p_4}{p_2} = \dfrac{p_5}{p_1}$.

But the assumption that there are no other, alternative, additive causes of

TABLE I

$CFD = G$	F	$\overline{F}$	
C	(a) p_3	(b) 0	$p_2 \cdot p_3 = p_4$
$\overline{C}$	(c) 0	(d) 0	0
	$p_1 \cdot p_3 = p_5$	0	$p_1 \cdot p_2 \cdot p_3 = p(G)$

C in our population is pretty strong. Let us abolish it now and assume that, besides being 'produced' by CDF, G may also be produced by an A which makes an alternative sufficient condition for G. Here again one should note that *A may be either one single 'natural' class of events or the whole alternative of such additive causes.* We have to examine now the possibility of predicting the probability of G for two interacting collectives C and F *as we did before, but under the circumstance that G* additionally may be produced by another, unknown additive cause A.

Let us assume that all the assumptions for the above model are valid here, i.e., all three variables, C, D, F, are independent in all possible combinations. Let us assume additionally that this is also true for the variable A, which has at the same time the probability of occurrence (both in the whole population and with respect to any single antecedent variable or any of their combinations) equal to p_6. If we now imagine a hypothetical fourfold table with the two *interacting causes C and F held constant*, the probability of G for the separate cells of our fourfold table will look as follows (Table II).

But our real situation will look a little different, because we don't know either $p(D) = p_3$ or $p(A) = p_6$ from the above model. Let us designate p_3 by x and p_6 by y. We have, then, the two following equations:

$$p_2 \cdot x + y - p_2 \cdot x \cdot y = p_4$$
$$p_1 \cdot x + y - p_1 \cdot x \cdot y = p_5.$$

From these equations we can compute the values of x (p_3) and y (p_6). We will then have:

$$x = \frac{p_4 - p_2 x}{p_1 \cdot (1 - p_4) - p_2 \cdot (1 - p_5)} \ .$$

TABLE II

$CFD \cup A = G$	F	$\bar{F}$	
C	(a) $p_3 + p_6 - p_3 \cdot p_6$	(b) p_6	$p_2 \cdot p_3 + p_6 - p_2 \cdot p_3 \cdot p_6 = p_4$
$\bar{C}$	(c) p_6	(d) p_6	p_6
	$p_1 \cdot p_3 + p_6 -$ $p_1 \cdot p_3 \cdot p_6 = p_5$	p_6	$p_1 \cdot p_2 \cdot p_3 + p_6 - p_1 \cdot p_2 \cdot p_3 \cdot p_6 =$ $p(G)$

And having found the value of x we can compute y, which is:

$$y = \frac{p_4 - p_2 x}{1 - p_2 x}$$

We see, then, that under the assumption of independence of all possible causes involved in 'producing' G *we can derive the predictions for two (or more) interacting collectives, whatever the number missing supplementary factors making them a sufficient condition for G and whatever the number of other unknown additive causes of G in this population.* All we need is the predictive power of each of these collectives for G separately, and the relative frequency of C and F (or of a greater number of controlled interacting factors in our population).

Let us examine now the situation in which C *and F are additive causes of* G. Since both have only a probabilistic predictive value for G, we have to assume that they constitute only necessary components of two alternative sufficient conditions for G and not the complete sufficient conditions of G. Suppose that the corresponding sufficient conditions (and the only existing ones in the given population) for G are CD and FH. According to the above reasoning the variables H and D have to be distributed randomly with respect to both F and C. Let us now designate:

$$p(C) = p_1$$
$$p(F) = p_2$$
$$p(D) = p_3$$
$$p(H) = p_4$$

Let us also designate:

$$p\,(G/C) = p_5$$
$$p\,(G/F) = p_6.$$

From these assumptions we obtain a model of relationships that can be presented in a fourfold table in *which two additive causes C an F are held constant:*

TABLE III

$CD \cup FH = G$	F	$\overline{F}$	
C	(a) $p_3 + p_4 - p_3 \cdot p_4$	(b) p_3	$p_3 + p_2 \cdot p_4 - p_3 \cdot p_2 \cdot p_4 = p_5$
$\overline{C}$	(c) p_4	(d) 0	$p_2 \cdot p_4$
	$p_4 + p_1 \cdot p_3 \cdot$ $- p_4 \cdot p_1 \cdot p_3 = p_6$	$p_1 \cdot p_3$	$p_1 \cdot p_3 + p_2 \cdot p_4 - p_1 \cdot p_2 \cdot$ $p_3 \cdot p_4 = p(G)$

Here again the unknown parameters of our model are the probabilities of the two unknown supplementary factors (D and H) which are denoted by p_3 and p_4. We can compute them in several ways. The most simple way is to start from the formula that $p(G/\overline{F})$ is equal to $p_1 \cdot p_3$. Knowing $p(G)$ in general and $p(G/F)$ we may easily find a $(G/\overline{F})$, and knowing $p(C) = p_1$, we may now find the value of $p(D) = p_3$. The other parameter of our model, p_4, can be found in a similar way. Now our prediction for the intersection of the two additive collectives is, as in our model:

$$p(G/CF) = p_3 + p_4 - p_3 \cdot p_4,$$

which sounds rather simple.

The assumption that the two additive collectives exhaust all possible alternatives causes of G is usually too strong. Let us therefore assume now that besides these two alternative sufficient conditions to which C and F

belong (and which are CD and FH correspondingly) there is in our population another additive cause of G, namely K, the role of which in producing G, or its probability is unknown. The only assumption we will make here is, as above, that K (being another sufficient condition for G) is independent of C, D, F, H, both pairwise and in all possible combinations. Here again it should be stressed that K may stand either for one naturally distinguishable causes of G or for their whole alternative. Let us now denote $p(K)$ in our population by p_7.

The corresponding fourfold table with additive causes C and F held constant then looks as follows:

TABLE IV

$CD \cup FH \cup K = G$	F	$\bar{F}$	
C	(a) $p_3+p_4+p_7 - p_3 \cdot p_4$ $-p_3 \cdot p_7-p_4 \cdot p_7$ $+p_3 \cdot p_4 \cdot p_7$	(b) $p_3+p_7-p_3 \cdot p_7$	$p_3+p_2 \cdot p_4+p_7-p_3 \cdot p_7$ $-p_3 \cdot p_2 \cdot p_4-p_2 \cdot p_4 \cdot p_7$ $+p_2 \cdot p_3 \cdot p_4 \cdot p_7 = p_5$
$\bar{C}$	(c) $p_4+p_7 - p_4 \cdot p_7$	(d) p_7	$p_2 \cdot p_4+p_7-p_2 \cdot p_4 \cdot p_7$
	$p_4+p_1 \cdot p_3+p_7-p_4 \cdot p_7$ $-p_1 \cdot p_3 \cdot p_4-p_1 \cdot p_3 \cdot p_7$ $-p_1 \cdot p_3 \cdot p_4 \cdot p_7 = p_6$	$p_1 \cdot p_3+p_7$ $-p_1 \cdot p_3 \cdot p_7$	$p_1 \cdot p_3+p_2 \cdot p_4+p_7$ $-p_1 \cdot p_3 \cdot p_7-p_2 \cdot p_4 \cdot p_7$ $-p_1 \cdot p_2 \cdot p_3 \cdot p_4$ $+p_1 \cdot p_2 \cdot p_3 \cdot p_4 \cdot p_7 = p(G)$

But our problem looks different here, because we don't know the three parameters from the above model, namely, $P(D) = p_3$, $p(H) = p_4$, and $p(k) = p_7$. If we denote $p_3 = x$, $p_4 = y$ and $p_7 = z$, then we can find these values from the following equations:

(1) $\quad x + p_2 \cdot y + z - x \cdot z - x \cdot p_2 \cdot y - p_2 \cdot y \cdot z. + p_2 \cdot x \cdot y \cdot z = p_5$

(2) $\quad y + p_1 \cdot x + z - y \cdot z - y \cdot p_1 \cdot x - p_1 \cdot x \cdot z \cdot + p_1 \cdot x \cdot y \cdot z \cdot = p_6$

(3) $\quad p_1 \cdot x + p_2 \cdot y + z - p_1 \cdot x \cdot z \cdot - p_2 y \cdot z \cdot - p_1 \cdot x \cdot p_2 \cdot y$
$\quad + p_1 \cdot x \cdot p_2 \cdot y \cdot z \cdot = p(G).$

And our prediction of $p(G/CF)$ is

$$p_3 + p_4 + p_7 - p_3 \cdot p_4 - p_3 \cdot p_7-p_4 \cdot p_7+p_3 \cdot p_4 \cdot p_7 = p(G/FC).$$

Let us look now at Tables I — IV from another perspective. In all four tables the singular cells have been denoted additionally by the letters (a), (b), (c), (d), so that the category $CF = a$, $C\overline{F} = b$, $\overline{C}F = c$ and $\overline{C}\overline{F} = d$. As we will recall, Tables I and II describe the *interaction* of two collectives CG and FG, whereas Tables III and IV describe their *additiveness*. What is characteristic for the first two tables is that the relative frequencies in $a > b=c=d$, when b, c, and d are either 0, or > 0.

We may generalize this by saying that if we make an *intersection of two interacting causal collectives* in a population in which all other alternative causes of the given effect are either absent or distributed randomly, we obtain a *pattern of relationships corresponding to the above formula*.

Comparison of probabilities in the cells of Tables III and IV reveals another pattern. It may be characterized as follows: $a > c$ and $b > d$, but also $a > b$ and $c > d$.

And this may again be referred to as a *general characteristic of the additiveness*[16] *of two causal collectives* in an otherwise random population.

So we see that for situations in which we can make the assumptions of randomness of the variables involved and can postulate the nature of the causal connections between the two antecedent variables and their effect, *we can derive the prediction of their joint probabilistic effect*, i.e. *we can find the probability of this effect for the intersection of two initial collectives*.

What is even more interesting, we do not have to make the assumptions about additiveness or interaction of the initial causal collectives a priori. Assuming the randomness of the variables involved we can test the relationships in our data and see whether they fit into the pattern of the additive or the interactive type, and if they do, whether we have additionally to postulate the existence of some other variables of the additive type absent in our analysis and which are the source of the random 'noise' in our relationships.

On the other hand, if we cannot assume the randomness of all the independent variables involved or postulate it when it does not actually exist, the above equations either will lead to wrong estimations and predictions of p (G/CF) or will not be solvable at all. In such situations (which, unfortunately, are pretty common in natural conditions) *the prediction of the probability of a certain effect for the class denoted by its several causes does not seem a priori solvable*. Neither can we do it for situations in which we cannot assume that both our 'predictors' are the causes of the predicted class of events.

Here we might distinguish two subcategories of such situations. In the first, the events C and F (or at least one of them) *precede G* in time but they *are not the genuine causes* of G. If this association (correlation) resulting in the spurious correlation between G and any of its antecedent (spurious) causes is lower than 1.0, then, of course, our models will not work.[17] In the other type of situation, at least one of our 'predictors' does not precede G in time and therefore we cannot use the above models, which assume that G is the common effect of all the variables involved in the prediction of $p(G)$. Unfortunately, as mentioned above, the case of the 'prediction' of Mr. Petersen's religion belongs to this last category.

NOTES

* Reprinted, with major modifications of the last section, from *Synthese* 23, (1972), where it was published under the title: 'Inductive Inconsistencies and Conditional Laws of Science'.
[1] C.G. Hempel, 'Inductive Inconsistencies', in his volume *Aspects of Scientific Explanation*, New York and London, 1965.
[2] Cf. also Hempel, *op. cit.*
[3] See the typology of general conditional causal relations and the analysis of statistical consequences of their conditionality in Chapter V.
[4] For analysis of the role of time-space coordinates in general scientific proportions, see Chapter IV.
[5] Cf. Hempel, *op. cit.*
[6] Cf.R. von Mises, *Probability, Statistics and Truth*, 1939. See also the analysis of this edition by C.D. Broad in his review of the German edition of von Mises' book, reprinted in the collection of essays by C.D. Broad, *Induction, Probability and Causation*, Dordrecht, 1968.
[7] Cf.G.H. von Wright, 'Broad on Probability and Induction', in the collection of essays by Broad cited above.
[8] The reasons for quotation marks are explained below.
[9] C.D. Broad, ' "Von Mises" Wahrscheinlichkeit, Statistik und Wahrheit', in the collection of essays cited above.
[10] G.H. von Wright, *Broad on Probability and Induction*.
[11] A better term might be 'interpolate'.
[12] See Chapter V.
[13] E.g., C.G. Hempel in his essay 'Inductive Inconsistencies', cited above.

[14] For analysis of the rules of deriving statistical relationships from corresponding causal assumptions, see Chapter V.

[15] I stress the causal independence of C and F, in as much as they might constitute links of the same causal chain $C \to F \to G$ and then, of course, the above reasoning wouldn't be valid.

[16] It should be noted here that the term 'additive' may be a little misleading with respect to probabilistic values of each of these collectives separately. What we are adding here are the predictive values of each of these collectives for G (i.e. $p(G/C)+ p(G/F)$ diminished by the probability that the effects of their action will occur jointly. Otherwise we would often have to admit the probability of two additive collectives higher than 1.

What should also be noted here is that within each of these collectives taken separately the *overall probability* of (G/C) and p $(G)F$ by no means expresses 'pure causal mechanisms' only. Let us say that by *causal components* of a certain probabilistic relation $p(G/C)$ we mean that value of probability which covers only those G's which *followed C and were caused by C*. It is clear that this causal component is equal to $p(D/C)$ or p_3 in our model. The rest of $p(G/C)$ is due to the fact that some C's occurred jointly with the other alternative cause FH when D didn't occur. This might be called the *spurious component* of this probabilistic relation p G/C. In Table III, e.g., this spurious component is equal to $p_2 \cdot p_3 - p_3 \cdot p_2 \cdot p_4$. In Table IV the spurious component of p_5 $(= p[G/C])$ corresponds to all the rest of the formula which follows p_3 after the first + sign.

[17] If C and F (or either of them) are correlated with some real causes of G and the correlation with these real causes is 1.0, *then* of course our models will work because the spurious causes are empirically (extensionally) equivalent to the real ones.

BIBLIOGRAPHY

Broad, C.D., ' "Von Mises" Wahrscheinlichkeit, Statistik und Wahrheit', in C.D. Broad (e.d.), *Induction, Probability, and Causation*, Dordrecht, 1968.

Hempel, C.G., 'Inductive Inconsistencies', in C.G. Hempel (ed.), *Aspects of Scientific Explanation*, New York and London, 1965.

von Mises, R., *Probability, Statistics and Truth*, 1939.

Nowak, S., 'Causal Interpretation of Statistical Relationships in Social Research', in this volume, p. 165.

von Wright, G.H., 'Broad on Probability and Induction', in C.D. Broad, (ed.), *Induction, Probability, and Causation*, Dordrecht, 1968.

CHAPTER VII

LOGICAL AND EMPIRICAL ASSUMPTIONS
OF VALIDITY OF INDUCTIONS*

1. THE ROLE AND NATURE OF EMPIRICAL
PRESUPPOSITIONS IN INDUCTIVE REASONING

The problem of the justifiability of induction has a rather long tradition in philosophy of science. On one side of the dispute we meet many advocates and codifiers of inductive thinking, and − not least − the practice of several centuries of scientific investigations based in their essential aspects upon the application of induction. On the other side, beginning with Hume, are many of its critics. Some of them − like Reichenbach − have demonstrated the lack of sufficient justifiability of inductive method, but have nevertheless sanctioned its use for pragmatic reasons as the best thing we have for trying to discover the general regularities of the world around us. Others have rejected induction in general, arguing − like Popper − for an approach whereby we should rather try to falsify the hypotheses than to confirm them inductively.

In all this dispute the most convincing argument for me is the history of modern science, which in a way has 'pragmatically' confirmed the usefulness of inductive thinking. It seems reasonable to assume that if something 'works' there must be some reason for that. On the other hand, if something 'works only sometimes' − as is notoriously the case with the results of inductive thinking − it may be that this reason is not valid everywhere. Therefore I would like to examine in this chapter the so-called *presupposition theory of induction*.

The problem is not a new one. The idea that we need certain assumptions about the nature of the reality studied by scientists was clearly formulated by J.S. Mill in his postulate of 'uniformity of nature' as a necessary requirement for the validity of demonstrative induction. Russell, Keynes and Reichenbach extended the argument to the area of statistical thinking about verification of hypotheses. This kind of argumentation was called on again by A.W. Burks in his 'presupposition theory of induction'. In one of his papers on the subject Burks recalled Hume's criticism of inductive method and wrote:

> One of the ways of treating his criticism is to say that the validity of inductive argument is in an important sense relative to some broad factual assumptions about the general nature of the universe and that these general assumptions are presupposed in an awkward way by the users of inductive arguments.[1]

I would be inclined to agree with this statement, but I don't think I would agree with another of Burks' conclusions, namely that these assumptions are basically unverifiable, because they constitute our ultimate reference in inductive argument. He wrote:

> Since according to postulate theory the validity of any inductive argument depends on these factual presuppositions it follows on this view that there can be no inductive verification of induction, and hence no non-circular inductive justification of the use of inductive method, but the claim remains that these factual propositions relative to which we use the studied inductive method may be validated and which in a certain way explain this method.[2]

The argument about 'circularity' of inductive justification of induction is a rather constant theme in discussions on this subject. I think that the basic source of misunderstanding here lies in the wrong understanding of the notion of *generality of inductive method*. I think we should distinguish clearly any particular case of inductive reasoning proceeding according to a certain logical pattern – e.g., the pattern of enumerative induction, the pattern of demonstrative induction, etc. – from *the logical structure of the pattern itself*. In this general logical pattern or 'model' of a particular type of induction we state what kinds of assumptions, described in a most general 'content-free' way, are necessary in order to come to some conclusions, described again in a formal, content-free way. These assumptions are, besides their formal nature, *strictly hypothetical*. The structure of the given inductive model tells us only what kinds of conclusions would be justified *if in any kind of reality these assumptions were satisfied*, but it is unable to say whether anywhere, and if so where – in what science and for what kind of phenomena – the assumptions are satisfied. On this general, formal level of analysis the validity of the conclusions of inductive argument of a given kind is therefore analytically 'guaranteed' by the structure of the given inductive model. The conclusions are certain if the given model permits certain conclusions; they are more or less probable in other kinds of formal inductive models, independent of whether there is any reality at all which corresponds to the given assumptions. If such reality exists, its study may be *conceived as a case of empirical interpretation of the given formal model.*

Now if we want to apply inductive reasoning in any empirical science, or in any life situation, *we have to assume that the phenomena we are studying satisfy the postulates of the given model.* This we usually know – better or worse – *from the theories existing in the given science* and based upon the evidence *collected before our inductive study.* The theory may be sometimes quite strong. A chemist who, on the basis of one single case of reaction of two or more elements, comes to the conclusion that any sample constituting the combination of these elements in proper proportions will always create the given compound has at his disposal quite a strong theory, which he uses as the presupposition permitting him to make such inductive generalization. His theory is based upon at least two centuries of previous empirical-inductive studies. Therefore he is 'entitled' to use the model of inductive reasoning that applies to this situation – we will discuss the structure of this and of some other models later – and to treat it as a logically consistent pattern of proper reasoning. He does not have to justify the induction in its general form, because its structure is deductive in its essence, and guarantees the conclusions in all situations where its assumptions are satisfied. The only possible error (besides the possibility of wrong assessment of diagnostic facts, errors of measurement, artifacts of experimental procedures, etc.) would come from his *wrongly assuming for the reality he is studying the empirical validity of the presuppositions of the given model of induction.*

Only if the induction were to be used for the validation of a 'theory' that claimed to constitute all knowledge about the world in general (whatever this might mean), and if such a theory were to be tested in all its complexity at one moment of time, could the assumptions of such induction would not validated empirically without the obvious danger of circularity of argument. Fortunately, for induction and for human knowledge, this is not what authentic sciences try to do.

Once we understand this, we can understand why induction works better in some sciences than in others. These are the sciences in which the researchers have at their disposal sufficient numbers of well-founded propositions to permit them to use the inductive method for situations 'where it will work' and not to use it in others. Even they are often corrected by the future course of their discipline, and although some drastic reformulations of the foundations of the given discipline are always possible, at least relative rarely in such sciences do we encounter naive generalizations which are rejected almost immediately by the next study.

Max Black came quite close to this view when he said that "inductions may be justified but not all of them at the same time."[3] By this he meant that the inductions applied in empirical sciences always use knowledge gathered and inductively generalized before. According to him the 'problem of induction' was invented by abstraction-loving philosophers who try to think: How do we know that 'all ravens are black', or that 'the sun will rise tomorrow'? — without taking into account the existing knowledge of biology or astronomy. When we look at what is actually being done in the sciences, we see how knowledge already established is used in generalizations and predictions. Black noted:

Skeptical critics of inductive procedure, from Hume to Russell, are always asking us how we know that the sun will rise to-morrow... But we are never in the situation of almost total ignorance if we are in the position to raise the problem of induction. We really know a good deal about the sun and anybody who can intelligently ask a question about its rising tomorrow also knows a good deal about empirical regularities.[4]

But I would not agree with Black that the problem of justification of induction in general does not exist at all, and that the justification is of a merely 'pragmatic' kind; we have to demonstrate that it has usually or often worked and on the basis of this extrapolate its value. *The problem of validity of induction in its general form exists, but it is not an empirical problem*. Its nature consists in defining sufficient numbers of models of inductive reasoning which, from their analytically specified assumptions, would lead to analytically guaranteed conclusions. *It is not the task of the philosopher to say anything about the nature of the world in general*, probably for the reason that nothing that would be sufficiently general to make any kind of induction universally valid would be true. It is even less his task to say anything about the validity of these assumptions in particular fields of empirical investigations. The specialists of the given discipline usually know this much better. Such a position puts induction into the field where it would appear to belong — i.e., into the area of logical analysis and formal methodological reflection — and leaves the decision concerning the use of particular methods to those who are capable of deciding whether the reality they study constitutes the exemplification, or at least sufficient approximation, of the given, formally defined, model.

Should future experience reveal the generalization or theory justified inductively within a particular science to be false, this does not invalidate induction in general, nor does it invalidate the given model of inductive reasoning in its formal structure. It only means that the scientists were wrong in believing that the presuppositions of the given model were satisfied in this particular case.

Now it may happen that the scientists using the given inductive method are aware of the possibility that its assumptions might not be fulfilled in the given case. They may even be able to specify a certain *probabilistic measure of the degree of this uncertainty*. Then, of course, even if the method leads to certain results, the conclusions made in the given case of reasoning would be of a probabilistic character, with the probability of a valid conclusion being equal to the probability of validity of the assumptions made. In cases where the given method in itself leads only to probabilistic conclusions and the assumptions of its applicability are also made in probability terms, the final conclusions would be the function of these two probabilities. But even then the two probabilistics should be stated separately, since one of them is the function of the general logical pattern of inductive reasoning and therefore is not 'adjustable' to our changing knowledge, whereas the other is strictly related to the state of knowledge at any given moment of time.

The task of this paper is to discuss certain models of inductive reasoning understood in the way presented above. *No claim will be made that induction as understood here is the only method by which new hypotheses or theories should be incorporated into the sciences*, since this would imply that no development of theories is possible unless the previous theories are sufficiently strong to justify it. A scientist may come to the formulation of a new hypothesis in a completely 'unjustifiable' way now and then, or may justify it on grounds of certain non-empirical premises. If he lacks premises for its inductive confirmation, he may accept his theory 'only because' he has not been able to find any contrary cases. This is the essence of the Popperian approach to the confirmation of theories.

The aim of this paper is rather to show the limits within which the inductive methods as understood above can be validly used, than to stress the omnipotence of induction. These limits are defined by the limits of validity of empirical assumptions of induction of the given type.

Moreover, nothing will be postulated about the degree to which any of these assumptions seem to be empirically valid in the world in general. For the use of inductive methods it is enough to assume that the conditions specified by the given assumptions do sometimes exist, and that the scientists may sometimes — with greater or smaller degree of certainty — be aware of their existence.

2. THE ASSUMPTION OF COMPLETE UNIFORMITY WITHIN A CLASS OF OBJECTS OR EVENTS

The most simple pattern of inductive reasoning is that based upon the assumption of complete, unconditional uniformity within a certain class of objects or events. If we assume that all the objects of a particular class are uniform with respect to a certain characteristic or set of characteristics, then of course one single observation is sufficient for the empirical justification of a generalization applying to the whole class of objects or events. One might think that in such a case we assume precisely what has to be empirically validated, and therefore no observation is needed at all as long as we rely on our assumption. This would be the case only if the assumed uniformity were identical with the assessed property of the objects of the given class. Such is not the case we have in mind: here the assessed uniformity refers to a certain *variable without the specification of the value of this variable;* the role of the empirical study is to specify the value of the variable for all the cases for which the assumption of uniformity was made.

Suppose that we have a set of objects $A_1, A_2 .... A_n$ (where n might be either a finite or an infinite number) which will be generally denoted by A. Suppose also that there is a variable characteristic (in short, a variable) B which may have as its values (quantitatively or qualitatively understood) the *properties* $B_1, B_2,... B_n$. Let us now say that $B_U(A)$ means that *all the objects A will have the same unspecified value of the variable B.* Now suppose that we have observed the object A_j from the class A and have assessed that it has the value B_i of the variable B. Then we will have, of course, a rather simple reasoning:

$$B_U(A)$$
$$\frac{B_i(A_j)}{B_i(A)}$$

The cases of reasoning corresponding to the above scheme are more frequent than one would at first suspect. A biologist observing under the microscope the eye structure of one single bee usually generalizes his conclusions to all bees, because he assumes (on the basis of his general biological theory) that the anatomical properties of all the examples of the same species are (within certain limits of variation) roughly the same. A nineteenth-century physicist who described the properties of all atoms of a new element on the basis of one study of a small sample of the atoms based his generalizations upon a presupposition (which actually turned out to be false after the discovery of isotopes) that all the atoms of that same chemical element are identical.

Most frequently, however, the assumption of uniformity has the character of a 'historical' proposition: we assume that all objects of class A which are delineated by certain time-space coordinates are uniform with respect to the characteristic B. For the justification of our generalization it is then, of course, necessary to assume that the observed object A occurs within the limits of the time-space area for which the assumption of uniformity has been made. If it is enough for certain practical purposes to measure the temperature of a human body in one particular spot (e.g., under the tongue) this is because the medical, or biological, theory says that at any given time the temperature of the entire body is sufficiently approximately the same.

The assumed uniformity can be of a still different kind. We may simply say that we expect (on the basis of previous knowledge or on the basis of our 'subjective' belief) that all the objects of class A are uniform (whether universally or 'historically' understood) with respect to one variable B. But if our assumption is drawn from a more comprehensive theory, it often happens that the variable B is an element of a set of variables and our theory says that the objects of class A are expected to be uniform with respect to the whole set of such variables, whatever the values of these variables might be. It may happen that our theory is more general in another sense, i.e. by stating the uniformity of a broader class K of subclasses of object when A is a subclass of K, distinguished by its special features; the theory says then that some other classes of objects mutually exclusive with A (e.g. all C's, D's, etc.) will also be uniform (although not always in the same way) with respect to variable B. Finally, our theory may be of such kind that it applies both to a broader class of objects and to a more comprehensive set of variables. One such theory says that all basic anatomical features of varieties of all species are, within each of these species, approximately the same.

Suppose now that the theory on which we base our assumptions of uniformity of all A's with respect to B says that not all properties of A's (i.e. a whole set of properties of which B is our element, but only a certain fraction of them — let us say 75%) are uniform for all A's. At the same time our theory is not specific about the uniformity of A's with respect to B particularly. Knowing this we will then expect that our assumption of validity has a certain probabilistic validity, and we will say that under this additional assumption the conclusions about all A's would be valid only at a 75% level of confidence. The same may happen when our theory says that not all mutually exclusive classes of objects (A being one such class) but only 75% of them have the tendency to be uniform with respect to B (when B is, for example, the color of the skin of a species of animal). Finally, when the relation of uniformity may be probabilistic both for the class of events and for the class of variables we have to take into account these two probabilities in order to obtain the probability of validity of our conclusions.

This does not mean, of course, that we expect, e.g. that about 75% of butterflies of a given species will have the color of wings we observe in our singular case. The estimation of probability of our hypothesis has as a reference class the frequency of the whole set of hypotheses concerning all anatomical properties of all species of butterfiles on which the probability of our hypothesis is based, and it means: 'From the whole set of such hypotheses, about 75% of them may be legitimately validated upon the observation of one single case of butterfly'.

The estimation of degree of certainty of our assumption of uniformity may be (and often is) based upon more subjective premises. Our knowledge is rarely of the kind that would permit the clear formulation of frequencies of uniform classes and uniform properties within a more general reference, categories of classes and sets of properties. The probabilistic value of this assumption might even be 'measured' by asking the scientists how much they would be willing to 'bet' on its correctness. But it seems to me that the logic of reasoning should then be exactly the same: We should ask them whether they expect that their estimation refers to the set of properties (variables) within the given reference class of objects, or to the uniformity of different classes of objects with respect to the given variable, or to both at the same time. In some cases the simple act of asking such questions might help to clarify the meaning of the 'bet' involved. We must also accept that sometimes the scientist will not be able to say anything more specific than that he is doubtful about the certainty of assumption of uniformity of all A's with respect to B only, and he feels that his degree of uncertainty should be expressed in numerical terms by the probability, let us say, 0.5. Accepting this, we should understand at the same time what uncertain and ambiguous inductive grounds we are on. The conclusion will always be the same: If the given assumption of uniformity permits certain conclusions but the assumption itself is 'probabilistic', the generalization should be made with the same level of probability.

3. THE ASSUMPTION OF RANDOMNESS IN STATISTICAL INDUCTION

Let us turn for a moment to statistical induction. Its basic feature lies in the procedures of estimation of certain parameters in certain populations on the basis of distribution of corresponding properties in the *random samples* drawn from these populations. This procedure, extremely useful in many practical situations, seems to be of relatively little value in the case of universal laws of science because these, by definition, apply to infinite populations, *which makes any procedure of artificial random sampling basically impossible*. The population may be so dispersed in space that we simply do not have practical access to all its members to assure them equal chance in the selection of our sample. Moreover, its unknown part may 'exist' in the past and in the future, which makes any sampling theoretically impossible. All this seems to rule out the application of standard procedures to such universal populations — unless they are... *naturally random*.

Let us understand the notion of randomness in the way defined by von Mises in his frequency theory of probability.[5] To say that the values of variable B occur at random with respect to the events or objects A is equivalent to saying that in an infinite series of A's there is a *limiting relative frequency* for the given (and any possible) value of the variable B_1 so that the greater the number of A's in a given finite, observed series the more unlikely that the proportion of those A's which possess, for example, the property B_1 among all A's observed will differ more than a small fraction from this limiting value of relative frequency. The mathematical properties of such series, which von Mises proposed to call 'collectives', are described by Bernouilly's Law of Great Numbers. What is essential for us here from the point of view of inductive analysis is the fact that a collective by definition *excludes the possibility of any gambling system* which might effect the limiting frequency of the series A and B_1 among all A's. There exist no systematic, predictable ways of changing the value of this probability (i.e. frequency limits) by any selection rule – either by a rule referring to the position of a particular A_i within the series, or by its relative position with respect to some or all A's preceding it or followed by it. What is also important, there is no way of changing the frequency limit by taking into account some space-time coordinates of occurrence of A's or the fact of their concurrence in the context of other events or properties.

Suppose now that we have a random collective of the cases A-B_1 in which $0 < p(B_1/A) < 1$. It is obvious that for such a collective *any natural series of observed events or any sample drawn from it in any arbitrary way constitutes a representative sample according to all rules we know from elementary statistics for random samples*, because – due to the notion of collective – there is no way of changing the limiting frequency of B_1 among A's. This frequency is 'entitled' to 'oscillate' around its limiting value in finite series of A's, but we assume that the longer the series, the smaller the probability that the frequency in the series will differ more than a certain small amount e from its limiting value. When increasing the length of the observed series we can decrease both the size of a probable 'error' and its probability in the way described by Bernouilly's theorem.

In empirical inductive studies we are faced by a situation in which we observe a certain finite series of A's and we see that only a certain proportion of them has the property B_1. Now, if we make the assumption that this is a finite series from a random collective, such a series constitutes for us – as said above – *a representative random sample for the entire infinite series*. The assumption of randomness without specified value of the frequency limit plays here the same role as the assumption of unspecified complete uniformity in inductive generalizations from one case to the whole general class; it permits us to estimate the value of a frequency limit in the whole infinite series in the way we estimate the values of the parameters in the finite populations on the basis of observation of statistics in a random sample drawn from it.

But due to the different nature of the assumptions our estimation is not exact. We extrapolate the observed frequency to the entire series ± a certain 'confidence interval' of other pobalities which cannot be excluded with a sufficiently high level of significance. We also know that the exclusion of the other possibilities has not been made in terms of certainty but in terms of certain probabilities, but these probabilities may be sufficiently high (or low) to give us almost complete certainty that the estimated frequency limit belongs to given confidence interval. If we toss a coin 10 000 times and come out which 8000 'heads' we still may 'admit the possibility that it was a 'fair coin', even though we know how small is the probability of obtaining the given result under the assumption that probability of heads in the entire series is 0.5.

In order to formulate such probabilistic conclusion from our finite series to the entire collective we must assume the *random character of the entire population*. Otherwise no inductive generalizations can be made. The question then arises, whether and when we can make such an assumption.

Hans Reichenbach, who stressed very strongly the necessity of attempting to estimate the frequency limits from the frequencies in the finite series, was rather skeptical about the possibility that we may know in advance that our series is random. He wrote:

In the analysis of Hume's problem we thus arrive at a preliminary result: If a limit of frequency exists, positing the persistence of the frequency is justified, because this method applied repeatedly must finally lead to true statements. We do not maintain the truth of every individual inductive conclusion, but we do not need an assumption of this kind, because the application of the rule presupposes only its qualification as a method of approximation.

This consideration bases the justification of induction on the assumption of a limiting frequency. It is obvious however that for such assumption no proof can be construction.[6]

As we know, Reichenbach found such justification in 'pragmatic' aspects of induction' "This is simply the best thing we can do; if limits exist, we will discover them – if not, then we will make errors in such cases both in our conclusion and in our actions based on those conclusions".

Here again, I think there are relatively few real situations in the sciences where no guesses about random or nonrandom observed series are possible, and usually one does not make any generalizations in such cases. A sociologist knows that the distribution of customs, values and institutions is nonrandom with respect to space and time; therefore he will not generalize the findings from his own nation or culture to mankind in general. On the other hand, there are quite a few situations in which the notion of randomness seems to be quite strongly rooted in the existing theory; therefore a physicist who has discovered a new radioactive isotope knows that the process of its decay will be random in such a way that half of its atoms in any sample will decay within a constant period of time for any sample possible: his only task here is to measure the period characterizing this whole, naturally random process. Even the insurance companies feel reasonably confident that the proportion of car accidents for a given category of drivers will be relatively stable and in a way random; therefore the only thing important for them is to estimate the probability of an accident for the given category of drivers and set their coverage rate correspondingly.

But there is one essential difference between the conclusions of a physicist studying radioactive decay and the conclusions of the insurance companies. The physicist makes the assumption of random uniformity *for the entire universally defined class of all atoms of the same radioactive element*, whereas the assumption of random uniformity made by the insurance companies applies to some *finite populations* even if not defined by number or in terms of definite space-time coordinates of their existence. As with the assumptions of complete uniformity, the assumptions of random uniformity may have the character either of universal laws of science or of 'historical generalizations' and, depending on the nature of the assumption, we extend our generalizations and expectations correspondingly. It is needless to stress that the extension of assumed random uniformity as well as its a priori probability depends strongly upon our previous knowledge.

The assumption of random uniformity of a certain infinite series may sometimes − on the basis of our previous knowledge − have a *more specific character*. We may feel entitled to believe that from the whole range of probabilities between 0 and 1 *only some are possible* for our collective $A{\rightarrow}B_1$. Let us assume that we admit only two such probabilities: P_1 and P_2. We may also believe that we know the *a priori likelihoods* of these two alternative probabilistic hypotheses. In such case − after assessing the frequency of B_1 in our estimation of the 'random sample' of A's − we may apply the Bayesian approach to probability that one of the two hypotheses in question is true. Our conclusion is made in terms of the probability of the hypothesis in question, but at least the frequency limit derived from the observed series is under these assumptions exact.

Unfortunately, real situations corresponding strictly to the Bayesian model with definite empirically justified a priori likelihoods are rather rare outside the area of artificial arrangements with randomly selected urns containing black and white balls. But fortunately, under the Bayesian theorem, the longer the series of evidence, the more the final decision depends upon the conformity of the observed series with one of the alternative hypotheses and the less on the a priori likelihood of the hypothesis in question. So – if our theory permits us to specify the number of hypotheses with exact frequency limits, their relative a priori likelihoods are relatively irrelevant. On the other hand, if the only thing we feel entitled to assume is the random character of the series in question, then our conclusions have to be made in terms of certain intervals of probability and not in terms of exact frequency limits. But if the series of evidence is sufficiently long, then these intervals may be so 'narrow' that they practically correspond to one value of p.

4. THE ASSUMPTION OF RANDOMNESS IN ENUMERATIVE INDUCTION

Let us return now to the situation in which a scientist tries to verify inductively a general (i.e. exceptionless) law of science. As we remember, under the assumption of complete uniformity of all the events or objects of the given class he is studying, one singular observation of the value of the variable in question is sufficient for the generalization referring to the whole universally defined class. *Any repetition of the observation or experiment is superfluous* here unless it is aimed toward the elimination of possible error of measurement or the check of correctness of experimental manipulations. On the other hand, we know that *in many cases scientists will require repetitions* of the observations, and the less they are certain of the validity of the uniformity assumption, the more they will tend to base their generalizations on a greater number of possible observations, tending more and more toward the use of the method of 'enumerative induction'. The structure of this method consists in inferring from the general descriptive statement 'of a great number of A's which have been observed, all had additionally the property B_1' the inductive generalization: 'all (existing or possible) A's are B_1',

Enumerative induction has not had a very 'good press' among philosophers of science, beginning with F. Bacon. On the other hand, the empirical scientists have tended to rely on it fairly strongly, especially in the sciences in which the theoretical foundations of presupposition about complete unity are rather weak. Does this mean that the use of this method is therefore unjustified?

I don't think so and I think we can find quite serious reasons for its use, as well as for the assumptions necessary for its application. These assumptions cannot postulate, in this case a complete uniformity, because this is what we want to 'prove' by enumerative induction. What we must assume here is that *our series of all events or objects A is naturally random* with respect to the occurrence of characteristics constituting the possible values of the variable B. Our *assumption of randomness is otherwise 'open*: it does not say which of all theoretically possible values of the variable B may occur jointly with A, nor what are their respective probabilities (limiting frequencies) in the whole series. It only implies that each of the values of the variable B will within the reach of our experience *tend to occur* with the frequencies corresponding to their probabilities in the entire infinite series.

If the variable B has only one value with respect to the objects A, (e.g. B_1), then of course in any natural series of observed events A, all of them will be also B_1. This would be analogous to the case of drawing balls from an urn in which all the balls are black. If some of the A's are B_1 and others B_2, the two will tend to occur within a naturally random finite series with frequencies determined by their probabilities in the entire series of A's.

The assumption of randomness of the studied series permits the scientist who uses the method of enumerative induction to come as close to general laws of science as the number of observed cases will permit him to do. He can apply the normal rules of statistical reasoning in estimating 'population parameters' from the 'statistics' of his 'random sample' by rejecting the alternative 'null hypotheses'.

Let us look at the famous case of generalization almost always occurring in the discussions on enumerative induction: 'All ravens are black', and let us assume that a scientist would like to confirm it inductively.

Suppose now that during extensive trips around the world he has observed 1000 ravens and has found that all of them were black. Assuming that the sample of ravens he observed was (due to the extended character of his trips and the natural movements of the members of the populations of ravens) *a random sample of all ravens*, he can ask himself the following question: "suppose that there is at least 1% of ravens which are non-black — what is the probability that none of them would be observed by me?" The answer is simple: $0.99^{1\text{-}000}$, which comes to approximately 0,00004.

On the basis of such data he may come to the conclusion that it is extremely unlikely that the proportion of non-black ravens among all ravens is greater than e.g. 1%, or in other terms: that *at least 99% of ravens or possibly all of them, are black*. The proposition he makes is not completely general, it is stated in terms of a certain *degree of generality for which he has sufficient empirical evidence*, but it leaves the question of greater and greater generality open for further observations and makes it dependent upon the increasing number of confirming cases. If the scientist will further observe only black ravens, and if he additionally takes into account all the observations about ravens made by all people in all periods of history, and finds that nobody has reported a case of a non-black raven, he may be *practically certain* that the generalization about ravens is *either completely general or the frequency limit of non-black ravens in the entire population of ravens is extremely close to 0*.

So we see that under a rather simple assumption of randomness of the observed series we can come *extremely close to a completely general hypothesis* and we are able to formulate our conclusions with any approximation to 1, excluding the values of complete generality and complete certainty. *In most practical cases such conclusions would be equivalent to complete certainty and complete generality*. But, we also know that the notion of randomness admits the possibility of realization of an extremely unlikely series. In other words, we know that it is still 'theoretically possible' that half of all the ravens are non-black and it 'just happened' by a random accident that none of them was observed by a human observer. Nevertheless, when we try to compute the probability of such an 'accident' we will exclude its 'possibility' without hesitation in our practical thinking, just as we reject practically the possibility that the water in the pot we are putting on the stove will freeze instead of boiling, even if it is 'theoretically possible' in physical theory.

The situation changes dramatically if the theory of given phenomena says that there is an a priori probability that the relationship in question is really *completely general*, because then we can apply Bayes' theorem for the estimation of the probability of this alternative. In general, the greater the a priori probability of a completely general relation and the smaller the probability of the given effect under the hypothesis alternative to our hypothesis, the more *a posteriori* probable is the hypothesis of complete generality of our series (on the basis of the given set of data correstponding to the scheme of successful enumerative induction).

But these conclusions depend also upon the assumption of the randomness of the series involved. What is especially important here is the postulation of the randomness of the series as predicted by the hypothesis alternative to our hypothesis (i.e. that which states that the relationship is not general but statistical only). If the series is not random, we cannot assume that any finite series of observations uniform with respect to the predicted property is able to reject our statistical null hypothesis. It might be, for example, that all consecutive 10,000 cases of A's are B_1 when the entire class of A's is by no means uniform with respect to B_1. It may turn out that all the other A's will be non-B_1, or differentiated on the variable B otherwise in a non-random way. In such case no series of any length be used for the rejection of an alternative hypothesis, and therefore for the confirmation of our hypothesis.

It may also happen that our general theory does not specify the a priori probability of our general hypothesis, but only says that it is possible (i.e. that its probability is higher than 0), and does not specify the mean value of the frequency limits of our property (evidence)under all alternative hypotheses, but only says that it is lower tha 1. Assuming the random nature of these series we can apply the Keynesian model of verification,[7] and say that under these assumptions the probability of our hypothesis tends to increase toward 1 as the number of confirming observations tends to increase to infinity. But, of course, we cannot give any numerical values of the probability of our hypothesis and thus cannot specify in Bayesian terms the length of the series, the uniformity of which is a condition for the acceptance of our hypothesis at a given confidence level.

Needless to say, the nature of our assumptions in all these situations depends upon our prior theoretical knowledge as well as upon an understanding of the conditions under which the observations are made. If we assume, for example, that the phenomena we are observing do not constitute a natural random sample from the entire − naturally random − universal population, because there are some natural boundaries of the 'free movements' of the objects in question, in the result of which only a certain subpopulation defined by certain time-space coordinates can appear within the reach of our observation, and at the same time there are no strong arguments in our theory in favor of randomness of the entire series – then of course we should formulate the assumption of random character of a *historical population* to which we have access and extrapolate the results of enumerative induction so that it would justify a 'historical generalization' only.

5. THE POSSIBILITY OF ESTIMATING THE DEGREE OF UNCONDITIONALITY OF GENERAL CAUSAL HYPOTHESES

Up to now we have considered the problems of inductive confirmation of hypotheses stating the joint or sequential occurrence of properties or events A and B (or B_1). We assumed that we obtain *uniformly confirming results* and tried to spell out under what kinds of assumptions certain conclusions may be made.

But it often happens in emirical research that induction of enumerative type fails after a certain series of confirming cases. This would imply differing conclusions:

(1) First, it might signalize that the concept A used for the description relationship $A{\rightarrow}B$ is useless for this hypothesis because there is no uniformity of A's with respect to B, and no valid hypothesis with respect to the relation between A and B may be formulated. This would imply the necessity of reconceptualization of our theoretical problem and that we have to search for some other class of phenomena D (somehow correlated either locally or generally with A) which might be suspected as generally (or almost generally) related to B.

(2) Second, it could mean that there is a certain uniformity of A's with respect to B but that this uniformity is of a statistical character with $p \ B/A < 1$, when B occurs at random with respect to A. This kind of problem was discussed in Section 3 above.

(3) Finally, it may mean that A is a useful concept for the formulation of the hypothesis stating the relationship between A and B in general terms, but that *this relation is of a conditional character*: A is accompanied (or followed) by B only when certain other conditions (e.g. C) also occur. This means of course that the relation $AC{\rightarrow}B$ is an unconditional, i.e. general, relation.

While the notion of conditionality applies to any type of relations, it seems to be especially important in the case of *causal relations*[8] It would be impossible to recount here the long dispute in philosophy of science over whether it makes any sense at all to say that among all the relations between events, some of which precede the others in time, we should distinguish as a subclass those which are causal. Let me just say that for the purpose of this chapter I mean by causal relations such necessary (i.e. general) sequences of pairs of events which possess one additional feature: *no matter how the antecedent of this relation were to be changed, whether it should change spontaneously or by deliberate action of a conscious agent, the consequent of the relation would 'behave' in the way described by the given 'causal law'.* Therefore by stating that a certain relation is causal (as distinguished from 'spurious') we imply additionally that *we are able to control the occurrence of the consequent of this relation by the purposeful 'manipulation' of its antecedent* – if the manipulation of the antecedent is practically possible for us. This notion of deliberate manipulation of consequences by the manipulation of the antecedent might be proposed as a partial, 'operational definition' of causal relations. It is obviously, however, much too narrow: In some sciences, like astronomy, we assume the existence of necessary, causal relations between such properties as the mass of the sun and the shapes of the orbits of its planets, even when we know that the antecedent of this variable is – at least at the moment – completely outside the limits of possibility of human 'manipulation'. But we might say that when stating the causal nature of such a relation we mean that no matter how its antecedent should be changed, whether it changed 'naturally' or due to the action of some 'conscious agent', the consequences of the change for the shapes of the planetary orbits would be, as described by our causal theory.

The principle of manipulable necessity seems to be the basic assumption of the 'controlled experiments' conducted within the rules of Mill's canons of demonstrative induction. And here we see an interesting phenomenon: Even the philosophers or scientists differ as to the theoretical notion of causality, if they accept it at all, they seem to agree that if there are causal relations at all, at least those which have been tested under the conditions of the controlled experiments are causal, which would imply they agree on the 'operational definition of causality' as specified above. (Those who do not believe that the observation of regularities under the conditions of 'manipulable' antecedents proves anything about their causal character must still agree that most of the regularities of this kind are conditional — i.e. the sequences are exceptionless when certain other conditions are also controlled or at least are observed as occurring with their antecedents. Therefore the analysis presented below seems to be independent of belief in the principle of causality.)

The conditions referred to may be of *positive* kind. A may be causally related to B (I assume that an experiment under the rules of the canon of only difference revealed that A is somehow related to B, i.e. that their relation is not a spurious one) but at the same time A may constitute only a necessary component of the sufficient condition for B, but not the complete sufficient condition. This means that there is at least one additional event or property (e.g. C) of such kind that both A and C are essential for the occurrence of B, and the joint occurrence of A and C constitutes the sufficient condition for the occurrence of B. The conditions may also be of a negative kind: A is followed by B (causes or 'produces' B) only when a certain *disturbing factor F does not occur*. Once we have stated on the basis of the experiment conducted within the rules of the canon of only difference that A is somehow related to B, *the discovery of these positive or negative conditions becomes the central problem of further inductive analysis.*

The general idea of this analysis is to test the relationship $A \rightarrow B$ under different combinations of antecedent conditions so that their relevance or irrelevance for this relation can be stated. The reader will find an excellent analysis of this problem by G.H. von Wright.[9] I would like now to discuss another aspect of the problem. Suppose that after the test of certain combinations of antecedent variables, and on the basis of observation of occurrence or non-occurrence of B, we come to the conclusion that the relation $A \rightarrow B$ is conditional with respect to a certain C while *other variables tested by us seem to be irrelevant.* We may be tempted to say that $AC \rightarrow B$ is really a general, i.e. unconditional, causal relation. On the other hand, we know that the number of variables we have tested in our study is fairly limited and that many others might be suspected as being not irrelevant for this relation. With respect to the relation $AC \rightarrow B$ they may either be positive factors which occured and acted in our experimental setting but we were not aware of their occurrence or of their importance for $AC \rightarrow B$, or they may be disturbing factors unknown to us which would 'cancel' $AC \rightarrow B$ had they occurred in our experimental setting. The question is: What should be done in order to be able to say that our hypothesis is really a general, i.e. unconditional, causal law?

The answer is simple but not very helpful. In order to obtain such certainty we should test all possible factors and thus prove their irrelevance. In a way our situation here is analogous to the case of enumerative induction. Since the number of possible factors (i.e. classes of events which we might define in general terms) is infinite, the proof of complete unconditionality is just as impossible as proof of the complete generality of a proposition referring to an infinite number of individual cases. But here again we can try to formulate certain assumptions which will make our task less hopeless. And this again depends to a high degree on the previous theoretical development of our discipline.

First, we might say that although the characteristics on which particular events may differ are supposedly infinite, the theoretical knowledge in many sciences limits their number quite drastically, *putting the phenomena studied by this science within a more or less definite conceptual scheme and enabling the assumption that the phenomena which cannot be defined in this conceptual scheme are for this science, and the phenomena studied by it, irrelevant.* If the conceptual scheme is really comprehensive, then of course it means that we are dealing with a certain closed system of variables which can be mutually dependent in different ways but cannot be dependent on phenomena outside this conceptual system. Even then the number of classes definable within the system may be quite large, but it may also be – at least in some sciences – finite.

Assumong now that the number of typologically defined events or properties which might be suspected as hypothetical conditions of the regularity $AC{\rightarrow}B$ is finite and not very large, and that we are able to test in proper inductive combinations the hypothetical impact of all of them, and finally that the results of all such inductive eleminations are negative (i.e. each of these variables turns out to be irrelevant for $AC{\rightarrow}B$) then of course we can say with *complete certainty* that the relation is absolutely unconditional in the typological sense, or – what is equivalent – that it is a true general relation in the theoretical sense.

The validity of this conclusion is again depended upon the validity of our assumption that our conceptual scheme is theoretically comprehensive for the kind of relationships we are studying, i.e. that it constitutes a sufficient frame of reference for a satisfactory theory of this class of phenomena. It is a rather strong assumption to say that we have a complete list of hypothetical conditions and that we are able to test the impact of all of them upon $AC \rightarrow B$ in inductive comparisons (with a negative result). It may seem justified in some areas of research when the conceptual apparatus is fairly simple as, e.g. in classical mechanics, but it is usually rather doubtful in most of the others, such as biology or the social sciences. We might then suspect that of all the possible variables which should have been studied in inductive comparisons, only some have been tested in our experimental studies (or by finding proper instances of naturally occurring cases) and that others which have not yet been tested in our research might eventually falsify the proposition about the unconditional character of our regularity. The question then arises a to whether we might be able to say anything in more definite terms about the confidence level of hypotheses, which states the *degree of unconditionality* of our regularity.

Let us concentrate for a moment on the discovery of hypothetical disturbing factors of a certain relationship, i.e. factors the occurrence of which would make the proportion $AC \rightarrow B$ false. Suppose that for the relation $AC \rightarrow B$ there is a finite but very large number of 'natural classes' of events of properties, where the notion of 'natural classes' is rooted in the conceptual scheme of the given theory and refers to the concepts definable in the comprehensive language of this theory. Suppose, moreover, that a great but finite number of such variables $D, E, F, G... O, P, Q...$ can be classified into two categories, so that the event denoted by definite values of dichotomous variables of one category are *relevant* for the sequence $AC \rightarrow B$ in such a way that the occurrence of the event denoted by the given value of such dichotomous variables may *cancel* the sequence $AC \rightarrow B$ when the events denoted by R. Suppose now that the relevant disturbing factors constitute a *relevant classes of events by R*. All the other variables (i.e. classes of events) which are *irrelevant* for the occurrence of the sequence $AC \rightarrow B$ will be denoted By $\bar{R}$. Suppose now that the relevant disturbing factors constitute a very small minority of all variables understood in the typological sense: say, that from the totality of variables distinguishable for the given type of analysis only 1% of them have the property of being disturbing factors for $AC \rightarrow B$. Let us say that in such case our regularity is *almost unconditional in the typological sense;* only a very small proportion of generally definable factors, when operating on the relation $AC \rightarrow B$, can disturb it.

Now suppose that, knowing the complete list of all hypothetically distinguishable factors and having no a priori hypotheses about the impact of any of them upon $AC{\rightarrow}B$, we start to test (e.g. experimentally) their relevance for $AC{\rightarrow}B$, *taking consecutive factors at random* from the entire population of variables which should be tested in inductive comparisons. Suppose moreover that after having tested the impact of a long series of generally defined factors in inductive eliminations we find none which would disturb the relation $AC{\rightarrow}B$. Then of course we can say, first, that with respect to the great number of the tested factors $AC{\rightarrow}B$ is unconditional, since the occurrence of events denoted by them is for this relation irrelevant. But we can say more, namely that in the entire population of factors the proportion of those which might eventually prove to disturb our regularity is, with a very high probability, very small, if there are any such disturbing factors at all. Assuming we tested them really at random, we can say that *with a given level of confidence our regularity is either completely unconditional or almost unconditional in the typological sense, where the term 'almost' denotes a certain degree of unconditionality understood in terms of the proportion of factors defined within the given conceptual scheme which are irrelevant for this regularity to all factors possible in the given frame of reference.*

Thus without testing the hypothetical impact of all the factors from our list (if the list is very long) we come fairly soon to the conclusion that, say, with a confidence level 0.95 *either none or not more than 1%* of all the generally definable factors might disturb the generality of our relation.

But here we should remember that a scientist almost never takes the hypothetical 'modifiers' of the tested relation at random even if he were to have a complete list of them. He usually begins by testing the impact of those variables which are most likely (on the basis of theoretical knowledge or of his scientific insight) to belong to the subclass of the relevant factors, and does not (at least primarily) study those which seem almost certainly to be the irrelevant ones. If he then finds that the tested factors happen nevertheless to be irrelevant, this *increases* considerably the *level of confidence* of the conclusion as to the unconditionality of the relation $AC{\rightarrow}B$ and increases the degree of unconditionality as compared with its value computed for the situation of really random selection of the factors tested. He is not able to judge numerically *how much more justifiable* his conclusions are and what should be the 'corrected' size of his inferred degree of unconditionality, but at least he can take the credibility and degree of unconditionality computed for the random choice of factors as the lower limit both of degree of unconditionality and of the level of confidence of his hypothesis. Here again the conclusions are the function both of the nature of the 'data' and of the validity of the conceptualization which defines the list of factors from which the sample of inductive tests was constructed.

Let us consider now a weaker assumption. Suppose now that a complete list of the hypothetical factors does not exist, but their number is postulated as finite. Assume moreover that the classes of relevant and irrelevant factors (if relevant, disturbing factors exist at all) are operating at random in the entire population AC. Suppose further that the scientist is studying their impact as they occur randomly in natural sequences of events of the type AC. Then, of course, due to the same kind of reasoning we conducted for generality in the frequency sense he can say, on the basis of observation of a long list of events AC differing in respect to a great many accompanying conditions and still leading to B that $AC{\rightarrow}B$ is unconditional or almost unconditional in the typological, i.e. theoretical sense, with a given degree of unconditionality and with a given probabilistic level of confidence.

Nothing changes in this reasoning if, instead of studying the naturally occurring cases, he 'produces' them by deliberate experimental manipulations. If we assume that from an 'unlisted' but finite number of factors he is able to choose factors at random he can again formulate his probabilistic conclusions about the degree of unconditionality of the given relation, understood in terms of the approximate highest proportion of the relevant factors to an unknown totality of all others, including of course the situation of complete unconditionality.

Here again, assuming that the scientist is not taking these factors at random but is able to concentrate primarily on those variables which have, due to prior knowledge, a higher a priori probability of belonging to the class of relevant factors, but which eventually turn out nonetheless to be irrelevant, both the level of confidence of his conclusions and the degree of unconditionality in the typological sense are 'better' than in the case of randomness of experimental trials.

The scientist who elects to concentrate in his inductive comparisons on those factors which 'a priori' are more likely to falsify his hypothesis rather than on factors chosen at random proceeds according to the rules prescribed by Karl Popper and other advocates of the 'falsificationist approach'. He tries to find those instances which would abolish his hypothesis. What is interesting is the fact that if the above reasoning is correct, the 'falsificationist' is inductively increasing both the degree and the probability of unconditionality of the tested hypothesis if this unconditionality is made in terms of a certain degree of unconditionality more or less close to 1. Therefore we see that *under the assumptions made above the falsificationist research strategy has the most powerful consequences for the inductive confirmation of the tested hypotheses*.

If we cannot make the assumption that we are testing primarily the relevant factors or at least the random ones, then of course the above conclusion is not valid. But this would mean – by implication – that in our experiments we are *testing non-randomly primarily the irrelevant variables*. Unless our theory says that the hypothetically relevant variables are outside the area of our access or experimental manipulations, such an assumption of non-random negative choice of the tested variables would seem a highly doubtful one.

Let us return to the situation of more or less random choice of the test variables. We have been considering instances in which the number of such variables was finite. The problem becomes more difficult, however, if we cannot assume the 'limited variety of the world' – to use Keynes' term. But even here one can postulate the assumption that if relevant 'disturbing factors exist', they occur at random among the infinite totality of all possible variables with respect to the sequences $AC \rightarrow B$ in either natural or experimental situations. Therefore, if after a great number of experiments our relation 'still holds', its degree of unconditionality and level of confidence can be estimated in the way presented above. Nevertheless, one should remember that we are speaking here not about an infinite number of events, a certain proportion of which might disturb our relation, but about an infinite number of classes of such events defined in general terms. This assumption seems also to pose certain ontological problems, because in some interpretations it would lead to obvious contradictions: If the total number of variations of a certain class of phenomena is infinite and the disturbing factors are randomly distributed, even in a small fraction, among all factors, these disturbing factors would still be infinite in any natural sample and this would make any regularity empirically impossible.

But here I would like to stress another aspect of this situation. If we see a certain class of phenomena as 'infinitely variable' from the point of view of possible categories of theoretical mechanisms underlying them, this usually means that we do not have any good theory of these phenomena, because any such theory usually limits the degree of theoretically relevant variations to a more or less clearly defined set of general concepts. We therefore would not have in such a situation a sufficient number of generalizations which have passed the test of generality in a certain number of conditions so that we might ask the question: *How really unconditional* are they, and how strongly can we believe in the given degree of unconditionality? Thus in disciplines where we are faced with the possibility of an infinite variety of the studied phenomena, the problem of verification of general hypotheses with respect to their degree of unconditionality is usually an academic one.

We discussed above the problem of estimating the degree of unconditionality with respect to the existence or nonexistence of hypothetical disturbing factors in the given relationship. Exactly the same kind of reasoning and argumentation can be applied to the hypotheses when certain relevant conditions might be *positive* in nature, i.e. when the occurrence of a certain event D (or E, F ...) is relevant for the regularity $AC \rightarrow B$, since we can say that its non-occurrence is then a disturbing factor for $AC \rightarrow B$. The corresponding inductive canons would be different, of course.

This section can be summarized by saying that demonstrative induction aiming toward the discovery of conditionality or unconditionality of the tested hypotheses can be justified in probabilistic terms in a way quite similar to enumerative induction. The propositions which we can infer validly (with a certain level of confidence) should be made *in terms of degrees of typological unconditionality instead of degrees of empirical generality*, and our inductive instances are here not singular events but typologically defined classes of such events, the structure of reasoning is much the same.

Let us look finally at the relation between the level of generality of a hypothesis (when this refers to the frequency-limit of cases $AC \rightarrow B$ *among all cases AC*) and the degree of its unconditionality as understood above. It is obvious that an unconditional hypothesis is by the same token a general one. But the reverse relation is more complex. Suppose that in our series of experiments we have discovered that among a certain set of potentially definable factors our hypothesis $AC{\rightarrow}B$ is negatively conditional upon 10% of them. This does not mean of course that when these disturbing factors are not controlled, $p\ B/AC$ = 0.1. It may happen that the individual events denoted by the full alternative of all possible disturbing factors are still extremely rare, so that the probability of occurrence of any of these *individual disturbing events* with respect to AC is, say, 0.01%. In such case, our hypothesis will be fairly *strongly conditional with respect to the theoretical framework in which it is located but empirically almost exceptionless* as far as its degree of generality is concerned. If we decided not to control the possible disturbing factors, we are still almost certain that $AC{\rightarrow}B$. And for practical reasons we may treat it as an almost general law of science in our predictions of future occurrence of B.

On the other hand, it may happen that from the whole set of generally definable factors only an extremely small proportion of them will disturb our relation $AC{\rightarrow}B$m. Suppose even that there is only one such factor F among the thousands of other, irrelevant factors, but at the same time that the probability of occurrence of events F with respect to AC is fairly high – let us say, 60%. In such case, of course, the relationship $AC{\rightarrow}B$ on the level of individual events will obviously be far from a general one, with $p(B/AC)$ = 0.4. Here it seems to be necessary to include the variable F into the set of antecedents of our theory and to say that $ACF{\rightarrow}B$, not only for reasons of theoretical correctness but also for practical use of our hypothesis in everyday explanations and predictions. The same applies of course to the situation where a variable is essential for the given relationship in its positive sense: when its occurrence is necessary for the regularity $AC{\rightarrow}B$. Suppose that D constitutes an additional necessary component of the sufficient condition for B, so that $ACD{\rightarrow}B$. If D occurs in practically every instance of the event AC, then the discovery of the causal relation with B is theoretically important but not so important from a practical point of view. The relationship $AC{\rightarrow}B$ will still be almost general even if we do not take D into account. On the other hand, if the probability of occurrence of D with respect to AC is fairly low, D has to be included into the antecedent of our law. Otherwise the relation $AC{\rightarrow}B$ will be fairly useless in practical predictions, or at best it may be used only for statistical predictions.

6. CONCLUSIONS

In this chapter I have tried to formulate the general assumptions which make inductive generalizations logically justified either in their strictly general (or unconditional) formulations or at least in terms of some approximations to such generality (or unconditionality), with certainty or with a possibly unambiguous probabilistic level of confidence. It is not assumed that these assumptions have to be generally valid. Inasmuch as they are in the nature of *empirical assumptions of a given type of inductive method*, their validity is to be assessed independently of the given model of inductive reasoning. Their validity has to be assessed on the basis of previous studies, and generalizations substantiated by them.

While the assumptions and the corresponding models of inductive reasoning are extremely simple, there is no basic obstacle to thus describing models of situations in which a greater number of variables are involved, the relations are of a more complex kind (e.g. the functional relations of quantitative variables), the nature of the randomness postulated is much more complicated, and where various kinds of assumtions imply the use of programs of inductive research of a more 'sophisticated' kind as a valid method for the study of the given phenomena. Construction of such models would go beyond the task of this paper, but it is basically possible,

There is nevertheless one class of situations which I would like to mention briefly, as I have discussed them at length in another paper.[10] These are situations where:

(a) the relation $A \rightarrow B$ is conditional with respect to a certain C (or a whole set of factors C);

(b) C occurs in such a way that entire 'historical' populations of A's (i.e. those A's which occur within a definite time-space area) occur jointly with C, while other such populations of A's occur jointly with non-C.

This requires a special scheme of inductive analysis in which both the 'individual characteristics' (A) and the 'population characteristics' (C) are controlled in such a way that their relevance for the occurrence of B might be discovered. These situations are especially frequent in the social sciences, which implies the importance of cross-cultural or cross-historical inductive comparisons, or the necessity of replication of many experimental studies in different cultural or national settings. And needless to say, such cross-population inductive studies may lead both to the formulation of general (almost general) proposition and to the formulation of statistical generalizations.

As I have already noted, whatever the nature of these assumptions of the given kind of inductive method for particular areas of research, their choice should be made by specialists of the given field of science, since their theoretical knowledge or even their intuitive guesses are usually much more competent than those of philosophers of science. They are able to know, or at least to guess with a fairly high credibility, what kind of uniformity, can be postulated for the area of their interest. They are also able (at least sometimes) to correct or to limit in their further studies the applicability of these assumptions if they turn out to be wrong. The only task of the methodologist is to specify as clearly as possible these assumptions and the corresponding inductive methods as internally consistent models of deductive reasoning from two sets of premises (assumptions and evidence) to the generalization justified by them.

At the close of this chapter I would like to stress again that I do not claim that induction is the only method on the basis of which generalizations or theories have been, are, or should be, 'admitted' into the sciences. Quite the contrary, I have tried to demonstrate that induction of the type presented above, i.e. which permits us to see to what degree our generalizations from the data are justified, requires that the scientist have some prior knowledge about the studied phenomena which permits him to make certain presuppositions necessary for the inductive confirmation of the next hypotheses. If he is unable to make these presuppositions on the basis of prior empirical knowledge, but is nevertheless willing to 'risk' them, he is of course 'fully entitled' to do so — provided he (and his readers) are aware of how strong the premises of the new generalizations are. Finally, if he does not feel entitled to make any assumptions at all, then he may apply the 'deductionist' i.e. the falsificationist, approach, as I have mentioned several times above. What I would like to stress here is rather the fact that the inductionist and deductionist approaches to the confirmation of hypotheses and theories are not incompatible philosophies, but rather complementary approaches, and that the application of either of them or their joint contribution to the development of science should always be related to the body of previous knowledge, or at least to our more or less credible and clearly stated guesses about the nature of the studied phenomena and their regularities.

NOTES

* This paper was written during the author's Fellowship at the Center for Advanced Study in the Behavioral Sciences at Stanford, California.
[1] Arthur W. Burks, 'The Presupposition Theory of Induction', *Philosophy of Science* **20**, (1953), p. 177.
[2] A.W. Burks, 'On the Presupposition Theory of Induction', *Review of Metaphisics* **8** (1954), p. 575.
[3] Max Black, 'The Justification of Induction', in: *Language and Philosophy,* Ithaca, 1949, p. 88.
[4] Max Black, ' "Pragmatic" Justification of Induction', in: *Problems of Analysis*, Ithaca, 1954, pp. 188–189.
[5] See R. von Mises, *Probability, Statistics and Truth*, London, 1957.
[6] Hans Reichenbach, *The Theory of Probability*, Berkeley, 1948, p. 472.
[7] See John Maynard Keynes, *A Treatise on Probability*, London, 1963, Ch. XX, 'The Value of Multiplication of Instances or True Induction.'
[8] See S. Nowak, 'Conditional Causal Relations and Their Approximations in the Social Sciences', in: P. Suppes *et al.* (eds.), *Logic, Methodology and Philosophy of Science*, IV, North-Holland Publishing Company, 1973, pp. 765–787.
[9] G.H. von Wright, *A Treatise on Induction and Probability*, New York 1951.
[10] See Chapter IV.

BIBLIOGRAPHY

Black, M., 'The Justification of Induction', in: *Language and Philosophy*, Ithaca, 1949.
Black, M., ' "Pragmatic" Justification of Induction', in: *Problems of Analysis*, Ithaca, 1954.
Burks, A.W., 'On the Presupposition Theory of Induction', *Review of Metaphysics* **8**, 1954.
Burks, A.W., 'The Presupposition Theory of Induction', *Philosophy of Science* **20**, 1953.
Keynes, J.M., *A Treatise on Probability*, London, 1963.
Mises, von R., *Probability, Statistics and Truth*, London, 1957.
Nowak, S., 'Comparative Social Research and Methodological Problems of Sociological Induction', in this volume, p. 133.
Nowak, S., 'Conditional Causal Relations and their Approximations in the Social Sciences', in: P. Suppes *et al.* (eds), *Logic, Methodology and Philosophy of Science*, IV, North-Holland Publ. Co., 1973.
Reichenbach, H., *The Theory of Probability*, Berkeley, 1948.
Wright, von G.H., *A Treatise on Induction and Probability*, New York, 1951.

CHAPTER VIII

EMPIRICAL KNOWLEDGE AND SOCIAL VALUES
IN THE CUMULATIVE DEVELOPMENT OF SOCIOLOGY

1. SYMPTOMS OF CRISIS IN SOCIOLOGY

The task of our Symposium is to answer the question whether there is a crisis in sociology.* I think the fact that such a question has been asked at an international meeting of sociologists constitutes at least a partial answer to it. It implies that there are a sufficient number of sociologists who *feel* that their discipline might be in a critical situation. Before trying to answer the question directly, let us consider what kind of crisis they might have in mind, because a crisis in a science may mean any of at least three different things:

(1) First, it can mean that the given science cannot develop or even cannot exist in a certain set of social conditions, because *these conditions do not, in the most simple and external sense, permit it to do so*. The means used can range from the pressure of public opinion not to develop it, through denial of the economic means necessary to the science, to simple bureaucratic decision blocking the development of the science as a whole or in certain essential sub-areas.

(2) Second, it can mean that the science, developing more or less 'correctly' from the point of view of certain internal standards, *does not fulfill some of its external social functions* in the way it should according to certain normative standards of its social functions. In such case we may be inclined to blame the situation on either the society for not using the science in the 'proper' way, or the scientists themselves for developing their discipline in such a way that it is of little social relevance.

(3) Finally, it can mean that the *crisis lies in the science itself*. Sometimes this may be a temporary phenomenon, as for example when the old 'paradigms' which were delivering the guiding principles for the development of the science do not work any longer, and the science has not yet been able to develop a new workable paradigm. In some other cases the crisis is of a more profound character, as when the representatives of the given science cannot agree on the basic assumptions of their discipline, its

nature and goals, and this disagreement is not solved for quite a long period.

There is no doubt that in some countries we have definitely a crisis of the first kind, and one of the papers in our Symposium discusses a typical case of this category.[1] There is also no doubt that there are various symptoms of crisis in sociology with respect to its social functions. According to some writers[2] contemporary sociology is performing the *wrong* social functions, being committed to conservative social values. In saying this they usually have in mind *ideological* functions of certain findings and theories, their real or possible impact upon human values and attitudes on a mass scale, as factors strengthening or legitimizing various 'Establishments'. But one can question as well the more practical, *instrumental* functions of contemporary sociology, the purposes for which it is being 'used' by different groups and agencies which need sociological information for guidance in their practical actions. These uses can also be judged from the standpoint of specific normative standards, and we might ask the question: Who is applying these social findings and theories and for what purposes, and what are the possible or actual instrumental social consequences of the various sociological orientations, i.e., what are their most likely practical uses?

But the question of the instrumental uses of sociology can be asked in more general terms: What are the *possibilities* of contemporary sociology in the area of guiding social actions, *to what degree is it really prepared to transform social reality in a scientifically justified way*? In order to see this problem more clearly let us imagine a kind of 'mental experiment'. Suppose there is a society which constitutes an ideal from the point of view of a sociologist's dream. It is guided by social values which sociologists accept (here we would have to assume a certain uniformity of values among sociologists) and it is eager to follow practical recommendations formulated by sociologists, transforming them into practical social actions. Now let us ask ourselves honestly: How much would we have to offer this society on the basis of our findings and theories, and especially *to what degree would we be able to agree on such recommendations* derived from our theoretical and diagnostic knowledge? The answer is – unfortunately – not too favorable to sociology, and one can easily predict that the number of recommendations for the solution of one specific social problem would not be much smaller than the number of sociologists involved.

If this is a correct 'prediction', then we have to assume that our discipline is also in a state of *internal crisis*. We could say, of course, that this means only that our discipline and especially our social theories have not been developed sufficiently to allow social scientists to make correct predictions and therefore to formulate scientifically valid rules of social action on which all are able to agree. This would entail admitting we are now at the stage of development medicine was before Pasteur, when every doctor was a representative of a specific 'school' and on the basis of the 'theory' of his school along with certain practical experiences he made his recommendations to patients. This is probably to some degree true, but I believe the situation is much more serious than a mere stage of development, and that contemporary sociology is in a state of profound crisis which has its roots in the philosophical foundations of our discipline.

As we know, some critics of sociology have not limited their criticisms to 'unmasking' the conservative functions of particular social theories or 'approaches'. They have attacked the very idea that sociology might be a science at all in the normal sense of the term. What do I mean by a 'normal' science? I mean by this a discipline in which a certain amount of intersubjective agreement is possible with respect to both its findings and its theories. I mean a discipline in which these findings are subject to a cumulative growth, in which the propositions established by one generation of researchers are modified or even rejected by others who on the basis of certain new findings are able to demonstrate that their position is right, in such a way that it would be accepted by a majority of other researchers. In general I mean a discipline which develops according to certain standards accepted by all those working in it, and in which we observe the *process of its cumulative development*.

It would not be fair to say that the process of cumulative development does not exist in our science at all; it is quite visible in some areas of sociology. But in sociology it has certain special features: The cumulativeness is usually limited to those scientists who are working within the same 'paradigm', or as we used to say, who accept the same 'approach' to study of social phenomena. One of the basic features of sociology is that each of the 'approaches' seems to have a dynamics of its own. Even if there are really paradigms in any sense of this highly ambiguous term, they lack the basic feature of the paradigm, i.e., they do not with the passage of time replace one another in a way that would obtain the general acceptance of all concerned. The rise, development and disappearance of different 'approaches' which we can observe in both the recent and the more distant history of our discipline has very little in common with the appearances of new 'paradigms' in the natural sciences, where the new paradigms become necessary when the old ones cannot account for newly emerging facts and theories. The occurrence and disappearance of 'paradigms' in sociology seems to have much more in common with changes in women's fashions — which can hardly be accepted as a satisfactory situation in science.

Moreover, this situation has been defined by some as a normal, not to say a desirable one. Predep Bandyopadhyay in his excellent criticism of 'radical methodology' presented this view in the following way:

Disagreements among sociologists are in principle irreducible, since they involve different paradigms, evaluative statements and different purposes... Value neutrality and objectivity are both unattainable... Therefore, there are and will be several sociologies, none of which are objectively true but merely different ways of looking at society, and — in the words of a fashionable school in sociology — different ways of constructing social reality.[3]

If one could say that all these attacks were merely the misunderstandings of 'outsiders' who, due to lack of competence in the philosophy of science, had misinterpreted some rather banal statements about the relations between social sciences and societies, there would be no reason to state the existence of an internal crisis of sociology. One is not necessarily responsible for the errors of one's incompetent critics. But unfortunately this has not been the case with sociology. Most of the arguments by which the scientific status of sociology has been questioned have come from 'academic' sociology. In some cases these arguments were presented as a caricature and reflected certain views which could be found in a more 'balanced' and at least partially acceptable form in sociology itself. In other cases no caricatured exaggeration was necessary.

Therefore I would like in this paper to discuss certain problems pertaining primarily to the existence of different 'schools' in sociology, and such related problems as 'objectivity' and 'value involvement', because in my opinion lack of agreement on these problems constitutes the basic obstacle in the cumulative development of our discipline, and therefore might be identified as a basic source of its internal crisis. The problem of the theoretical consequences of 'value involvement' deserves special consideration because the radicals have attacked the methodological assumptions of 'traditional sociology' in the name of certain social values: equality, freedom and human dignity. But many sociologists share these values with their critics, and it is perhaps for this reason that rebuttals have not been freely forthcoming. Thus polemics revealing the spurious relationship between radical social values and 'radical methodology' have been rather rare.[4]

However, these problems are related to crisis in the first and second senses as well: Wrong notions about the relations between sociology and society, about its functions for the society, create expectations which sociology can never – at least as a scientific discipline – satisfy. At the same time, by contributing to the internal crisis in our discipline such notions indirectly diminish the possible applications of sociology to the transformation of society in the direction corresponding to our value systems. Moreover, we should also remember that any element of internal crisis in sociology, and especially those arguments which question the possibility of its existence as a normal science, may be used as an argument against sociology by social forces which, from the point of view of their interests, perceive the existence of sociology as dangerous or at least inconvenient. Therefore, those for whom crisis in the first sense distinguished above is of primary importance should remember that the elimination of internal crisis from sociology may be relevant in this respect as well.

In general, I agree with Franco Ferrarotti that it is "important to rediscover the nature of sociological enterprise, its necessarily unfinished character, its problematic disposition and its ambiguity",[5] provided the main goal of rediscovery and reassessment of sociological enterprise will be accompanied by the elimination of ambiguities, wherever this is possible. The present paper is an attempt in that direction.

2. CUMULATIVE CHARACTER OF EMPIRICALLY TESTED PROPOSITIONS AND THEORIES

When speaking about the possibility of cumulative development in sociology it seems reasonable to start by looking for areas in which an unquestionable agreement can be reached. It seems that the easiest thing to agree on is on the *truth* of certain propositions.

Let us say that a proposition is true when the facts really are as it states. In empirical sciences it is more convenient to say that a proposition is empirically justified to the degree that we are entitled – on the basis of certain data – to assert with greater or smaller probability its full or approximate truth. Now, without going much into a philosophical discussion on the meaning of the term 'empirical proposition' (some of these problems will, nevertheless, have to be mentioned below), let us say that any two sociologists should be able to agree on the empirical validity of such statements as 'New York City has a given number of inhabitants'. They should also be able to agree on the truth of the statement that in a certain sample of persons studied by a social mobility researcher, those subjects whose parents belonged to the lowest paid group have the smallest probability of getting into highly paid occupations or of finishing university studies.

One can also say that most sociologists should be able to agree on some generalizations from such observational data, even if the intended validity of the conclusions overreaches the extent of the observed reality, especially if the chance of inductive errors in generalizing procedures can be more or less rigorously estimated. This can be done most easily where the studied population constitutes a random sample of the larger population, because in such situations the rules of statistical inference define the probability of making errors of different size, but we could expect a certain amount of agreement also in less standardized research situations. Thus, e.g., most sociologists would probably agree that the amount of social inequality in contemporary Sweden is smaller than in contemporary France, or (even without too rigorous data) that the amount of illiteracy in Africa is much larger than in Europe. In other situations, where the risk of inductive error cannot be so easily estimated, different scientists may *differ in their belief* as to the empirical validity of certain generalizations, but at least they should agree that their empirical validity may be questioned.

Nothing is changed when these propositions (especially the empirically validated generalizations) are related to each other in such a way that we would be inclined to call them 'theories'. The term 'theory' is used in many different ways and there is not enough space in this paper to mention all of them. The meaning to which I would like to subscribe defines theory as a set or system of empirically testable general laws or law-like statements which can be used for the explanation and/or prediction of the phenomena within the realm of their applicability.

The theory is therefore not identical with certain problem orientations, although the kinds of problems of interest to a scientist may co-determine

the theory he will formulate and the theory will be composed of hypothetical solutions to these problems. Nor are the concepts identical with
theories. Concepts constitute a better or worse language in which we might
eventually formulate a theory. Proper conceptual language determines to a
high degree the value of a theory, but it does not replace it. What is clear is:
We should choose such concepts as will be *most useful* on the most general
scale; i.e., with the use of which we can formulate the best possible laws and
their whole systems.

There is no place in this paper for reviewing the characteristics of good
theoretical concepts; their optimal shape depends upon the nature of the
studied phenomena, and on the kind of questions to which the theory
should constitute an answer. We can say, e.g., that if they are classificatory
concepts denoting certain sub-areas of reality, they should denote classes of
phenomena whereby one can formulate a relatively large number of true,
i.e., testable and tested, and at the same time possibly general propositions.
If they refer to more abstract properties or 'dimensions', they should be
defined so that they can occur *in the same meaning* in many different
theoretical propositions, etc. What is more important is that having these
criteria (and many others omitted here) in mind, *one can rationally argue
which of two (or more) proposed sets of concepts is better* for the formulation of the theory in question.

The elementary requirement of any theory is that the hypotheses, once
they have been conceptualized, are (directly or indirectly) empirically
testable by the theory's consequences. Therefore, all that has been said
regarding testability – and consequently about the possibility of getting
inter-subjective agreement on the truth, plausibility, and finally the hypothetical character of propositions – applies to the same degree to theories, if
they are understood in the propositional sense.

However, all tested theories are at least potentially cumulative. Each of
them constitutes a contribution to the totality of social knowledge quite
independently of the value assumptions or 'approaches' which led to the
formulation of the research problems to which the theory constitutes the
answer. Cumulativeness means also that each such previously tested theory
(as we know, the verification of a universal proposition is never complete)
may be rejected or modified by new empirical evidence. Its conceptualization may be challenged by demonstrating that reformulating its propositions
with the use of other concepts will increase the generality of the theory or

its explanatory or predictive power. And this is again independent of any value judgments or prior assumptions. Such theories may be cumulative in a more general sense as well, inasmuch as they can sometimes be meaningfully related to each other. By relating different theories to each other (or by finding that they are apparently unrelated, at least at the given stage of our knowledge) we obtain a more and more comprehensive structure of theoretical knowledge about social phenomena.

Here one should mention one problem, namely, the old dream of systematizing all relevant theoretical knowledge about society into one 'unique' and all-inclusive theory. In sociology this dream takes usually one of two forms. The first starts from reductionist assumptions and the attempt is made to construct a general behavioral theory from which all other social theories can be derived. The other approach to realizing the dream starts from the other end and aims to construct a theory of a macrosociological type. Both approaches to an all-inclusive theory tend to end with declarations that they eventually will unify all sociological theories. But unfortunately no theory can be simply declared; it has to be formulated and tested, and the same applies to an all-inclusive social theory.

The dream that there will be one theory for one science does not seem to be justified by the development of most of the sciences, when we study them historically. This does not mean we should not work toward a continual integration of our theories by building more and more comprehensive theoretical structures, of either a reductive[6] or a systemic type. But we have also to assume that whatever our progress in this direction we will have to live for a long time with many *partial theories* – mutually complimentary, and cumulative in different senses of the term, applicable for different aspects of social reality, answering to different theoretical questions, and useful for different practical social purposes.

In saying that different theories can be potentially cumulative, I do not mean to say that all theories we may find in contemporary (or classical) sociology constitute empirically valid systems of generalizations formulated with the use of optimal sets of concepts, and therefore that they should be included in their present form. I am of quite the opposite opinion. I think they are often badly formulated and poorly conceptualized, and their empirical evidence is of questionable value. What I only want to say is that we should work on the improvement of those theories and criticize them in the way one does in any other science, i.e., on the basis of relevant empirical evidence and not by revealing their usually 'spurious contradiction' to our own 'theoretical approach'.

3. EMPIRICAL AND NORMATIVE COMPONENTS IN THE DIVERGENT INTERPRETATIONS OF FINDINGS

Certainly not all the differences of opinion and divergent interpretations of the same data in sociology are of spurious character. Let us now look at those where the differences are real in order to see which of them can be solved on an empirical basis or more generally within an area of argumentation which allows resolution in terms of strictly scientific discourse, and which cannot be so solved because they are related to differences in scientists' focus of interests and directly or indirectly to differences in their value systems.

(a)   The most simple case would be when one theoretician says that a certain general relation exists between two or more variables while the other says that is not true.

In this situation it can be said that at least one of them is wrong and that it is up to empirical studies to demonstrate who is right. Actually what is usually the case in such situations is that both scientists are partially right and partially wrong, because they both omit in their theoretical formulations certain additional conditions which determine whether this relation occurs. In such case both views require reformulation: Adding these necessary conditions we obtain a more general theory of which the former two are elements, valid for different sets of conditions omitted before.

(b)   A second kind of situation is when two theories or at least two laws are incompatible because one of them is more general than the other, and it turns out that the more general theory logically implies the falsity of the less general one in spite of our insistence that this less general theory is true.

Here again, one would think that we have to reject at least one of the theories. But sometimes the impression persists that both theories have sufficient empirical support. The only solution in such case is to reformulate one of the theories in such a way as to eliminate the logical contradiction. This we usually do either by adding some modifiers or qualifiers to the more general theory so that these exclude from the limits of its validity the cases described by the less general theory, or by redefining the concepts of the less general theory so that they do not imply it is derivable from the more general one.

      (c)   The third and most interesting case is when two scientists agree on certain empirical generalizations but disagree on which concepts to use and on which theories should be applied for the interpretation of the findings and generalizations.

In this kind of controversy in sociology we can easily distinguish two types, although they often occur jointly. In the first type the scientists, agreeing upon the validity of certain data, disagree in that each would like to use for the description and interpretation of these findings certain *terms* and concepts which are *different primarily in their evaluative function*, i.e., their meanings have different emotional and evaluative components. In the second type two sociologists insist on using different terms, because they want to treat the observed variables as indicators of different theoretical concepts.

It is well known that most of the concepts we use in sociology not only denote something in reality but also express certain emotions we associate with this reality, convey certain evaluations of it. Two terms may be empirically equivalent, but may at the same time convey quite different emotions and evaluations. Thus, someone who has negative feelings about the term 'social class' may readily accept certain empirical findings about his own society if they are described in terms of 'social strata' but will reject them if interpreted as 'class differentiations'. It may happen that a sociologist will reject the notion of 'nationalism' as describing the attitudes of his co-citizens but will readily accept the term 'patriotism' even though the empirical meaning of it would be equivalent. The reason for his resistance is the same as indicated above, the unwillingness to accept a proposition which, although empirically true, conveys emotions which he cannot accept.

Terminological differences which involve only the evaluative components of the concepts used are of minor importance for the cognitive functions of sociology (although they may be of serious importance for the groups whose social or political interests have been threatened by application of a certain evaluative label). As long as extensions of the terms are identical, they are 'translatable' one into another. Even if two statements formulated by two scientists sound different, the users of the corresponding concepts — even though they cannot agree on the evaluative implications — can agree on the empirical, non-evaluative content of factual statements and theoretical generalizations, and these empirical results would constitute a contribution to the cumulative development of sociology.

This will be possible, however, only if the concepts we use in our empirical studies and in the formulation of our theories are defined with such preciseness that we can distinguish the *denotative* (i.e., empirical) from the *expressive* (i.e., evaluative) components of their meaning. Only then will we be able to say which components denote something that exists in social reality and which express our evaluation of this reality. Needless to say, we are rather far from this ideal in many areas of sociological enquiry: Many of our concepts are defined with the use of terms which in themselves are too vague to secure for the given concept a sufficiently precise meaning. In many other cases we have *terms* lacking any definitions at all, and the correctness of interpretation of the author's findings and theories depends completely upon an assumed community of meanings as between the author and all of his audience. In such a situation the postulate of a clear distinction of the evaluative and non-evaluative components of the meaning of the concepts used cannot of course, be realized.

Here we come to the next point: What constitutes the empirical, denotative components of sociological concepts? Until now I have assumed for simplicity's sake that we are dealing with phenomena the existence or occurrence of which can be assessed by observation only. As we know, things are not so simple in sociology. While there are, of course, certain phenomena of a strictly observational nature − like the number of inhabitants in a city or the sex of a given individual − most of the phenomena we are interested in are of a more or less *inferential* nature. We know that certain apparently similar behavioral sequences may have quite different *meanings* for the behaving persons and for those who are interacting with them, and for this reason we would in our analysis classify into separate conceptual categories a fight between two hooligans and a boxing match between two professionals. We also know that two observationally different situations may, due to the similarity of their meanings, belong to the same conceptual category, e.g., as when we classify both the behavior of Wall Street brokers and the behavior of some scientists during an international conference into one conceptual category of 'competitive behavior', even when from a strictly behavioral point of view the two situations are rather different.

When we do this we are not really very different from natural scientists, who also have to postulate the existence of some unobservable entities and to use certain *hypothetical constructs* which denote them. The science of 'genetics' wouldn't be possible without the assumption that there exists some unobservable property called a 'gene'; without the postulation of such unobservable entities as 'atoms' or 'elementary particles' the construction of physical and chemical theories wouldn't be possible. The whole area of social reality which we can grasp by the method of 'Verstehen'[7] — to use the old term introduced by Dilthey and developed for theoretical use by Weber — is made up of phenomena which are directly unobservable but which exist in human minds and so determine both individual behavior and social interactions and relations. Therefore we have to take them into account in our conceptual apparatus if we want our concepts to represent social reality with sufficient adequacy and to be useful for explanations and predictions. We have to reflect in our concepts both the meanings of individual behaviors and the whole complex configurations of such meanings, which constitute (jointly with their observable counterparts) the structures and cultures of whole societies.

This means that we will have in the conceptual apparatus of sociology (1) Concepts which are defined in strictly observational terms — like sex, age or number of interacting persons; (2) Concepts which are 'behaviorally open' and defined strictly in the language of the individual or social meanings of the given behavior or situation — such as the concept of 'competitive behavior', (3) Concepts whose contents combine both meanings and observable behaviors — such as 'cooperation of a group of workers in a factory'.

Whatever the character of the concept, it is useless to empirical science until its content has been somehow related to strictly observable phenomena so that we are able to say on the basis of some observable 'indicators' when the phenomena denoted by it exist or occur. This implies that in social sciences (as well as in natural ones) we are dealing with two levels: The level of observable phenomena which we may use in our research as the 'indicators' of our study, and the level of hypothetical, inferential phenomena which also belong to the studied reality but the existence of which are assessed indirectly by inference from the observable ones and under some additional theoretical assumptions which cannot be discussed here in detail.[8] Nevertheless the problem of whether the hypothetical correspondents of the given observable phenomena exist in a given case is solvable within strictly scientific discourse, just as in everyday practice we understand other people's behaviors and the complex social situations we observe or participate in. If we are wrong in our assessments, future study or future personal experience may correct our mistake. A man may incorrectly interpret someone's behavior as friendly, until some future personal experience informs him of his mistake. A doctor may make an incorrect diagnosis of a patient's illness on the basis of observing its first symptoms, but future observation of the progress of the illness or of the patient's reaction to medical treatment is able in many cases to correct the mistake. We should remember that the relation between the 'indicators' and the concepts they indicate constitutes in itself an important part of social theory in exactly the same way was knowledge of the symptoms of different illnesses in medicine.

Both the phenomena of the observable level and the meaningful components of social reality are usually rather *complex syndromes*. Thus *scientific concepts never 'reflect' them in all their complexity*, but have to concentrate upon certain of the aspects or components only. The same syndrome of 'meaningful behavior' may be an example of 'consumer behavior' for one scientist and an example of 'keeping up with the Joneses' for another, if they concentrate their attention on its different aspects. And the indicator of the two corresponding concepts in this situation may be exactly the same: Buying a certain expensive object. This does not mean of course that any case of different inferential interpretations of the same behavior is necessarily valid (actually both may be wrong); it means only that the same empirical situation (whether of a strictly observational or of an inferential nature) can often quite justifiably be categorized into different conceptual categories.

Suppose now that in a certain area of social studies the conceptual apparatus has been defined so that the content of the corresponding concepts is clear to us and we are able to distinguish its denotative and evaluative components. Suppose also that two scientists studying the phenomena within this area agree on the validity of certain strictly observable data, but disagree on the theoretical way they should be interpreted — thus their differences are not primarily related to differences in the evaluative components of the concepts. In such case what the scientists are seeking is to interpret the variables they observe as indicators of different concepts. But here we need to distinguish three kinds of situations.

In the first situation both scientists are interested in treating the indicator (I) as indicating a certain variable (V) (on the meaning of which they agree), but one of them believes that the assessment of I indicates V and the other does not. Two survey analysts disagreeing on the frankness of questionnaire answers and therefore on the validity of some survey data in the study of 'attitudes' would be an example.

For the second situation, suppose two scientists agree that a certain indicator I potentially indicates a complex, meaningful-behavioral syndrome of social reality. Suppose additionally that they agree on the validity of a certain problem formulation – e.g., they would like to explain a certain behavioral pattern by reference to the values and knowledge of the behaving persons, with the indicator I being used for assessing the existence of such a complex syndrome as 'values and knowledge about social reality possessed by a member of the social group', the culture of which is known to the scientists from other sources. The indicator for the possible explanatory concept is then simple: membership in the given group. But the question of how the indicated concept should be defined for the purpose of the explanation is equivalent to the question: Which of the totality of values and information possessed by the members of the given group are 'responsible' for the explained behavior, and which are in this theoretical context irrelevant? In other words, the issue is: Which of the great number of concepts that might be used for the denotation of the given indicated situation is *theoretically fruitful*? It may not be easy to solve such a problem, especially if we take into account the actual situation in sociological theory, but in general this is again a category of problems that can be solved within scientific discourse.

Finally, the scientists may agree on the applicability of a given set of indicators for different concepts, and may also agree that for the given problem formulation the given theoretical interpretation of the indicators would be optimal, but still be in disagreement of a more profound nature having to do with the *kinds of questions* they are interested in. In such case we are at a level where empirical differences are not involved, and the answers given by different scientists are different because they are determined by the scientists' various individual interests.

This kind of disagreement may also occur before the data have been collected. The two scientists may then be inclined to ask different questions, and consequently to use quite different concepts when asking the questions *important for them*. Usually they also find that these concepts have to be operationalized in different ways, by the use of different indicators. Then, of course, they will not face the situation of divergent interpretations of the same data because their data will be quite different. In any case they will present different pictures of social reality; sometimes these pictures will refer to roughly the same social phenomena, or may even be based upon different interpretations of the same data.

Now it may happen that one scientist will not accept the findings and generalizations of another research as contributions to the development of sociology, not because these are – in his opinion – false in any of the senses distinguished above, but *because he does not view the questions to which these propositions constitute answers as relevant questions of sociology*. Differences in understanding as to what constitutes (or better, should constitute) relevant problem areas of sociology produce as many obstacles in the process of its cumulative development as do differences about the empirical validity of answers to these problems.

4. NORMATIVE AND EMPIRICAL ASSUMPTIONS OF PARTICULAR 'APPROACHES'

Whatever our conclusion on the empirical validity and cumulative nature of particular findings or particular theories, the fact remains that in contemporary sociology these findings and theories can be located in (or rooted in) one of many 'approaches' which exist in our discipline, in one of several 'schools' of sociological thinking. Without trying to define what makes an 'approach' in sociology, we can cite several examples: functionalism, historicism, evolutionism, behaviorism, phenomenology, struturalism – in any of the several meanings of these terms. As I noted earlier, the existence of different approaches as they now stand and what is much more important, the claims of some of them to 'superiority' or 'unique validity' constitutes the basic obstacles in the transformation of sociology into a 'normal science'.

Without entering into a detailed analysis of the content of the various approaches we can say that they may be regarded and compared from several different points of view:

What is the particular *problem area* defined by the particular approach,

i.e. what kinds of questions are of special importance for the representatives
of the approach? From this standpoint the different approaches are at least
potentially complementary, either when they ask quite different questions,
the answers to which lead to completely unrelated theories, or when they
deal with similar problem areas and lead to related theories.

Once the list (or system) of problems typical for an approach is relatively
clear we may ask some other questions, and especially:

What are the normative assumptions of the kinds of questions asked – or
in other words, what might be the functions of finding the true answers to
these questions in the course of scientific investigation?

As we know, when undertaking a certain study we may be motivated by
the expectation that the results of the study will have one or more of the
following consequences or functions:

(1) They may increase, or even basically change, our knowledge about
the phenomena in question – let us call this the *cognitive function*.

(2) This may lead to transformation of the reality the science is dealing
with, by enabling people to find the proper means leading to their goals –
what might be called the *instrumental functions* of science – or by influencing their world outlooks, attitudes, values and motivations to undertake
certain actions, what might be called the *ideological function*.

It should also be noted, however, that even a strictly cognitive function
involves a certain value judgment – namely, that one should (or that we as
sociologists should) contribute to a better understanding of the mechanisms
governing social phenomena, and that this might be sufficient reason to
undertake a study. In other words it gives to curiosity the status of a
legitimate human value.

The *intended functions* of or reasons for the undertaking of a study are
not, of course, always identical with the real consequences of the study:
someone may have in mind 'increase of pure knowledge' only to have it turn
out that as a more or less direct consequence he has contributed to the
production of the atomic bomb (as was the case with the theory of
relativity) or has changed in an essential way the value systems and world
outlook of the whole of mankind (as in the case of Copernicus). In other
instances, studies which were undertaken with the aim or realizing strictly
instrumental goals have led to discoveries of great cognitive importance
(radar being an example). The same holds true in both the natural and the
social sciences. Banal as it may be, it must nevertheless be noted that in

some discussions of values in sociology, the real consequences of a certain study are often cited as 'unmasking' the value system of its authors or of a representative of an 'approach'. It would seem that we need to distinguish clearing two things: the actual intentions of a scientist when he is formulating a certain research problem, and the actual function of his work — if we can assess these empirically, or at least the foreseeable possible consequence and 'uses' of his work.

The analysis of the real or foreseeable, possible social consequences of specific social theories or findings is important, not only as an essential part of sociology itself (namely: 'sociology of sociology') but it is important also on a meta-theoretical level, because:

(a) It makes each individual social scientist sensitive to the different kinds of possible functions of his study, permitting him to maximize the intended functions and thus to increase the social or theoretical importance of his research.

(b) It produces premises for an inter-subjective agreement on the question of whether with respect to the given set of social values the study of the given problem, or the solution of a certain hypothesis is of primary importance, or whether it is desirable at all. Eventually this permits us to distinguish which of the differences in the formulation of our research problems may be 'justified' by the scientist's convictions about the cognitive importance of certain concepts, hypotheses or theories, and which of them explicitly or implicitly reflect differences in value orientations of the scientists and therefore, cannot be solved within the area of strictly scientific discourse.

A *second category of assumptions* is of a different character. These are involved in certain *analytic procedures* characteristic for the given approach and in certain concepts which are analytically necessary for them. If someone is interested in determining a typical pattern of behavior in a certain society, statistics will provide him with a series of alternative but precise ways of understanding what might be 'typical'. If he seeks an estimated risk of error in making his generalizations, the theory of inductive inference constitutes a set of analytic procedures which he should use.

The third and the least clear category of assumptions of different approaches understood as some problem areas are the *assumptions which refer to some empirical features of the studied reality*. The precise formulation of any question about reality actually implies that we assume something about this reality — only under this assumption asking the given question may make sense. If I ask someone, "Did you stop beating your wife every morning?" I assume that the asked person did beat his wife before. Otherwise neither the answer "yes", nor the answer "no" makes any sense to me.[9]

If we do not study centaurs in contemporary zoology but study elephants, this is because we assume that elephants exist and centaurs do not. If we study the functions of institutions for societies, this is because we assume that the institutions may have some functions for the societies to which they belong. If we study the social causes of institutional change, we must first assume that the institutions do (or at least may) change, and additionally that these changes have or may have social causes. In order to ask: "Which is the basic factor in the integration of a society, its system of social control or the community of values among its citizens?" I need to have a long list of assumptions which make the formulation of this question sensible. I have to assume that the society is integrated, that it has a system of social control, and that the values of its members are similar; additionally, I have to assume that these two factors are contributing to the integration of the society, and that one of them is the more important one. Only then will a *direct answer* to the question of such a study make sense.[10]

Such empirical assumptions are sometimes hidden behind certain conceptualizations typical for certain approaches, and they are sometimes implied by the research techniques which characterize some of the approaches. In order to study attitudes, we must assume that the attitudes exists; in order to study them by questionnaire surveys, we have to assume they can be expressed in the form of questionnaire answers, etc. In the case of abstract conceptualizations, the assumption is often reducible to the assertion that the concept proposed for denoting social phenomena — and the processes and realtions among them — and are not empirically empty classes. Otherwise it would not male sense to try to study them.[11]

What is important is that as long as these assumptions are of the nature of empirical propositions, one can discuss the problem of their validity and also the *problem of the limits of their empirical validity*. Since most societies change, it makes sense to study them for answers to the question of what contributes to their change; some societies, however, do not change, at least for certain periods in their history, and then the assumption of the existence of change is not valid for them. Human aggregates may be 'internally functionalized' to various degrees, from a task group working within the strict pattern of division of labor to the amorphic agglomeration of an audience at a theatre performance. The assumptions which make certain theoretical questions sensible for groups or societies of the first kind are not true for the other.

One of the important problems in sociology is the analysis of empirical assumptions of different 'approaches' in order to see to what degree they seem to be justified and *what are the empirical limits of their validity*. Such an analysis would reveal the empirical 'legitimacy' and limits of applicability of various approaches, as long as they are understood as problem areas.

Different approaches often use different *concepts* in the study of social reality. These concepts should be evaluated from two standpoints. First, they may be regarded as components of the questions which define the problem area of a particular approach, in which case all that was said above about the validity of empirical assumptions of questions directly applies to the concepts too. But these concepts may occur also in explanatory hypotheses and theories, and in such context we apply to them the criteria of theoretical fruitfulness. If it turns out that two approaches try to answer the same kind of questions, we may legitimately ask which of the proposed sets of explanatory concepts is better for that purpose. The fact that they were proposed by an approach does not make them 'untouchable' in terms of strictly theoretical evaluative standards.

To give an example: In the context of whatever approach whereby we come to the question of the influence of 'attitudes' by 'class membership', there is probably only one optimal way of defining for this purpose the concept of 'class' and the concept of 'attitude', and the final acceptance of whichever conceptualizations should be independent of the author's approach and determined only by comparison of the theoretical usefulness of various conceptualizations.

Many approaches include not only certain problems and concepts but also specific *theoretical hypotheses* and quite developed theories about certain social phenomena, or society in general, its change, etc. Needless to say, each of these theories has to be judged on its own merits, according to normal procedure of empirical verification. And all that has been said about the cumulativeness of theories applies as well to the hypotheses and theories which constitute parts of certain 'approaches'.

Finally, an 'approach' may include a certain set of research procedures, typical operationalization of research variables, and data collection methods — as in the case of, for example, the survey approach, the experimental approach, or the phenomenological approach (with its stress on introspectional data or participant observation).

Now it may be that these indicators *define completely* for the given approach the content of the corresponding concept, being its 'operational definitions'. In such case the different approaches would be dealing *by definition* with different areas of social reality and their findings and generalizations would obviously be unrelated and non-comparable. But, fortunately enough, the period of radical operationalism seems to be over in the methodology of social sciences, and more and more scientists understand that our theoretical concepts should be defined in such a way that the indicators used in the research do not exhaust all their meanings. In some cases they may not even belong to the content of the given concept, as e.g., when we take the 'place of inhabitance' as an indicator of a respondent's 'income'. In such cases different indicators and different research techniques may be regarded as *alternative operationalizations* of the same theoretical concepts, suitable for different research conditions or for different kinds of studied populations. Then we can of course relate the findings made within different approaches to the same theoretical hypothesis, formulated in a language independent of the different conditional operationalizations of the concepts of the hypothesis. Then we could have, for instance, a situation in which the 'phenomenological analysis' of groups or institutions in natural conditions and the sociometric data would lead to independent tests of the same theory of interaction. In any case one should aim toward the direction in which the problems, hypotheses and concepts would determine the research tools, and not the reverse.

Thus we see that from a strictly cognitive point of view the so called approaches are, at least potentially, complementary. At first they constitute different although often overlapping problem areas, and thus *increase the diversity of questions in sociology*. Then they formulate different hypotheses aimed toward answering these questions, defining the concepts for the hypotheses and making more or less precise the relations between the phenomena denoted by the concepts. And the hypotheses may be true or false quite independently of which approach was their source, because their truth depends only upon the nature of the reality they refer to. Finally, they may be complementary also in terms of different operationalizations of the same concepts according to the research procedures employed, provided we see clearly that the indicators and techniques which are optimal for one set of conditions are not necessarily optimal for another kind of research situation.

In saying this I do not mean that everything we find in the works of representatives of different 'schools' constitutes a piece of 'partial truth', and that all we have to do is 'edit' these works, one by one, in a 'cumulative way' and the totality of social knowledge will arise as the sum of these complementary partial views. Quite the contrary. What I propose here is to develop sociology in such direction that the term 'school' will mean the same as it does in any other science: a group of people interested in a common problem area, using similar research techniques and equipment, believing in the hypothetical fruitfulness of certain research hypotheses, and *presenting both these hypotheses and the data in support of them for the evaluation of all other scientists*, whatever other school they might happen to belong to. From the point of view of this maybe not so distant goal, the existing schools and approaches in sociology constitute only a starting point, and their problems, hypotheses, and research techniques should come to be evaluated from one standard: The degree to which they seem useful for the construction of what might be called an inter-subjectively testable and acceptable body of theoretical knowledge about social phenomena. The *first step in that direction would be that the different approaches stop claiming their uniqueness and universal validity, and admit the partial nature of their problem perspectives, concepts, theories, and research techniques*, trying at the same time to define that kind of social situations and that kind of research conditions their 'approach' seems to be most suitable for.

5. INSTRUMENTAL FUNCTIONS OF SOCIOLOGY

Let us now look a little closer at the instrumental functions of sociology, i.e., it's applications for the scientifically valid transformation of social reality in the direction prescribed by a certain value system. For this purpose we may use certain more or less descriptive findings of social reality which lead to the diagnosis that some elements or fragments (institutions, patterns of behavior, etc) do not correspond to normative standards accepted by the scientist or by the group or institution.

Sometimes a correct diagnosis of the state of affairs closes the task of applied social research. This is the case when the means leading to transformation of the reality in the desired direction are more or less obvious. In more complex cases, sociologists may also be asked what should be done in order to eliminate the gap between the reality and that which is normatively postulated. Technically, this means that the sociologist is asked whether in his theoretical knowledge there are any propositions which

- have as dependent variables conditions that are more or less close to the 'goals' prescribed by the given value system;
- have as independent variables conditions that could be practically realized by those who postulate these goals or by someone else ready to contribute to their realization;
- are valid for the situation in which the social action is intended or postulated.

Such theoretical propositions (assuming causal connections between the 'goals' and the 'means' may or may not already exist in social theory. But if we do not know of a theory (or a more limited generalization) that would satisfy these requirements, then, for purely practical reasons, we would have to develop one. We can therefore say that in many cases the strictly instrumental social application of sociology will have to be accompanied by the development of objective, i.e., true, social theories.

From a strictly instrumental point of view, only empirically valid generalizations and theories have any value at all, because only such theories permit us to make the proper choice of means and therefore to formulate recommendations for practical actions. The same applies to diagnostic studies – if they are not empirically valid they can lead to action that is unnecessary, or to not undertaking action where it would be desirable. Therefore we can say that *the instrumental functions of sociology are a powerful factor toward the 'objectivization' of its findings and theories*. And, needless to say, the various facts and theories used for the implementation of various societal goals are from a strictly cognitive point of view, cumulative insofar as they are true. This does not mean however, that theories developed for strictly instrumental needs are the same as those developed from more cognitive orientations. A theory is a set of general propositions that can be used for explanations and predictions. But the theories which (due to their generality and level of abstraction) explain a lot are not always the best theories for the purpose of practical prediction in concrete social situations. For this purpose it is often better to have generalizations of limited validity but which within the limits of their validity (e.g., within one society) are able to predict events with great accuracy, than to try to resort to very abstract universally valid general theories. By the same token, generalizations which are of great practical use are not always the best formulation of theories constructed for more cognitive reasons. This is why it is so important that the scientist be cognizant of the reasons for his study.

It is obvious that scientists, since they start their studies from the points of view of diverse value systems, diverse goals, and diverse normative standards will concentrate their attention as well on diverse aspects of their societies when undertaking diagnosis, and on diverse theories needed for the proper choice of means leading to the given set of social goals. Their different pictures of society will be by necessity partial — which does not necessarily mean false — and the different theoretical mechanisms will be complementary, assuming they correctly describe the corresponding causal relations. The real instrumentality of any findings or theories assumes their empirical validity, or simply, their truth.

In general one might say that all social theories, whether they be formulated for cognitive or for practical instrumental reasons may have certain instrumental functions if:

- the phenomena constituting the dependent variables of the theory are the objects of certain value systems and are positively or negatively evaluated;
- the phenomena constituting the independent variables of the theory are within the possible range of 'practical manipulation'.

Since almost any phenomena in the society is likely to be somehow (positively or negatively) evaluated by some groups, the first criterion will eliminate very few theories from having some potential practical applications. But the second criterion imposes limitations, and recognizes significant differences between the groups and institutions in any society; the phenomena which are 'manipulable' by the government or any 'power elite' are not identical with those which might be influenced by the 'underdogs' of the given society. This has nothing to do with 'objectivity' in the sense of the empirical validity of particular theories. But it justifies the assertion that different theories may have different practical social implications.

On the other hand, I believe we usually underestimate the degree of flexibility of possible 'uses' of social theories. Even when the manipulation of a certain variable is limited to one group only, and the manipulation used in the interest of the members of this (let us say, privileged) group, knowledge of the relevant theory may motivate members of other groups toward certain counter-actions. In some cases the simple 'unmasking' of a particular theoretical mechanism may play a positive social function, making its use against the values of the majority practically impossible. Actually it is difficult to find any theory which in different social conditions or milieux could not be applicable for the implementation of different – often quite opposite – values, even if only within the area of strictly instrumental uses of knowledge. The same applies to descriptive findings. Nevertheless one should clearly distinguish the *hypothetical* uses of the results of sociological studies in various thinkable social conditions from their *probable* uses which can be more or less easily foreseen in a particular society at the given moment of its history. It seems it is these highly probable uses that should be taken primarily into consideration, because they delimit the use of the immediate social consequences of our studies. For instance, we are now witnessing an amazing development of certain branches of neurophysiology; but it is hard for many people to envision uses of findings or theories on the neurophysiology of the brain other than those which could serve for the 'physical control of mind'.[12] It is difficult in most contemporary societies to contemplate the uses of 'subliminal propaganda' without fearing that it might be abused. Indeed, such possible uses of sociology are frightening some sociologists to the degree that they are ready to abandon the goal of developing theoretical social knowledge altogether, and to limit the tasks of sociology to strictly idiological functions. Let me quote again from the paper by P. Bandyopadhyay a fragment which in my opinion constitutes an answer to these fears:

Conservatives since the Russian Revolution have burned countless gallons of midnight oil trying to refute the claims of social sciences correctly to analyze the societies of today and to provide rational foundations for socialist development. Indeed they have often denied the claims of social science itself. They have usually argued this in terms of the impossibility of objectivity, the necessary intrusion of values, the falsity of determinist analysis and therefore of predictions. They have inveighed against 'scientism' and the necessity for revolutionary times. It has been the radicals among the makers of social science who have championed and labored to prove the connection between the objective social sciences and the necessity for social science and socialism. Today, in North America at least, the positions seem reversed.

Some conservatives proclaim the possibility of objective predictive social sciences, and radicals deny this possibility. In doing so, however, the radicals are in strange company: that of the conservatives of yesterday, still toiling to complete objective social science, bury Marx yet again and invalidate the laws of social development.

This would be a matter for laughter were it not that the stakes are so high: the scope of reason, the comprehensibility of history and society, and the transformation of society in the direction of the fullest possible realization of equal freedom of all. It is, therefore, a great disservice to radicalism to adopt positions that conservatives have adopted in defense of the status quo on the basis of the none-too-convincing arguments.[13]

To this admonition one can only add that positions which reject the possibility of the use of social theories for good purposes are extremely pessimistic, whereas radical values have traditionally been the elements of an optimistic attitude toward the world.

Some writers propose a special solution to this dilemma by the undertaking of so-called 'action research' which develops certain theories of very low generality, the intended applicability of which is limited to a specific set of social goals. I think this combination of research and action toward the improvement of social reality is an extremely interesting phenomenon in contemporary sociology. It is interesting for many reasons. First of all, it acknowledges the need for direct involvement in socially meaningful actions on the part of sociology. And it also indicates that some important problem areas of social theory have not been sufficiently studied by sociology, so that the hypotheses needed for the guidance of action have to be tested in the course of the action itself. While the first aspect of action research is very encouraging, the second should be viewed as an unavoidable necessity only. The situation here is similar to that of a doctor who has to cure a patient when neither the diagnosis nor the remedies are to be found in the medical textbooks but must be found in the course of the medical 'action research'. Such action research may be very stimulating for the development of medical knowledge, but any patient who makes an 'interesting case' from this point of view knows very well that he has no particular reason to be happy about it. It is much safer for the patient to belong to a category of cases which are described by current medical theory. It is – in exactly the same way – much safer for a social group or society to have the kinds of social problems that have been analyzed before, and the nature and theoretical mechanisms of which are relatively well known, than for its problem to have to be solved by a theory the validity of which is being tested in the course of the action itself.

This does not mean I am against involvement of sociologists in socially meaningful practical actions aimed toward the improvement or change of social reality. Nor am I against 'action research' when the existing theory is not sufficient for guiding practical social actions. I do not accept however, an attitude which regards action research as a substitute for so-called 'traditional research' aimed toward the verification of social theories. *Action research is not a substitute for but a supplement to 'regular' research*, and both these kinds of research can and should contribute to the development of sociological theories. And the better our theory the more efficient are its applications, and the more beneficial it will be for the societies, provided its applications are guided by proper social values.

Moreover, if the theoretical conclusions of any piece of action research are really empirically valid, we do not have any quarantee that they will not be used in the service of other — even quite opposite — social values, as is possible (at least theoretically) with any true finding or generalization.

While I disagree with those who propose to take sociology out of theorizing altogether, or to limit it to action research situations, I would like at the same time to strongly oppose any form of social or moral careless- ness. I think in the first place that any sociologist who undertakes an applied social research task ordered by a 'client' (whether this be a social group, an institution or a person), and which is aimed toward the realization of certain goals, *by implication accepts these goals and assumes moral co-responsibility for their eventual realization*. Secondly, I think it is a sociologist's obligation to try to foresee the possible consequences or 'uses' of his work according to his best assessment of the social forces operating in the given society — the value systems, etc. — and to formulate his research problem with the clearest possible picture of both its cognitive and its social significance vis-à-vis his own value system. *If it should happen in some instance that he is able to foresee only an application of his theory that he cannot approve morally, I think he should abandon this particular avenue of inquiry. While it is our task to develop social knowledge, it is an even more important task not to develop it in situations where its misuse can be easily foreseen and would have dangerous social consequences*. The medical prin- ciple 'primum non nocere' — 'above all, not to do harm' — should be printed on the first pages of all our textbooks.

The postulate of 'social awareness' is necessary but not sufficient in the case of sociology. We should encourage and develop within sociology a branch of sociological reflections and studies which could help us to understand better the possible instrumental utility of sociological findings and theories in different social contexts and for different value orientations — and thus to better foresee who is likely to use 'our sociology' and for what purposes. The development of just such a branch of sociology as this is for me of the greatest social importance.

6. IDEOLOGICAL FUNCTIONS OF SOCIOLOGY

By an ideology we usually mean a set of fairly general values, some more specific evaluations of certain areas of reality when the knowledge about them is confronted with the more general evaluative standards, often some philosophical assumptions and beliefs, some more or less general descriptive and theoretical propositions referring to this reality and — last but not least — certain prescribed patterns of action aimed toward the conservation or change of reality. Some of the components of ideology have the character of empirically testable propositions, whereas others are of a normative character. But when we speak of the ideological functions of scientific — i.e., empirically testable — sociological propositions we usually mean the impact of these propositions or of some other empirical proposition which do not belong to ideology upon evaluative and motivational areas or ideologies, and especially the degree and direction in which they may influence human evaluations and motivations to specific actions, as well as the degree to which they may be able to integrate these actions in a collective effort toward a commonly accepted goal. Let us look first at the motivational functions of social sciences, at their role in 'triggering' individual and collective actions.

To begin with, we can say that when diagnostic knowledge is able to 'trigger' some motivational mechanisms aiming toward social actions, theoretical knowledge can orient such actions channeling them toward the realization of means necessary for the implementation of the goals. Sometimes discovery of a theory may itself play the role of a motivational 'trigger mechanism'. When the existence of a problem has been known for a certain time but no means were known which might be used for its solution, theoretical discoveries which reveal the *possibility* of a solution can start the motivational mechanisms and integrate individual motivations into one collective action.

In some situations the theoretic knowledge can even have a strong impact upon values. There might be many things we would like to have, frustrations we would like to eliminate, goals we would not object to realizing – but since we do not believe these things are possible, they exist as dreams somewhere on the pulphery of our value system, scarcely belonging to it, and certainly not motivating our behaviors. If it were to be discovered by a theoretician that they could be realized, that the means leading to them are such and such, and if additionally it should turn out that these means are within our range, it is very likely that this would lead to a basic change in the structure of our value system. The dreams would become real social goals with a powerful motivational force. On the other hand, if it could be proven that something we were aiming at is beyond the range of practical possibilities, this would probably lead to a decrease in the attractiveness of such a goal. In most instances, however, our theoretical analysis will be of a somewhat different kind. Our theories will say that realization, more or less probably and that the means necessary for that are more or less 'expensive' in the broad sense of the term. Such statements, of course, may also have some impact upon the attractiveness of the goal and may to different degrees influence human motivations.

One should remember that we are speaking here about certain *sociopsychological mechanisms determining the impact of knowledge upon human evaluations and motivations*. Some of these mechanisms correspond more or less closely to the pattern that is described in psychology as *rational reasoning and rational behavior*, but this does not mean that other psychological mechanisms cannot operate here. Thus for instance we observe in the formation of evaluations certain simple mechanisms described by theories of learning: e.g., when we discover in our theoretical analysis that a certain state of affairs is instrumental for some our goals, we see that a positive value is being attributed to this mean too. It is even likely that due to the functional autonomization of instrumental values, the phenomena which were primarily seen as the 'means only' will become positive values in themselves quite independently of the goals to which they were supposed to lead. Their instrumentality is often forgotten and ultimately be treated as independent 'goals'.

In other cases we see the operation of mechanisms that are fairly distant from the ideal of human rationality. According to elementary logic the same object when regarded from the standpoint of its various properties, and confronted with an identical set of evaluative standards, may be good from the standpoint of some of its properties and bad from the standpoint of others. We may arrive of course at some general evaluation of this object, taking into account its diverse partial evaluations without violation of the rules of rationality. But we also know that on a mass scale people feel uneasy when confronted with such instance of 'cognitive dissonance' and tend to accept only some of the evaluations – those which go in only one emotional direction – and not to accept the others. This usually leads to the rejection of all 'dissonant' information. Finally, from a strictly rational viewpoint, the concrete evaluations should be functions of our knowledge and our general normative standards, and the rules for actions should be derived from these premises jointly. We know, however, that many people behave otherwise. They will adapt the general evaluative standard to the requirement of a concrete evaluation if this concrete evaluation is somehow important to them, or they will be unwilling to admit any information that would make the acceptance of both the general evaluative standard and the concrete evaluation impossible. They may also be inclined to look for 'proper' information necessary as 'rationalizations' of action, rather than true information for use as a guide in their behavior.

Here I would like to mention one type of conclusion that is often encountered in discussions about the ideological implications of certain theories. If theory is understood as a set of general propositions describing the relations between the variables, such theory may be 'used instrumentally' for the implementation of certain social goals – as I discussed above – but it cannot justify, at least within the rules of rational thinking, any such goals without additional value judgment. By the same token, it cannot be used without such normative assumptions for either the 'justifications' of the status quo or the recommendation of social change. If the theory is believed to have such implications it is because these value assumptions are – usually unconsciously – taken for granted.

This can be especially clearly seen in recent discussions about the 'conservative assumptions' of functionalism, where the critics of functionalism failed to perceive that it has such conservative implications – for conservatives only. In criticizing the critics I do not have in mind, of course specific evaluations of certain institutions made by representatives of the functionalist school or their general view of 'social equilibrium' as a desirable social goal. What I do defend is the *general problem area of functio-*

nalism, which stresses the importance of looking for the social functions of different cultural patterns and institutions, and leaves (or should leave) open the problem of acceptance of these functions and of various kinds of social equilibrium. The conclusions to be drawn us to the appropriate *social actions* must be based on the confrontation of these empirically assessed functions with different human values.

The psychological mechanisms which determine the formation of evaluations may, when acting on a mass scale, influence the reception of sociological knowledge in such a way that the ideological conclusions formed from the sociologist's work are more or less distant from the ideal of human rationality. These mechanisms may also be 'used' by various groups or institutions which, more or less deliberately, will try to derive from the sociologist's work conclusions which it does not justify, in order to influence attitudes and motivations on a mass scale. Finally – and which is most dangerous here – the need to influence attitudes by means of certain sociological information may lead to greater or lesser pressure upon sociologists to 'deliver' independently of its truth information necessary from the standpoint of its specific ideological function, and to suppress certain other information even if it is true. I noted earlier that for strictly instrumental purposes, false information is useless. But – unfortunately – this is not the case with the ideological functions of sociology, because any information which is *believed* to be true by its 'receiver' will have an impact upon his attitudes and actions, either within the scheme of strictly rational thinking or according to any other kind of psychological mechanism. Therefore, if a group or institution is more interested (consciously or unconsciously) in influencing these attitudes in an a priori direction than in shaping them in an empirically justified, rational way, and at the same time possesses means of moral, economic or more direct pressure upon the sociologist, it may try to use these means to influence his enquiry in the desired direction.

Of course it may happen that such a group will 'profit' from such 'misuse' of sociology in the short term. But in the long run such behavior is usually irrational even from the standpoint of the values which underlay the pressure, because short-term 'gain' in shaping human attitudes cannot compare to the costs of acting under a false diagnosis of the situation, and guided by the wrong theoretical assumptions. In practice this means that actions undertaken under such circumstances are either unnecessary or inefficient or both. It may also mean not undertaking action when action is necessary. It is also my firm belief that a sociologist who, sharing the values of some group 'yields' to its pressure with respect to the truth of his results does not do a service but an injury to the cause he believes in.

When these pressures are perceived by the scientist as coming from 'outside' he is in a relatively better position to resist them. But when they act on him in a situation where he is a member of the pressuring group, and thus presumably accepts its values, he may 'yield' more or less unconsciously. In such case the danger of diminishing the cognitive and instrumental value of his study – even from the standpoint of precisely the values he accepts – is especially serious. It seems that *the more we are engaged in a value system, and the more we are involved in the realization of the social goals determined by these values, and the more important the goal of our social action is to us, the more we should be alert to possible disturbances of the objective validity of our perceptions, analyses, and interpretations.* Above all, we should remind ourselves that nothing is more harmful for the realization of social goals than wishful thinking. Therefore, any increase of our value-involvement in the realization of certain social goals should be accompanied by any increase in our willingness to realize these goals by means of all the possibilities offered by objective scientific investigation.

Remembering that whatever we do may have an impact upon the thinking of those who are not specialists in our field and that only some of their conclusions may be rational, we should also try to stress the rules of correct inference in the presentation of our findings. One of the important tasks of the sociologist is to eliminate as much as he possibly can misinterpretations of his work, taking into account all the known mechanisms of the sociology and psychology of knowledge and trying to counteract them if they would lead in the wrong direction. Another equally important task is to make 'premedited ideological misuses' of his study as difficult as possible.

All of this does not mean the sociologist is not entitled to evaluate the societies he is studying. Any human being has a 'right' to his values, and the sociologist is in no way obliged to be 'unhuman' here. For any human being – including the sociologist – it is quite 'legitimate' to undertake actions aimed toward the realization of his values. And he is also 'entitled' to communicate his values to all who are inclined to listen, as well as to those whose interest has not been aroused. But as a scientist he has some obligations which I would not hesitate to call *moral* ones. He must present as empirically valid only such facts and theories as have been discovered or tested in empirical investigations, *adjusting the strength of his assertions to the strength of the empirical evidence supporting them*. Whenever he enters the area of 'loose' theorizing or hypothetical speculations – both of which are necessary in the development of any science – he should not pretend that these constitute tested theories. Whenever he evaluates an area of social reality he should present his values distinctly from the assessed facts, leaving it to the reader or listener to accept the general values and concrete evaluations apart from the facts, or possibly to come to different evaluations on the basis of the same facts. Finally, when proposing some course of social action he should specify both the value assumptions which underlie his proposing or accepting certain goals and the theoretical and diagnostic findings which legitimize the proposed course of action. *Such a sociology would be at the same time both objective and value involved*. Its strictly cognitive components would develop in a cumulative process, in which we would obtain better and better understanding of this part of empirical reality that we call 'society'. This in itself might constitute a legitimate goal of study for those scientists guided by predominately cognitive motivations, provided they do not overlook the possible 'misuses' of their science. Such an objective yet value involved sociology could also serve as an intellectual tool toward the transformation of our world into one such as we would like to live in. We will need for all of this a great deal of the *theoretical and diagnostic knowledge* which sociology can eventually give us, assuming its optimal scientific development. And we need as well some additional knowledge about the relations of sociology to society, its possible instrumental and ideological functions, in order to *guide the development of our discipline in a wise way* – so that it can be properly used and hardly ever abused. The faster we develop our discipline with respect to these criteria, the better for our social causes and for mankind in general.

NOTES

* Paper prepared for the Round Table 'Is There a Crisis in Sociology?' of the VIIIth World Congress of Sociology, Toronto, August 1974. The present extended and modified version of this paper was written during the author's fellowship at the Center for Advanced Study in the Behavioral Sciences at Stanford, California. Reprinted from the volume Crisis and Contention in Sociology including all the papers from the Round Table (Tom Bottomore, ed.) published by Sage Publications; with the kind permission of the Publishers.

[1] Armando De Miguel: 'Undertaking Sociology in Authoritarian Countries: The Case of Spain. A Pessimistic Reflection'. Paper prepared for the Round Table mentioned above. (Mimeographed).

[2] See, e.g., A.W. Gouldner, *The Coming Crisis of Western Sociology*, New York, 1970.

[3] P. Bandyopadhyay, 'One Sociology or Many', *Science and Society* Vol. XXXV, No. 11 Spring, 1971, pp. 1 – 2.

[4] See, e.g., the paper by P. Bandyopadhyay cited above. See also H.S. Becker and J.L. Horowitz, 'Radical Politics and Social Research, Observations on Methodology and Ideology', *American Journal of Sociology* 78, No. 1, July, 1972.

[5] F. Ferrarotti, 'Some Preliminary Remarks on "Is There a Crisis in Sociology",' paper presented at the Round Table mentioned above. (Mimeographed).

[6] For the analysis of some problems related to the formulation of reductive theories in sociology see Chapter XI.

[7] See Chapter I.

[8] For a detailed discussion of logical and empirical relations between the indicators and the indicated phenomena denoted by the given theoretical concept, see Chapter I.

[9] For a closer analysis of the empirical assumption, involved in the formulation of questions of different kinds, see J. Gedymin, *Problemy zalozenia, rozstrzygniecia* [*Problems, assumptions, solutions*], Poznan, 1964. See also S. Nowak, 'Methodology of Sociological Research', Ch. I., *The Formulation of the Research Problem and the Choice of the Right Methods*, D. Reidel, Dordrecht, Holland, in press.

[10] By a *direct* answer I mean our answer which does not abolish the validity of the assumptions of the given question. In the case of the above example, an *indirect* answer might be: "Neither the system of social control, nor the community of values of ist members contributes to the integration of this society. In fact this society is not integrated at all'. Such an answer would abolish the validity of assumptions of our question. In some cases, when we conduct our study stepwise, the facts established at its first stage are used as the assumptions for the next question. The most simple example would be a series of interview questions: "Do you have children?" "If yes, how many?" The weakest form of assumption is that phenomena in question at least might exist. Therefore we ask people about their children, but we do not ask them how many horns they have. (See: J. Gedymin, *op. cit.*).

[11] Unless we define our concept as referring to an 'ideal type', which does not exist in reality, but the postulation of which is useful heuristically in the formulation of our theory. In such a case we assume that the empirical phenomena may be ordered as different approximations to our ideal type and that such ordering is fruitful from the theoretical point of view.

[12] See: J.R.M. Delgado, *Physical Control of the Mind: Toward a Psychocivilized Society,* Harper and Row, New York, 1969.

[13] P. Bandyopadhyay, *op. cit.*, pp. 21 – 22.

BIBLIOGRAPHY

Bandyopadhyay, P., 'One Sociology or Many', *Science and Society* **XXXV**, 11, 1971.

Becker, H.S., and J.L. Horowitz, 'Radical Politics and Social Research, Observations on Methodology and Ideology', *American Journal of Sociology* 78, 1, 1972.

Delgado, J.R.M., *Physical Control of the Mind: Toward a Psychocivilized Society*, Harper and Row, New York, 1969.

Ferrarotti, F., Introductory paper to the Round Table, 'Is there a Crisis in Sociology', presented at the VIIIth World Congress of Sociology, Toronto, 1974. (Mimeographed).

Gedymin, J., *Problemy załozenia, rozstrzygniecia* (Problems, Assumptions, Solutions), Poznan, 1964.

Gouldner, A.W., *The Coming Crisis of Western Sociology*, New York, 1970.

Miguel, A. de, 'Undertaking Sociology in Authoritarian Countries: The Case of Spain. A Pessimistic Reflection'. Paper presented at the VIII World Congress of Sociology. (Mimeographed).

Nowak, S., 'Concepts and Indicators in Humanistic Sociology', in this volume, p.   .

Nowak, S., 'The Logic of Reductive Systematizations of Social and Behavioral Theories'.

Nowak, S., *The methodology of Sociological Investigations* (English Edition, in print).

CHAPTER IX

CULTURAL NORMS AS EXPLANATORY CONSTRUCTS IN THEORIES OF SOCIAL BEHAVIOR*

1. TWO TYPES OF SOCIOLOGICAL EXPLANATIONS AND PREDICTIONS

The more and more rapid development of theoretically oriented sociology, social psychology and other sciences dealing with man's social behavior and with the functioning of human societies puts at the social scientist's disposal every day more numerous and also more reliable sets of scientific generalizations. These generalizations describe regularities of events within such spheres of phenomena as, e.g., 'political behavior', 'perception of social phenomena', 'interactions in small groups', as well as in other fields of the social behavior of men. The statements are usually formulated in a language much like the theoretical language employed in the natural sciences, and they have, as a rule, the form of an implicational relation between two sets of 'qualitative' phenomena, or the form of a functional relation between one or more independent quantitative variables on the one side and the quantitative dependent variable on the other. The relations that have been established are often of a statistical nature such as when one says that the occurrence of one phenomenon increases the probability of occurrence of another phenomenon, or that a positive correlation exists between two variables.

The scheme of explanations and prognoses looks much like that in the natural sciences. It is assumed that knowledge of a relevant universal law or historical generalization i.e., a proposition the validity of which is limited to a certain time-space area[1] along with establishment of the initial conditions described in this law or historical generalization enables us — provided certain additional assumptions have been taken into account (e.g. the relative isolation of the system from intruding counterfactors) — to predict the occurrence and the intensity of the dependent variables. It is also assumed that in order to explain a certain phenomenon it should be established that both the antecedent and the consequent of a given generalization have occurred. This deductive-nomological way of approaching ex-

planation and prediction of social phenomena[2] seems to be achieving more
and more successes.

At the same time, however, an old principle that has been successfully
applied for some thousands of years has not lost its validity. In accordance
with that principle, if we want to foresee how the members of a particular
society would behave in a certain situation we would first of all seek to
learn the *normative rules and prescriptions* governing the behavior of members of that community. For instance, if we want to foresee how an
American soldier will behave himself under certain circumstances, or how he
will – let us say – conduct himself at a particular time of day, we would
find the most efficient 'predictor' in the U.S.A. Army regulations. Similar
knowledge of norms creating the 'social role' of the foreman in a motor-car
factory will explain to us his behavior toward his superiors as well as his
subordinates, at least within a more or less definite range of variability. Just
to know the 'ethos' of the feudal knight in Western Europe will elucidate to
a certain, fairly high degree the manner of behavior shown by representatives of the *feudal* elite toward each other on the one side and toward
members of other social classes on the other. Even within the sphere of
those phenomena that have been relatively thoroughly analyzed by means
of 'nomothetic' theoretical reflection – as e.g. the theory of small groups – it
is impossible not to get the impression that knowledge of the regulations
and rules of summer scouting will be more helpful and useful in our
predicting how a certain scouting group leader will behave in a certain
particular situation than the reading of works written by any number of
small group psychologists.

In a rather peculiar way we have to apply in our explanations of social
behavior two kinds of premises:

(a) certain 'nomothetic' generalizations (assuming that their antecedents
are satisfied); and/or

(b) certain 'idiographically' described patterns of normative regulations
of human behavior which, if described with sufficient preciseness, may be
unique for the studied population.

In some situations one of these premises is sufficient for explanation and
(or) prediction. If we have knowledge of certain rituals accepted by members of a given group, we are able to foresee with great accuracy how, say, a
group of ladies will behave during a tea party, or how a military parade will
be carried out. On the other hand, we will use our nomological knowledge if

we want to explain why the 'underdogs' of a given society are fairly likely to be dissatisfied with that society. In some other cases we have to apply both of these approaches jointly for the explanation and prediction of *the same phenomena*. Suppose that we know the norms controlling the behaviors of most of the members of a certain society, and at the same time our general nomothetic theory tells us, what kind of members of this society are most likely to obey their norms. Then of course we will use both these premises in order to predict the *exact patterns* of behaviors of those members of the given society, which are specified by the antecedent of the general theory and which according to this theory are likely to obey these norms in an exact way. In some other cases nomothetic theories and normative regulators explain different aspects of the same behaviors. Even in very ritualized situations there may still be a place for 'dominant' and 'submissive' behavior, and which individuals will behave in the given way may sometimes be predicted by certain personality theories. On the ground of adequate laws one could even predict which people even when complying with the norms of their culture or behaving in accordance with expectations connected with their social roles will additionally shape those norms of those roles according to a specific personal pattern, or who while adhering to the standards of thinking which have been imposed upon them by the culture will in one way or another reinterpret the values of their society.

The same predictive capability of different approaches applies to many problems on a more macro-social level. When studying human behavior within an organized social system, we sometimes concentrate our attention on those aspects of behavior which retain a certain number of 'degrees of freedom' from the normative rules shaping the structure of the system and the relations between the persons functioning within it. Then we look for certain nomothetic regularities which might explain why different people behave differently within the 'limits of freedom' left by the rules of the system as we look for general regularities explaining the behavior of deviants.

In my view, essential part of sociological (and socio-psychological) theories should be concentrated on the problem: How do the normative regulators work, how are they able to shape the behavior of human individuals in a way that is approximately *isomorphic* to the rules. But if we want to avoid the methodological 'dualism' signalized above, we will have to do this in such a way as to permit us to interpret the action of normative regulations of human behavior within the broader context of the 'traditional' nomological way of thinking.

2. CULTURAL NORMS AS 'MATRICES' OF HUMAN BEHAVIORS

I would like to suggest that the only solution that may, at least on the level of methodological principles, do away with the above mentioned duality of explanatory and prognostic procedures employed in sociology, and that will foster — at least in a more distant future — some hope for a theoretical integration of these two types of analyses, is the acceptance of the assumption that *social norms are antecedents of sociological laws* (to say it more exactly: that social norms are necessary components of some sufficient conditions occurring in these laws), when the dependent variables of these laws describe human behaviors from a special point of view: *whether and to what degree these behaviors correspond — are 'isomorphic' — to a given normative pattern.*

Under this assumption all explanations and prognoses in which we do not refer to a corresponding general law, but only mention the existence of an appropriate norm as the cause of a particular sort of behavior, would be regarded as *incomplete explanations*, as so frequently are met with in other sciences. But we should be aware that there are two possible kinds of such incomplete explanations.[3] In the first case we omit reference to the corresponding law only for reasons of brevity. The law is so obvious that it does not warrant stating. But in other cases of incomplete explanations, we have to omit the law according to which an event occurs simply because we don't know such a law, i.e. we are unable to indicate the complete set of conditions which constitute the antecedent of a general or probabilistic regularity entailing the occurence of this event. In many natural situations, even if we don't know the general causal laws relating to classes of phenomena we still often 'feel' that they are somehow causally related.[4] The relations between the cultural norms of societies and the corresponding behaviors apparently belong in this last category.

Let us analyze more deeply some methodological features of the *potential laws* which would include cultural norms as antecedents and conforming patterns of behavior as consequents.

I think that among the various propositions describing the regularities in the occurrence of events it is worthwhile to distinguish a category of laws in which the relation between the antecedent (or, as a rule, one of the elements of the antecedent) and the consequent is that of 'structural isomorphism' such that we may consider the event explained by that law as a sort of 'photocopy' of one component of the antecedent, as a kind of 'faithful reproduction' in the consequent, of a definite configuration of the features in the antecedent. Let us call that element of the antecedent which is photocopied or reproduced the 'matrix' of the relevant consequent so as to underline more firmly the analogy to the process of photography or projecting a picture on a screen.

The analogy between the matrix and its isomorphic product need not be of a spatial nature, as in the photographic examples. 'Structural' correspondence of a spatial nature can still be discerned, for example, between the pattern of a dress illustrated in a fashion magazine and the dress made according to that model, but there is also a 'matrix relation' between what one wears at a formal dinner party and fashion edicts detailing the proper apparel for given social situations. Other examples of such a 'matrix relation' would be the relation between the text of 'Hamlet' and the sequence of words uttered by the actors on stage, that between 'No parking' sign and the behavior of drivers, etc. Needless to say, matrix-relationships can be found in other sciences as well. In genetics the study of matrix relations between the genotype and the phenotype is one of many such examples of isomorphic relations in the natural sciences.[5] We might call such generalizations 'the matrix laws' or the 'laws of isomorphism'. The matrix in these laws is this component of their antecedent which is able to shape their consequent in the isomorphic way. The statements relating to the efficiency of norms do not exhaust all the vategories of the laws of isomorphism in the social sciences. Since 'motive' is sometimes defined as 'perception of the aim of the action', motive can well be considered the matrix of the state of affairs for which we are aiming (when the program of the action towards its realization of the goal is the matrix of the course of our action). Thus the laws that describe the circumstances under which motives efficiently control human behavior are also matrix laws. For those who say that the superego is the result of internalization of cultural norms and values, obviously the group culture is the matrix of the individual's superego.

Let us symbolically denote the relation between the 'matrix' M and the behavior B isomorphic to it as $M \rightarrow B_M$. But to say that there is a matrix relationship between certain norms and the behaviors isomorphic to them is no more than to propose a *new term* for a class of relationships which seem to be distinct from others; it does not yet make such relations laws of the social sciences.

3. METHODOLOGICAL FEATURES OF THE LAWS OF ISOMORPHISM

We could call the matrix relations laws of the social sciences only if we could say that they really 'work' in their general form, i.e. that whenever any such 'matrix' exists the consequent behavior will be isomorphic to it. But this is obviously not the case in any science. In genetics certain 'matrices' are able to produce a working mechanism, while other combinations of genes lead to an obvious biological failure. Moreover, we know that the success of certain genetic matrices is codetermined by the external conditions in which the organism has to live.

The same is true for the matrix regularities governing the social behavior of human beings. The proposition that any normative matrix (whatever the term 'norm' should mean; see below) will act successfully under any conditions is obviously false. In some situations certain moral norms control the behavior of people quite well, in some others they clearly fail. Organizational rules sometimes work quite smoothly, in other cases they do not work at all and the organization disintegrates. Therefore the proposition that 'human behavior is determined by the norms influencing the person' is obviously false if we try to interpret it in general sense and literally, i.e. at the level of each particular norm and all persons under all conditions. At best we can only say that such a process occurs 'sometimes' or 'often', but propositions including terms like these cannot be called laws of science.

Therefore we have to be more specific and ask about *additional conditions* which are involved in such matrix-relationships, assuming that the statement 'under conditions C_1, human behaviors are successfully shaped by the normative regulations existing in the given society' constitutes a general logical scheme of such law. Our scheme of the matrix law would then be:

$$(C_1 \cdot M) \rightarrow B_M.$$

But here again it seems doubtful whether the specification of conditions only is enough to make our theory *valid for all possible kinds of norms*, independently of their content or the area of human behavior to which they apply.

Many theories in the social sciences which involve the 'matrix relationship' — such as theories of learning, theories of socialization, theories of interpersonal influence etc. — *are obviously too general* with respect to the matrices they are dealing with. Having found a relation of isomorphism between *some* pattern of behavior we would like to reinforce and the consequent behavior of the rat in the maze, we should then ask the question of whether such regularity applies to all possible 'matrices'. The answer has to be a resounding negative. This is also the case, if not more so, with more natural social situations. We do not seriously think that the conditions facilitating the actions of such matrices as youth fashions will also facilitate the actions of such matrices as those which determine the behavior of soldiers during battle. The efficiency of the organizational rules getting the time of lunch for workers in a factory is co-determined by a different set of conditions etc., than the efficient action of the rules establishing the minimum daily production per person.

Therefore the norms in our matrix relations have to be specified in the same way as do the conditions, so that the propositions described by our matrix laws would be empirically valid:

$$(C_1 \cdot M_1) \rightarrow B_{M_1}.$$

If we had to formulate a separate law for each and every matrix — for each and every normative regulator of human behavior — it would probably make the formulation of such theory practically impossible. But we usually can assume that for a given set of conditions C_1, *a whole class of different matrices* M_1, M_2 .... M_n *can 'work successfully'* toward shaping the consequents of the corresponding matrix relationships in an isomorphic way. This is true for matrices of various kinds. For instance, we know very well that the range of variability of printing plates for a given printing technique is in general fairly wide, and various different photographs or snapshots may be exchanged with and replaced by one another without interfering with the adequacy relation between the matrix and its product. We are also aware that if we want by means of newspaper plates to reproduce a detailed plan of city or a map of a mountain range with all its details, we will nevertheless eventually obtain a gray spot if the shape of the matrix under the given system of conditions has exceeded a *permissible range of variation*. It is also well known that with a proper type and a given intensity of sanctions a community can be induced to obey a considerable number of norms, but by no means to comply with all the thinkable norms. Teachers know very well that there is a fairly firm connection between the normative contents to be imparted to pupils and the conditions for and ways of transmitting them. They are aware that any attempt to ingrain by means of 'stick and carrot' the norm that dictates 'you should love your teachers' has no chance to succeed for it exceeds the permissible range of variability of the 'educational matrices' under given system of conditions.

Consequently, when trying to formulate a theory of the efficiency of social norms, our attention should be drawn to the problem of the possible range of norms which can efficiently control human behavior under a given set of accompanying conditions. For some definite set of conditions (let us say, C_1) there is a certain range of norms (let us designate them here as N_1, N_2, N_3), for which the processes of normative control of behavior are almost complete. Moreover, it should be borne in mind that under the given system of conditions C_1 a certain range of norms (e.g. N_4, N_5, N_6) still has a chance to exert some influence on human behavior, and that eventually will be a broad range of norms say, N_7, N_8, $N_9 ... N_n$) which have no chance for realization under C_1.

But here we must also deal with another peculiarity of the matrix laws. It is clear that a given matrix N_1 itself may sometimes be 'duplicated' or 'photocopied' under different systems of accompanying conditions. Therefore we might begin our systematization by taking a certain matrix N_1 as the focus of our attention in theory construction.

A drawing of a certain design may be printed either by hand operated press, by foot-lever, or even by platen or rotary press, while the color of the printing ink may be variously changed without a change in the matrix. The motive 'I want to eat' can be realized in several very differentiated (but by no means all possible) systems of accompanying conditions. The norm 'Thou shalt not kill' can be ingrained by means of very different systems of instructive conditions, as well as by different systems of enforcing this rule against persons who have defied it.

Here again we should take into account the *range of various conditions* within which the given norm may operate in general as well as the range of those conditions within which it may operate in the most efficient way. It doesn't matter whether we approach the problem starting from the conditions or from the matrices themselves. From whichever end we begin our systematization, at the final stage we should obtain a two-dimensional scheme (a 'matrix' in the mathematical sense) of interrelated variables that will enable us to see their interactions in 'producing' the behaviors isomorphic to the given normative matrix. This scheme might take the form set out in Table I, if we assume that $C_1...C_p$ constitute a range of conditions, $N_1...N_n$ are the norms regarded by our theory, and + denotes the configurations of norms and conditions which produce behavior isomorphic to the given norm.

TABLE I

Conditions for isomorphic relations	Normative matrices of human behavior								
	N_1	N_2	N_3	N_4	N_5	N_6	N_7	...	N_h
C_1	+	+	−	−	−	+	−	...	+
C_2	+	+	+	−	−	−	+	...	−
C_3	−	−	−	−	+	+	−	...	−
C_4	−	−	+	+	−	+	−	...	+
C_5	−	−	+	+	−	−	+	...	−
...	...	...	...	...	...	...	...	...	...
C_p	+	−	+	−	+	−	−	...	+

Such a table could present all the relations we find in our analysis. It is of course open for any additional norms we would like to include, as well as any other conditions. The table does deal with the theoretical findings at a fairly low theoretical level, because it presents the norms in all their concreteness and the conditions are specified in a rather narrow way. Thus the next step should probably be toward *finding more general mechanisms underlying the relations between the ranges of conditions and the ranges of norms.* It is difficult to say to what degree such general classificational schemes (let us denote this broader class of norms by N_{G_1}) would have to take into account more general elements of the *contents* of the norms, and elements of some socio-psychological processes involved in their operation under the given condition. Also the conditions should be, insofar as possible, defined in a general way (C_{G_2}), so that we could say, eg.:

$$(N_{G_1} . C_{G_2}) \rightarrow B_N.$$

This means that *whenever a certain norm N possesses the more general characteristics G_1 and it works under the generally definable conditions C_{G_2} it will produce the behavior isomorphic to its concrete contents.*

One should be aware that a theory of the above design *does not explain why norms of a given type exist* in the given society. *It takes the existing norms for granted,* as well as the conditions facilitating their impact upon consequent behavior. In order to explain the existence of the given norms we would have to apply quite different theoretical approaches; we might for example explain their existence by pointing to certain *'functional requirements'* of the society, which has produced the normative regulations necessary for its existence and functioning. We might, in other cases, look into the more or less distant past of this society to see how these norms have been formed and transmitted over several generations, and so on. Our theory would permit us to explain only why some norms existing in the society are able to shape in an isomorphic way the behaviors of its members, while others are obviously inefficient.

It is obvious that the sphere of phenomena that has been so far defined as 'conditions' indispensable for the efficient operation of a matrix, and denoted by the letter C, is not homogenenous one. It therefore seems advisable to distinguish two categories of its components, namely:

(a) properties of the subjects whose behavior the norm tries to shape. (We may denote these by P);

(b) conditions external to the 'shaped' subject, these being the elements of *the situations* that are essential to making the subject conform to the matrix (denoted by S).

A physical analogue to this would be that the exactitude and precision of the design on a medal impressed by means of a proper punch depends upon both the force of the pressure (denoted above by S) and the hardness and malleability of the metal (denoted by P) from which the medal is molded. On the socio-psychological level, we know very well that the efficiency of an educational process is dependent to a great extent on the personal characteristics of the individual including both their personality traits and the norms and values that have been previously acquired (P), and on the 'external' condition of the educational process (S).

Such being the case the general scheme of the law of social isomorphism can be presented as follows:

$$(S \cdot P) \quad \rightarrow \quad (N \quad \rightarrow \quad B_N)$$

or briefly

$$B = (f)\, NSP.$$

All of the problems I have already mentioned with respect to combinatorial relations between C and N remain valid here but they need to be translated now into the language of possible combinatorial relations among the three variables NSP simultaneously, which would make Table I a three-dimensional classificatory scheme. Thus we say that we should consider the range of variability of norms (N_1 ..., N_n) which can efficiently operate in the given external situation S_1 and for people with the definite individual features P_1. Or we may state what is the permissible range of the variability of personality types (P_1 ..., P_n) if in the given situational conditions S_1 we want the norms of the given content (N_1) to operate. Finally – and this is the central problem from the standpoint of the efficiency of social control – we can seek to establish the range of variability of external conditions (S_1, ..., S_4) that will guarantee the norms of the given content N_1) efficiently operating upon people with the given set of cultural features and personality characteristics P.

4. DIFFERENT MEANINGS OF THE TERM 'CULTURAL NORM'

If our theory is to describe the influence of the relevant normative matrices on human behavior, the dependent variables in such generalizations will refer to *human behavior*. As we will recall from another essay in this volume,[7] the explained behaviors can be understood either in a strictly behavioristic sense (when the norm refers to their external course only), or in terms of certain 'meaningful-behavioral syndromes' including in the definition of the given behavior both its external aspects and the 'subjective meanings' related to them, or they may be understood in a meaningful way only, leaving the external, observable aspects of the behavior definitionally 'open'. Analysis of the contents of the given normative matrix usually enables to discover the meaning of the 'behavior' (in the sense of the above classification) to which the given norm refers. Norms prescribing a certain ritual at the Royal Court refer primarily to the external behaviors of the persons to which they apply. Norms specifying the social role of a doctor refer both to his overt behavior and to his attitudes toward patients on the assumption that these attitudes will be involved in the doctor's behavior vis-à-vis his patients. Norms specifying the rule of solidarity among members of a clan may place the primary emphasis upon clan members' attitudes toward each other, leaving open the behavioral aspects of the rule as depending upon the circumstances. Thus we see that the content of the normative matrix usually specifies the nature of the *dependent variable* in the laws of social isomorphism.

But what is the nature of the normative matrix itself, constituting the *independent variable* of such a law? In general one may say that we have here in mind certain *cultural norms*, i.e. norms which, depending upon the definition of 'culture', either belong to the culture of the given group or society, or comprise this culture. Some definitions of the term stress very strongly that culture consists (primarily or exclusively) of *patterns* of thinking and behaving, and that these patterns are able to *govern* individual thinking and behavior. To quote Gillin and Gillin:

The customs, traditions, attitudes, ideas and symbols which govern social behavior show a wide variety... These common patterns we call culture. [8]

Or the definition of culture by Kluckhohn and Kelly:

By culture we mean all those historically created designs for living, explicit or implicit, rational and nonrational, which exist at any given time as potential guides for the behavior of man.[9]

The 'normative' definitions of culture therefore stress that it is a set or system of 'matrices' of human behavior and thinking.

But still the meaning of the term 'norm' and therefore the character of the antecedent of laws of social isomorphism is not clear enough. One of the ways of clarifying the meaning of such concepts may be -- as suggested by Maria Ossowska[10] to define the corresponding term in a *contextual way*. Instead of asking what we mean by a social norm, it may be better to ask in what situation and under what conditions do we use in the social sciences the statement: 'In society S there exists the norm N_1'. But when we look at the usages of the term 'norm' after they have been 'retranslated' in the contextual way we see that they may refer to quite various phenomena. Sometimes they refer to certain *generalized patterns of behavior* of the members of societies, on the assumption that these patterns are able to influence the behavior of other members. Sometimes they refer to certain *expectations* the members of the group direct toward each other, and the tendency to control the behavior of others according to these expectations. Finally, they sometimes refer to *internalized values* of the group members, on the assumption that these values are able to control behavior. The antecedent of the matrix relations is in each of such cases different, of course, and accordingly the corresponding law of social isomorphism will have quite different theoretical character and will belong to (or be explainable by) quite different theories of sociology, social psychology or psychology. And of course, depending upon the meaning of the term 'norm' in its given sense, the conditions facilitating its action may also be different. Therefore when systematizing norms as factors of human behavior we should take into account not only their content (i.e. the pattern of the behavior such a 'norm' determines) but also the nature of the phenomena which are *definitionally equivalent to the existence of the given norm*.

Let us consider more closely the various meanings of such terms as 'social norm' or 'normative cultural pattern'. In saying that a certain norm 'exists' in a certain society as a potential factor determining human behavior, we may have in mind any of the following aspects of 'normative phenomena':

(i) the content of an external command, expectation or prescription directed to all or some members of a given society with the deliberate intention of shaping their behavior in a given way. I would call this a *norm in the coercive sense*.

(ii) the existence of a certain opinion or conviction in the minds of members of the given society; this opinion or conviction refers to *how they themselves ought to behave* in a certain situation. Since this normative conviction constitutes at least a potential motive for subsequent behavior, I would call it a *norm in the motivational sense*.

(iii) an actual pattern of behavior that is either a typical (i.e. modal) way of behavior in some given society or possesses other special features (e.g. it is the behavior of especially prestigious members in a given field of social life). We might denote this type as a *norm in the behavioral sense*.

It appears that when one says a certain, specified norm exists in a particular society, he usually means norm in one of these three senses. Sometimes, however, he may have in mind a more *complex sense*, viz. he wants to state that in a given community there is an adequacy of the norm in more than one sense. The external expectation and prescription is in agreement with the internal motive, or the motive with behavior, or the behavior with command; or it may even be that there is a full adequacy of the norm in all three of the above-named meanings. Thus in speaking of the existing norm, we may have in mind

(iv) the adequacy of the norm in two or all three of these meanings. I would call this a *norm in the complex sense*. Depending on which phenomena constitute the complex, we would also be able to speak of norm in the coercive-behavioral sense, the motivational-behavioral sense, etc.

Let us now proceed to a more detailed analysis of the meanings of these terms, whereby I shall propose a typology of matrix regularities in the action of norms in each of the meanings.

5. NORM IN THE COERCIVE SENSE

The use of the term 'norm' in its coercive sense seems to be most frequent in the social sciences. According to this meaning, to say that norm N exists in a given society is to say that a relatively constant command or expectation or prescription of contents N is being directed toward the individuals of a certain category, aimed at molding their behavior in a way B_N.

Thus statements concerning the contents of norms in the coercive sense, and characterizing the conditions of efficiency of operation of social pressures and social control. In sociological literature the word 'prescription' is very often replaced by some other more or less synonymous terms such as e.g.: order, demand, requirement, expectation, etc. All of these statements can be presented in any of the following formulas: 'you should B_N', 'it is an imperative that you B_N', 'you ought to B_N', etc.

Relations between external expections and the behavior they are able to enforce can be studied at the strictly behavioral level: we simply try to find out what kind of expectations of which 'norm-senders' can under these conditions be isomorphically related to the behaviors of the 'norm-recei-vers'. But we usually obtain a much clearer picture of the mechanism of the action of norms in the coercive sense when we look at the nature of the mediating psychological processes involded. One such approach is to look at the *reasons* for the conformity to external normative pressures, because when people obey commands of this kind they may do it for a variety of different reasons:

(i) on the ground of their conviction that they should obey the comman-ding person, for he or she has the right to give them orders, which is to say on account of the *authority of the norm-sender;*

(ii) on account of the conviction that it is much better to obey the person giving orders because otherwise you may get into serious trouble, which is to say they wish to avoid sanctions safeguarding the norms or they expect certain *rewards* for their obedience.

(iii) on the ground of the conviction that the norm is *right* and *just* is in agreement with the rest of their belief system, which is to say they believe in the rightness and *fairness of the norm itself.*

(iv) because the behavior demanded by the given norm somehow satisfies certain of their *needs*. These needs may sometimes be quite conscious and the person knows why he conforms to the expectations of others; or he may be quite unaware of these particular needs and therefore unable to explain why the external expectations are so appealing to him. Whenever the conformer is unaware of his real reasons, he may accept (i), (ii) or (iii) as rationalizations of his obedience to the given norm.

It should be mentioned here that in these four reasons for obeying coercive norms I have distinguished types of regularities which are studied by quite different social or psychological theories. Some special personality traits and certain situational conditions, for instance, will play an essential role in securing obedience to an order when this order has been reinforced by the authority of the norm-sender. Other personality factors will come into the picture when the norm is accepted as valid and in force owing to safeguarding sanctions and rewards, and still different factors may come into operation when the main element of efficacy of the given order is its congruity whith already present opinions of the norm-receiver or his conscious or unconscious needs.

Speaking in most general terms, the matrix of the contents B_N may be involved in regularities described by various theories in which any number of factors may appear as accompanying conditions. Moreover, a basic condition for the efficacy of operation of a definite matrix regularity, in which the norm N_1 operating in a given society is involved, is the existence and the direction of operation of *other normative pressures*, their concordance, indifference or counteraction with regard to the functioning of our norm N_1.

Any individual in any society that has a relatively complex structure may face the situation of normative cross-pressures; other groups or individual 'norm-senders' may try to shape his behavior in a way incompatible with N_1 or contrary to it. The acknowledged authority of the competing 'senders', the sanctions they can administer, and the relation of the contents of their expectations to the individual's normative convictions and more personal needs all play a decisive role here.[11]

6. NORM IN THE MOTIVATIONAL SENSE

In saying that in a particular community the norm N exists, or that for a given community the norm N is characteristic, we sometimes mean the 'normative conviction' that 'I ought to behave in the way B_N'. In accord with this meaning, which I have called 'norm in the motivational sense', people 'address' the content of the norm to themselves and to their own behavior. Thus the norms *we accept* as just and valid determine our actions on the ground of certain motivational mechanism which make us to obey our own convictions how we ought to behave. Therefore the theory which will relate the influence of normative matrices upon subsequent behaviors will be a part of more general theory of motivation. Here the problem arises: Why should we distinguish from the whole set of motivational factors of human activities only those motives that are of a 'normative' nature, i.e. norms in the motivational sense? Would it not be better to search for regularities on an even higher level of generality, to tend to the theory of the efficacy of human motives in general, seeking conditions unfar which the motives are turned into their corresponding behaviors?

Beyond any doubt, at least some of the statements of the theory of motivation will turn out to be valid as to both normative and nonnormative motives, and looking at the problem from this standpoint the normative motives will by no means be specific ones. But I think that at least a part of the regularities will be rather specific for those, and only those, motives that apart from whether they have arisen as a result of the socializing processes of external normative coercions or whether they have been formed by the given person in the course of its own reflection, are marked by a sort of emotional loading connected with a *sense of duty*.[12] There will be some specific external conditions under which people will resign from fulfilling the norm but not from pursuing their non-normative as a corollary 'wishes', and various individual types will more or less easily yield in one or another area. Some writers have also stressed that there may be also different effects for the individual depending on whether he disobeys to his normative conviction, or he resign from satisfaction of his 'wishes'. In the first case his 'punishment' will be the feeling of guilt with all its implications for his self-esteeme etc. In the other case the main psychological effect will be the feeling of frustration with all its known consequences. All this seems to suggest that normative motives deserve special attention within the broader theory of motivation.

7. NORM IN THE BEHAVIORAL SENSE

A third meaning of the statement 'the norm N exists in society S'. I have called 'norm in the behavioral sense'. In saying that certain norms exist in a given society, we often have in mind neither external prescriptions reinforced by sanctions or authority nor 'normative convictions' but simply patterns of *actual human behavior*, the sheer fact that within some particular community in certain situations people behave alike or at least in a sufficiently similar manner that we can speak of uniformities of behavior. According to this meaning we can say that there exists in Poland the norm of kissing ladies' hands when we want to state that most Poles do actually do this. Similarly we could say that there exists in France the norm of drinking wine with dinner, and in Italy that of eating a great amount of spaghetti. The description of culture as a system of norms very often constitutes an account of the *manners of behavior* common for most members of a particular community and which sometimes are characteristic for that community only.

But it is not sufficient to state that the norm is here equivalent to a certain uniformity of human behavior. We would not rank among social norms the fact people assume the vertical position when they walk or that they are omnivorous creatures, for these things are *biological regularities* characterizing the human species. We would not reckon either among 'normative uniformities of behaviors' quite a number of behaviors of a clearly social nature. We would not number among the 'norms' characterizing the groups located at the bottom of the social ladder typical for these groups low self-esteeme, or common occurrence of acts of agression by means of which people in general react to frustrating situations, because these behaviors are the effects of regularities of non-matrix character, although undoubtfully produced by the social structure.

I would say that when we assume the existence of norms in the behavioral sense we mean a relatively uniform pattern of behavior that in addition is isomorphic with norms in the coercive or the motivational sense, or both. Therefore norms in the behavioral sense are those uniformities of behavior in a given group which either are required from the group's members or are accepted by them as 'right', or both.

When we use the concept of norm in the behavioral sense as the *antecedent of a regularity*, the consequent of which is in the behavior isomorphic to it, we must limit the focus of our attention in such a way that we *avoid an obvious tautology*. Assuming that here norm means 'the pattern of behavior characteristicfor most people in the group' (i.e. the modal pattern of behavior) it should be pointed out that above all we must avoid any formulation of our theoretical 'matrix hypotheses' in which *the same pattern of behavior of the same people* would appear in both the antecedent and the consequent. The fact that Poles kiss ladies' hands should not be explained by the fact that Poles act this way. Such tautologies can be avoided if the dependent variable of our hypothesis concerns the patterns of behavior of people *other than those whose behavior is described by the independent variable* of the hypothesis. The dependent variable will then concern:

(i) behavior of persons who have been members of the group for quite a long time but so far have behaved themselves in a manner inconsistent with the group's pattern — thus our hypothesis would constitute a peculiar case of a somewhat more general hypothesis reading: 'modal patterns of behavior tend toward being turned into general patterns'; or

(ii) newcomers who have joined the group either those who have entered it due to social mobility, or new generations which, in the case of the more durable groups, usually conform in the result of socialization, to characteristic patterns of behavior.

In both these cases we have the relationship of isomorphism between the pattern of behavior of a majority of the group members and the subsequent behavior of those who did not behave in the 'modal' way before. But the norm in the behavioral sense may also refer to the patterns of behaviors of a certain *minority* in the given society, if these patterns, due either to the special prestige of the 'norm-sending group' or due to some other mechanisms may influence and transform the behaviors of the majority.

Many of these mechanisms were discussed already. But one other mechanism not mentioned above seems also to play quite an important role in the process of conformance of behavior to certain modal patterns or to certain important 'normative models': the 'imitation' of behavior in the absence either of external pressures or of internal convictions to support it. Contemporary learning theory with its sheers upon the role of 'models' in the learning process seems to have rediscovered the importance of the processes of imitation, described by Gabriel Tarde many decades ago.

8. NORM IN THE COMPLEX SENSE AND PATTERNS OF NORMATIVE INTEGRATION

I have attempted to indicate various meanings of the term 'social norm' that are specific enough to enable us to interpret statements on the efficiency of their operations within the context of certain more general theories of sociology or social psychology. But when sociologists find within a particular community a set of norms of definite content they very often understand the term 'norm' in such a way that it includes more than one category as listed above. When we say that within our civilization there exists the norm 'thou shalt not kill', we have in mind not only that the command (upheld by the appropriate sanctions) has been directed to the

members of the community by the surrounding society and by its various
specialized agencies exercising social control of behavior, but also that on a
large scale the content of this command has been recognized as being right
by almost all people, and that the overwhelming majority behave in com-
pliance with that norm.

Whenever the sociologist or anthropologist in speaking of the existence
of a social norm means to include more than one of these categories as being
related to the content of this particular norm we shall use the term 'norm in
the complex sense'. Depending on the type of phenomena included in the
normative complex, we might speak either of a coercive-motivational
complex, a motivational-behavioral complex, or a coercive-behavioral complex, or finally a coercive-motivational-behavioral complex.

Such a complex may refer only to the joint existence of external
pressures, internal convictions and behavior corresponding to them. But
when describing it we may also be interested in the *direction of causal
relations* between the components of the given 'normative syndrome'. In
respect to norms in the coercive sense and to the problem of their efficacy
we have been concerned with the conditions under which they turn into
motivational norms. In respect to norms in the motivational sense we have
assumed that the phenomenon of their being transmuted into human
behavior is by no means a rare event, although we cannot characterize
adequately the necessary conditions. Thus a normative syndrome might, for
example, be conceived as a causal chain of the following structure:

Norms in the sense of
coercive   →   motivational   →   behavioral.

At the present stage of social science, to be sure, the relations existing
between the links of this chain are at best of a statistical nature, and usually
we can formulate only existential propositions about them since we are not
able to spell out the conditions which together with our matrices would
determine the occurrence of the next lin link, or at least entail its occurrence with a certain probability. Even so, the matrix relations existing between
particular elements of the normative complex do not necessarily have to
proceed in accordance with the scheme shown above. First, there can be
direct relations between the coercive and behavioral components of this
syndrome, as happens whenever we obey a command or conform to an

expectation without having faith in its 'fairness' and only because we wish
to avoid sanctions for its non-observance. But what is no less important, it is
also possible for the causal relations to run in the *opposite direction* from
that indicated above. If a particular norm has become the motive of human
behavior on a mass scale — i.e. if people believe they themselves ought to
behave in this particular way — it is highly probable that this will result in
others being coerced into an analogous pattern of behavior, for people
usually don't like others to behave in a way inconsistent with their own
values. It is also well-known that a habitual pattern of behavior or external
conformity to the expectations of others may turn into conviction of the
rightness of this behavior; in other words, produce a norm in the motivational sense.

Thus the *potential scheme* for all possible connections within a norm in
the complex sense should be designed:

Norm in the sense of

It is potential because it takes into account all the possible combinations of
causal connections within the normative complex. An appropriate theoretical analysis of a given normative situation might bring us to the conclusion
that:

(i) the given normative syndrome is composed of fewer than the three
postulated elements; and,

(ii) moreover, apart from how many elements are present, the system of
causal interrelations represents only some of the possibilities listed above.

In examining the set of theoretical problems related to norm in the
complex sense I have dealt only with those regularities that would explain
for us the structure of a relevant complex of a particular norm: its completeness, and assessment of causal influences between elements.

If we say that a certain 'normative syndrome' is *incomplete* with respect
to the norm N_1 we mean either

(a) that norm N_1 is represented in the given society only by certain external pressures and possibly by behaviors conforming to them, and that there are no normative convictions in the minds of the group members as to how they ought to behave in situations to which N_1 applies;

(b) or that at the same time there exists another norm in the minds of the group members that they *ought not* to behave in the prescribed way.

The same may apply to all other possible *inconsistencies* within the normative syndrome. Thus we may say for example that the memebers of a certain group in general believe that they ought to behave in accordance with a certain norm N_1, and send similar expectations towards the behavior of others, but it turns out, that the real behavior of the members of this group is either 'random' with respect to their norm in coercive and motivational sense, or clearly opposite to it. Finally it may happen, that the behaviors are on the mass scale consistent with the normative convictions of the group members, but they either don't take into account the expectations of some — otherwise 'legitimate' norm — senders or are clearly opposed to these expectations.

These two situations — where one of the elements of the normative syndrome is simply *missing* and there is — instead of it — an opposite norm in the given sense — should be conceptually distinguished, because the first refers only to a certain *incompleteness* of the normative syndrome, while the other refers to an internal *inconsistency*, to what might be called the disintegration of the normative syndrome.

When we say that in a given group the norm N_1 exists in the complex sense, we mean that the pressures, the motives and the behaviors of most people in that group correspond to the matrix N_1. But this is only one of many possible situations. From the standpoint of a certain matrix-pattern of behavior with the content N_1 in a given society:

(i) the norm N_1 may exist in the form of an external pressure, or internal motive or overt common behavior — corresponding in its shape to the matrix N_1;

(ii) one or more elements of this syndrome — i.e. pressures, motives, or behaviors — may be shaped by a matrix which in its content is directly or indirectly contrary or contradictory to the norm N_1, a situation we can designate as non-N_1;

(iii) the norm N_1 may be non-existent in one or more senses but we cannot state the existence of 'norm non-N_1', a situation we can designate as N_0.

All of these situations are presented in Table II. A full 'normative integration', in other words the existence of norm N_1 in the complex sense, appears in the table at cell number 1, when the full 'normative integration' with respect to the norm non-N_1 is in the cell 27. But the other 25 cells present quite a range of possibilities for theoretical analysis. Thus cell 5 (N_1 – N_o, N_o) seems to represent a classic case of complete lack of socialization with respect to N_1, while cell 7 signifies 'obedience to overt pressure with internal opposition to that pressure'. Cell 19 stands for 'obedience to the voice of conscience against external pressures', and so on.

The scheme of Table II presents only a *typology* of patterns of integration or normative disintegration. But we may also ask certain questions concerning the *dynamics of such syndromes* and especially the *transformation of one of these patterns into another*. These questions refer both to

TABLE II

Norms in the coercive sense	Patterns of normative integration and desintegration			
	Norms in the motivational sence	Norms in the behavioral sense		
		N_1	N_o	non-N_1
N_1	N_1	1	2	3
	N_o	4	5	6
	non-N_1	7	8	9
N_o	N_1	10	11	12
	N_o	13	14	15
	non-N_1	16	17	18
non-N_1	N_1	19	20	21
	N_o	22	23	24
	non-N_1	25	26	27

the conditions under which previously well-integrated normative patterns lose one or more of their components and which of them are most likely to start the process of normative disintegration, and to the conditions under which the incomplete or internally inconsistent configuration tends to restore its full consistency around either N_1 or non-N_1. There are many

interesting theoretical comments in sociological theory which are pertinent
to this problem, and the typology proposed here might well be used for
their more coherent systematization as well as for the formulation of some
new problems.

The problem becomes even more interesting if we follow R.K. Merton's
approach[13] and — within the given area of reality to which the norm applies
— look separately at the norms specifying certain *goals* and those which
prescribe the means toward realization of these goals. Accordingly, we
should apply the above typology separately to the goals and to the means,
and classify our social types taking into account their places *in both these
classificatory schemes*. Then we can say — taking as an example one type
from Merton's typology — that 'innovators' would be in cell 1 with respect
to goals but in cell 9 with respect to means, and can classify other of
Merton's types accordingly. But here again by strictly combinatorial analysis
we would probably find many other types deserving theoreticians' attention.

9. MATRIX LAWS AND MACROSOCIOLOGICAL THEORY

The notion of matrix law relating certain normative regulators of human
behavior to their isomorphic behavioral consequences was proposed in this
paper as the solution to an obvious 'dualism' in our approach to the
explanation and prediction of social behaviors. But at the same time we
could see that, depending upon our understanding of the term 'norm', the
matrix generalizations relating human behavior to 'norm' in the given sense
should be treated as elements of quite different, more inclusive, substantive
theories. The main point of this paper was to point out that *within the
existing ways of theorizing about the phenomenon of conformity of human
behavior to social norms, one should take the content of the corresponding
normative regulators much more into account than is usually the case*.
Otherwise we formulate in theories of 'socialization', 'social control', or
'human motivation' propositions which are obviously both too general and
too specific. They are too general because they often refer to *any* normative
antecedent, whereas under the conditions specified by the given theory they
should apply only to *some* of them. On the other hand, they are too specific
because they do not take into account that with respect to norms of the

given shape, different conditions may constitute the 'functional equivalents' for their efficient action.

To systematize these propositions within different approaches to theory construction in the behavioral sciences may be of course an important goal in itself. But one should also point out the usefulness of such generalizations in explaining the phenomenon of the existence of complex normative syndromes, discussed in the last section of this paper, and also *inexplaining the functioning of whole societies in all their complexity*.

When regarded from a special perspective the functioning of each society may be to a high degree described in terms of the efficient action of a complex configuration of normative patterns which either support each other in their action or at least do not act against each other. Some of these normative patterns are traditionally classified as belonging to the 'culture' of the studied population; others, defining the social roles of the society's members, the relative positions of different groups, and the mutual relations among these groups usually characterizing 'social structure'.

We can *describe* the culture and structure of the given society by concentrating on its normative regulators and assuming that these regulators operate more or less successfully. We can also explain the existence of some of these regulators at the 'macrosociological' level by referring to their *functions* for other normative regulators. The Marxist theory of ideology points out that many ideological norms can be explained in terms of their *instrumentality* for those norms which specify more basic features of the social structure. Similar kinds of explanations can also be found in many 'functionalist' writings.

Nevertheless it seems that relatively little attention has been paid in social theories to the problem of *how these complex normative configurations are maintained at the level of behaviors of the members of the given society*. Of course every student in a course of elementary sociology is taugh that a society and its structure and cultural values functions and persists due to the processes of socialization and social control; but if we agree with what was said above about the obvious incompleteness of these 'theories', then we have to admit that the *basic mechanisms through which societies exist and persist are still rather unclear to us*. The attention of theorists has been focused on the mechanisms of *deviance* from the prescribed patterns of behavior, rather than on those mechanisms through which the society functions in a 'normal way'. Therefore theories of 'abnormal' social behavior seem to be much more developed than theories of 'normal' behavior explaining why a society functions the way most of its members wish to function. A detailed theory which could explain the normal functioning of societies by as complete as possible a specification of mechanisms at the behavioral level which are involved in this seems still to be rather far in the future.

One can point to several reasons for the necessity of such a theory. One is of course strictly theoretical: We should try to know as much as we can about our societies. But more practical reasons can also be cited. Suppose we observe the spontaneous development of new processes or that we would like to change some elements in our society, but at the same time we would like to preserve certain of its basic values and norms or certain elements of its social structure. As long as we do not know whether the spontaneous or planned change will eliminate certain conditions necessary for the functioning of these norms, or will bring into existence their functional equivalents, we are unable to predict anything. Similar arguments for the need of such theories may be rooted in quite different value orientations. Suppose we would like to create a basically new society with altogether different value orientations for its members, with some basically new structural features and patterns of mutual obligations. It may turn out that the planned social structure is under the conditions in which we plan to realize it, theoretically impossible. In any case, knowledge of the proper matrix laws relating these norms to the conditions of their efficient operation would permit us to design a detailed 'blueprint' for social action based on a sound social theory, or to choose the best of several possible alternatives for this action.

I wouldn't like to be understood as proposing to *replace* macrosociological theories describing the relations between the macroparameters of societies and their dynamic interrelationships by a microtheory analyzing them at the strictly behavioral level. As I point out in another chapter in this volume,[14] each level of theorizing is necessary and theoretically 'legitimate'. What I would only like to stress here is the *necessity of interrelating the theories from the different levels* in such a way that each can check the validity of the other, each revealing the weaknesses of the other and delivering the premises for its improvement.

In order to have a proper frame of *reductive reference* for a macrosociology dealing with the existence and functioning and the dynamics of societies as seen from their normative side, we need proper and as complete as possible theories of normative control of individual human behavior. On the other hand, when developing such theories the more we take into account all the regularities heretofore discovered at the macro-level, the more efficient our theory will be explaining both the phenomena of individual conformity (or deviance) to social norms and in the reductive explanation of the laws of functioning of societies in all their complexity.

NOTES

* Modified version of a paper presented at the VIth World World Congress of Sociology, Evian, 1966. The original paper was published under the title 'Cultural Norms as Elements of Prognostic and Explanatory Models in Sociological Theory' in *The Polish Sociological Bulletin*, 2(14), 1966.

[1] For a discussion of functions of 'historical generalizations' in social theories, see Chapter IV.

[2] See C.G. Hempel, *Aspects of Scientific Explanation*, New York, 1965.

[3] See C.G. Hempel, *op.cit.*

[4] For an analysis of this kind of interpretations of certain sequences as causal, see M. Schlick, 'Causality in Everyday Life and in Recent Science', in: H. Feigl and W. Sellars (eds.), *Readings on Philosophical Analysis*, New York, 1959.

[5] On the other hand, it is obvious that the relation between the parent's genotype and the genotype of the child is a relation of partial isomorphism only and its strength depends upon the number of genes the parent and the child have in common.

[6] However, if we find *ex post* the relation of isomorphism between a certain normative pattern existing in the given society and the behavior of a person, we have quite a strong argument that this relation is not 'accidental', even if we don't know the proper matrix law. The more complex are two isomorphic patterns the less likely it is that they occurred together in all their complexity at random. The degree of complexity of patterns related to each other isomorphically plays here role similar to that of the number of instances in the case of regular inductive analysis. But it may be used only as support of the proposition that the isomorphic relation is nonaccidental; it cannot take the place of the formulation of the complete law of isomorphism specifying the conditions necessary for the given matrix regularity.

[7] Chapter I.

[8] J.L. Gillin and J.P. Gillin, *Introduction to Sociology*, New York, 1942, quoted from L.A. Kroeber and C. Kluckhohn, *Culture, a Critical Review of Concepts and Definitions*, Cambridge, Mass., 1952, p. , .

[9] C. Kluckhohn and W.H. Kelly, 'The Concept of Culture', in R. Linton (ed.) *The Science of Man in the World Crisis*, New York, 1945, quoted from Kroeber and Kluckhohn, *op. cit.*, p. 50.

[10] M. Ossowska, 'Fictitious Beings in Sociological Definitions', *The Polish Sociological Bulletin* 1-2, 1961.

[11] For an analysis of the mechanisms of interaction of the given norm with other norms directed toward the same 'norm-receiver' see R.K. Merton's discussion of 'role-set' and other patterns of multiple group membership in 'Continuities in the Theory of Reference Groups and Social Structure', in: R.K. Merton, *Social Theory and Social Structure*, New York, 1968.

[12] For an extended analysis of the role of norms in the motivational sense in the control of human behavior, see J. Aronfreed, *Conduct and Conscience, The Socialization of Internalized Control over Behavior*, New York, 1968.

[13] R.K. Merton, 'Social Structure and Anomie', in: *Social Theory and Social Structure,* New York, 1968.

[14] See Chapter XI.

BIBLIOGRAPHY

Aronfreed, J., *Conduct and Conscience, The Socialization of Internalized Social Control Over Behavior*, New York, 1968.

Hempel, C.G., *Aspects of Scientific Explanation*, New York, 1965.

Kluckhohn and W.H. Kelly, *Culture – A Critical Review of Concepts and Definitions*, Cambridge, Mass., 1952.

Merton, R.K., 'Continuities in Theory of Reference Groups and social Structure', in: *Social Theory and Social Structure*, New York, 1968.

Merton, R.K., 'Social Structure and Anomie' in: *Social Theory and Social Structure*, New York, 1968.

Nowak, S., 'Concepts and Indicators in Humanistic Sociology', in this volume, p. 1.

Nowak, S., 'The Logic of Reductive Systematizations of Social and Behavioral Theories', in this volume, p. 376.

Ossowska, M., 'Fictitious Beings in Sociological Definitions', *The Polish Sociological Bulletin* 1-2, 1961.

Schlick, M., 'Causality in Everyday Life and in Recent Science', in: H. Feigl and W. Sellars (eds.), *Readings in Philosophical Analysis,* New York, 1949.

CHAPTER X

ROLE AND LIMITS OF THE 'FUNCTIONAL APPROACH'
IN FORMULATION OF THEORIES OF ATTITUDES*

1. INTERNAL AND EXTERNAL FUNCTIONALITY
OF ATTITUDES

In his survey of various theories of attitudes, Daniel Katz[1] classified the theories into two groups. One group comprises those assuming that man is a *rational being* who thinks and considers thoughtfully what is around him so as to behave accordingly. The others do not assume rationality as the basic machanism of attitude formation, but rather tend to account for the shaping of attitudes in terms of such mechanisms as conditioning, or strivings towards homeostasis, or cognitive consonance, etc. Katz remarks pointedly:

The major difficulty with these conflicting approaches is their lack of specification of condition under which men do act as theory would predict. For the facts are that people do act at times as if they had been decorticated and at times with intelligence and comprehension. And people themselves do recognize they have behaved blindly, impulsively and thoughtlessly.[2]

According to Katz, this conflict of contradictory models of human personality can be reconciled for the purpose of attitude theory 'in a functional approach', by which he means "primary attention to the motivational bases of attitude and attitude change".[3] He sees three basic reasons for the superiority of such an approach. It permits us to understand that 'the same' external situation may have different functions for different individuals, producing different kinds of attitudes. Secondly, it reminds us that there are different sources for the same attitude in different persons' minds. Finally, it enables us to specify the conditions under which the predictions formulated by different attitude theories will hold true.

Stated simply, the functional approach is the attempt to understand the reasons people hold the attitudes they do. The reasons are, however, at the level of psychological motivations and not of the accidents of external events and circumstances.[4]

In this chapter I shall analyze the structure of reasoning behind the 'functional approach' to attitude in order to evaluate its validity for the formation of an empirically testable and predictively useful theory of attitudes. But before I do this, let me specify what will be understood by 'attitude'. It would be useless to attempt even a brief review of the various definitions of this concept since the time W.I. Thomas and F. Znaniecki defined it[5] in their study of the relation between human beings and those social objects that constitute a certain 'value' for them. Many such historical reviews can be found in the literature.[6] I would like to stress only that these definitions also can be classified into two categories. The first, starting with the more 'behaviorist' assumptions, treats attitudes either as strictly behavioral, internally consistent syndromes, or at the least as a set of predispositions to such behavioral syndromes, without specifying the psychological character of these predispositions in terms of certain 'mental states and processes'.[7] The second, represented among others by Thomas and Znaniecki in their definition of attitude, refers primarily or exclusively to mental prosesses and states as determinants of the given consistency of the behavioral patterns.

For reasons I have explained in another chapter of this volume[8] the second approach to the definition of most sociological concepts seems to be more theoretically fruitful. But even within this tradition, which seems more and more popular in recent times, we can find two different orientations or approaches toward the definition of attitude. The earlier approaches, fascinated by the newly discovered techniques of measurement of attitudes, tried to define the concept as a *unidimensional construct* with the main stress upon the 'positiveness vs negativeness' of attitudes.[9] The other orientation assumes that attitudes are *complex phenomena, with different components* and with possibly different structural patterns of relations between these components. This implies that attitudes can be classified in a multidimensional analytical scheme, which would take into account both their *evaluative aspects* and their *cognitive aspects*, and finally their *behavioral components* (understood either as actual behavior toward objects of the attitude or as a predisposition to such behavior). These three components of attitudes, distinguished by M. Brewster Smith about thirty years ago,[10] seem to be accepted by most theoreticians of attitudes who accept the non-behaviorist approach to the study of attitudes.[11]

Accepting the structural approach to the definition of attitude as well as the three components of attitudes distinguished by Smith, I would like to propose the following formulation as a definition of attitude.

The attitude of a certain person toward a certain object (or class of objects) is the totality of relatively enduring predispositions of this person toward evaluating this object in the given way or experiencing certain emotions in connection with this object, which may additionally be accompanied by certain more or less articulated knowledge about this object and a certain relatively stable predisposition to behave toward it in the given way.

As we see, the above definition takes the evaluative-emotional component of attitudes as the *definitional necessity of their existence*, but admits that this component *may* also be accompanied by a more or less developed *cognitive component* and by a more or less precise *program of behavior* toward the object of the attitude. Since these latter two components are not necessary but only possible, we will have four different kinds of attitudes: (1) those having only the evaluative-emotional component (which Katz and Stotland propose to call 'affective associations'[12]); (2) those having affective and cognitive components but lacking the behavioral one; (3) those having affective and behavioral, but not cognitive, components; (4) those in which all three components are present and more or less clearly distinguishable. In this chapter I intend to discuss only the last category of attitudes, i.e. the most complex ones, assuming that my comments will in the main also be valid (in more simplified form) for those attitudes in which one or both of the definitionally non-essential components are missing.

Now let us see what it might mean when we say that attitudes are *functional* or – as I prefer to say – that they are *instrumental*. It seems we can distinguish at least two different kinds of such functionality or instrumentality. First, attitudes may be instrumental in the strictly *internal* sense; that is, some of their components are functionally adapted to 'serving' the other components, or in other words they are instrumental to the other components. To give a simple example: I know that Ingmar Bergman produces films of a certain kind (knowledge) which I like very much (evaluation), and I 'adjust' the behavioral component of my attitude toward Bergman's films by being ready to see them as often as possible. In this kind of instrumentality the behavioral disposition 'serves' the other two components of my attitude toward this class of object *without necessarily involving in the whole causal mechanism any other phenomena which do not belong to my attitude* toward Bergman's films. In another case: I may learn by my own experience that 'dogs bite', and adjust both my evaluation of dogs and my behavior toward them in a way that will be instrumental to my knowledge of this empirical fact, and my evaluation of it.

When by attitude we understand a complex 'organization of beliefs' concerning a certain object, then the instrumentality is internal, whereby *some of the beliefs are able to create, to change others* so that there will be as little 'cognitive dissonance'[13] among them as possible or so they will be internally consistent.[14] In all these cases our field of theoretical interest may stay within the area of phenomena belonging to attitude only, but if we do try to discover external factors which may influence the internal consistency of attitudes or the instrumental relations between their components we do not assume the relation of instrumentality between them. For suppose that we understand the elimination of 'cognitive dissonance' to mean eliminating the logical inconsistency between beliefs. There is no doubt that such a predisposition as 'intelligence' plays a role in this process, but we would not say that the process of eliminating dissonance (in the specified sense of this term) is instrumental to human intelligence — we would rather say that intelligence is instrumental to reducing dissonance. This is what I mean by saying that the mechanisms of instrumentality are limited to the internal relations between the components of the attitude.

In speaking of the *external instrumentality* of attitudes I have in mind a category of situations in which the attitude as a whole 'serves' something that is *conceptually external to it*. We might say, for example, that a certain political attitude we often encounter among the very wealthy in a given society *is instrumental to their class interest*: it gives them a rationalization for their privilege, or may make them work even harder toward increasing their wealth. We might say that a racist attitude is instrumental to the needs of persons who are characterized by 'authoritarian personality' as understood by Adorno and others, meaning that the possession of this set of attitudes satisfies some (usually unconsious) personality needs of these persons and especially the need to reduce their anxiety in their confrontation with the social world.

In the above instances the *criterion of instrumentality* lies outside the area of the given attitude, *inhering in the psychological or social needs of the given person*. These needs are usually more general than the attitude itself, and from person to person may be (at least potentially) satisfied by different attitudes as well as by different patterns of behaviors more or less related to these attitudes. The notion of instrumentality (functionality) discussed by Katz in the paper cited above belongs to this category, i.e. he is interested in the external instrumentality of attitudes and treats their internal structure as the function of their external instrumentality. He emphasizes that the functional approach is a search for psychological motives related to definite human needs. Apparently, what is looked for are not momentary states of motivation, attitudes usually being conceived as more or less stable, but rather certain more continuous motivational structures determining the formation of attitudes and their continuity for some reasonable period of time.

In considering such stable motivational structures, what should be stressed is that we have to understand the needs as individual predispositions of persons, but not necessarily only those which are discussed and distinguished in psychological theories of personality. From the standpoint of the problems discussed here, *needs of a definitely social origin* and corresponding to the social values or social norms accepted by the given person, or needs related to his social position, social role, etc., can play the same role in the formation of attitudes as 'strictly psychological' needs: they can shape human attitudes and behavior in such a way as to make them instrumental − functional − to the social needs of the individual, contributing to the satisfaction of these needs in exactly the same way they do to the satisfaction of his 'personality needs'. Thus it is sufficient to assume here that needs and values are manifested in certain relatively stable motivational mechanisms which in turn determine the formation of attitude.

2. THEORY OR A HEURISTIC DIRECTIVE

The basic assumption underlying the approach advocated by Katz − as well as by many other authors cited by him − apparently is the belief that human behavior is *determined* by motives, by needs and values, both conscious and unconscious, and that the processes of attitude formation are a peculiar class of such behavior. If we want to explain the origin of attitudes, we should try to define the needs or values with respect to which the attitudes are functional. But then the question arises: How can such a 'principle of instrumentality' serve as a directive for constructing a theory of attitudes?

It can be remarked that the principle of instrumentality in its general formulation as given above fails to be a theoretical proposition, or more strictly that it fails to constitute a law of science, for we can hardly accept as a scientific law the proposition that attitudes, consciously or not, serve to satisfy certain needs or values — in other words, that the former are *functional* to the latter. Such a proposition can be, at best, a *heuristic directive* of research, postulating that if we analyze attitudes having certain specified contents, *we should seek to determine their functions* as related to personal or social needs and strivings. If and when we discover the 'functional relations' between some specific need and some concrete attitude, then we may claim to have discovered a law of science.

The heuristic directive makes the researcher sensitive too a certain type of problem, as it postulates a search for functional references to the analyzed attitudes, but it remains to be found toward what (if anything) the given attitude is functional. Theoretical statements concerning the functions of attitudes must be formulated on as concrete a level as possible.

A double typology of such statements is possible, one to deal with the *type of needs* (motives) satisfied by the analyzed attitudes and the other with the *types of attitudes* we attempt to explain. That is to say, our typology can be based on either the independent or the dependent variables of a 'functional theory of attitudes'. In his paper on the functional approach, Katz proposed an initial typology of the relevant mechanisms with respect to the character of needs that can be satisfied by attitudes, discussing briefly the following:

(1) the adjustment function of attitudes
(2) the ego-defense function
(3) the value expressive function
(4) the knowledge function.[15]

Obviously, we could as well attempt a typology of mechanisms based on a *substantive classification of attitudes*, classifying them with respect to their content and their positive and negative signs and trying to determine what sorts of functions — and under what conditions — can be served by positive or negative attitudes toward religion, toward certain features of the socio-political system, toward particular people and institutions, etc.

We would expect from a functional theory of attitudes that it be a typological arrangement of propositions alternatively relating to each other various motives, needs or values on the one hand, and on the other various attitudes concretely defined by their contents so far as possible, and with appropriate situational and psychological modifiers, determining which of the alternative connections are realized in the given conditions.

3. THE DANGER OF TELEOLOGISM

The utility of the heuristic directive depends, in the first place and quite obviously, on the soundness and the degree of universal validity of the principle of instrumentality, or functionality, of attitudes. But however this may be, a number of more general problems arise relating to the application of functional thinking in science. For in the functional frame of reference it is assumed not only that attitudes are subordinated to needs, values, and strivings, but moreover that by understanding the function of an attitude, by identifying the need to which it is instrumental, we are able to apprehend certain *causal mechanisms and thereby to explain the origin of the studied attitudes*. It is assumed that the need to which the attitude is instrumentally subordinated is an essential factor conditioning the presence of the attituda in the mind of an individual, and thus that need is an important *cause of attitude*. But such a manner of explaining the origin of attitudes may become teleological, and several traps lie in wait which are by no means always safely avoided by those who make use of the functional, and particularly of the teleological, type of explanatory pattern. We shall now consider some of those traps.[16]

The first of the lurking dangers is that of explaining the cause in terms of its effect, i.e. explaining an earlier event in terms of a subsequent one. We would be committing this fallacy if we explained the origin of attitudes by the fact that they *satisfy* specific needs, since the fact of satisfying a need is ipso facto subsequent to the existence of that need. Thus it must be made clear that when we explain the origin of an attitude, we do it by reference not to the satisfaction of a need, but to a fact that is prior to the emergence of the attitude, i.e. to the *existence* of a need that demands satisfaction. Since the need referred to precedes in time the attitude which is instrumental to it, we thereby avoid the first and most dangerous trap of teleologism. The existence of attitudes is explained by the existence of needs, and the next link in our chain of cause and effect is the satisfaction of the need by the attitude, in accordance with the scheme:

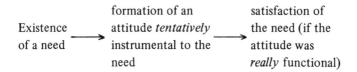

	formation of an	satisfaction of
Existence	attitude *tentatively*	the need (if the
of a need	instrumental to the	attitude was
	need	*really* functional)

This scheme is applicable for a certain special category of attitudes which are with respect to objects satisfying our *longitudinal needs* and which elicit behavior that cannot be evinced repeatedly (like for instance our attitude toward a change in the political system). In other situations, where the object of the attitude appears fairly often within the area of our experience and our behavior toward it may be repeated many times, the scheme of functional determination usually looks different. On the basis of our needs we form a kind of 'tentative attitude' toward the object and then acting in accord with this 'attitude' we try to ascertain whether it really leads to satisfaction of the given need, and we adjust our initial attitude according to the relevant experience. In such case one could say in a special sence that the satisfaction of the need determines or modifies the structure of the attitude, *without assuming that the effect determines the cause.* We need only to be clearly aware of the *time sequence* of the events and to understand that our *past* satisfactions of needs may determine our *future* attitudes.

However, if we are to recognize the existence of a need or its satisfaction as causing the emergence of a phenomenon which is functional with respect to it — in this case, an attitude instrumental for the need — we have to assume besides that the whole system in which the need is present functions in a manner that is *purposeful* from the standpoint of that need. Purposeful systems can be found both in nature and among man-made objects (like homeostatic machines). Considering what we know about the human mind, it can safely be treated as a purposefully acting system. A person indeed frequently behaves, both 'inwardly' and 'outwardly', in purposeful ways. But this raises a series of new problems.

The first problem involves the fact that all *homeostatic systems* (either stable or directionally changing) *are limited in their capacity for homeostatic adaptation.* This is also the case in respect to the instrumental character of the processes of attitude formation. In the first place, even if we consider an isolated need, people can build up (or take over from their surrounding cultural environments) attitudes which are *in various degrees instrumental to this need.* Some attitudes are actually, 'actively' functional, while others might have been shaped as a result of inefficient 'homeostatic mechanisms'. This can best be illustrated by attitudes which serve mainly, according to Katz, the purpose of outward adjustment. Different people, depending on the scope and validity of their knowledge about reality, or depending on their ability to draw valid consequences from observed facts, evaluate differently the object of their attitude, even if they are motivated by the same drives, desires and values; their attitudes can vary as to the *objective* (distinguished from *subjective*) *degree of their instrumentality* in satisfying a need that is common to all of them. As a simple example, some members of an oppressed group are apt to be politically active and work toward changing the political system, while others will build up a purely cognitive attitude of passive, though intensive, dissatisfaction with their social position, or even with the whole social system. Which of them turns out to be the more adaptive and therefore 'more functional' with respects to the needs of the given person depends upon various additional factors — primarily such factors as the rigidity of the given social system, its strenth, its reactions toward those who oppose it, etc.

By evaluating certain attitudes as 'functional' in the last example we use the notion of functionality in its *objective* sence, i.e. we judge the instrumentality of an attitude (or behavior caused by it) by the objective consequence of the corresponding action evaluated finally in terms of the value system of the given individual. But — as we know — humans are 'rational' (or inclined to behave 'instrumentally' with respect to their needs and values) at best in the *subjective* sence of the terms rationality and instrumentality. They react, evaluate things and behave toward them in a way which *seems* to them to be instrumental to their needs and values. If their knowledge is incorrect, their actions may fail in leading to satisfaction of their needs or values.

Therefore when dealing with *conscious* actions, decisions and processes of attitude formation, social scientists always relate them not to any possible 'perfect knowledge' about the situation but to the knowledge

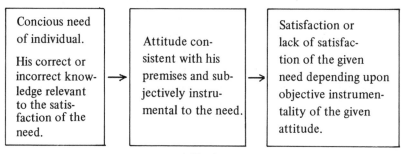

Fig. 1.

actually possessed by the given individual. If his evaluative or behavioral conclusions are *consistent* with his premises, we classify such an attitude or behavior as at least *subjectively rational* or instrumental, even if it fails to satisfy the corresponding needs objectively. But now our functional scheme of causal determination of attitudes by needs is as shown in Figure 1.

Then in a particular case we may say that a certain individual did not develop attitudes which would be even subjectively instrumental to his given need that he has, when in other cases we find attitudes at least subjectively (if not objectively) instrumental to this need. The statement about the relationship between the need and attitude is here accessible to empirical control.

The question becomes more complicated if we take into account the unconscious needs. As long as the attitudes and behavior of the individual are objectively functional to some need (which we assess by indirect, e.g. projective, techniques or assume on the basis of theoretical premises) the situation is fairly simple. We postulate (or assess by projective tests) that the given person has a certain need, and then we observe that his behavior (whether he is aware of it or not) really satisfies this need. The relation of instrumentality is here more or less visible. But suppose we would like to extend also the notion of objectively unsuccessful instrumentality to needs that operate on the unconscious level?

We fase here two possibilities. First we may assume that other premises necessary for the formation of an attitude, especially the individual's knowledge and beliefs about the object of the attitude, are of a *conscious character*, and may therefore somehow be assessed empirically by the researcher. If we then come to the conclusion that the given attitude is consistent with the set of premises we postulate, i.e. with his conscious knowledge and unconscious needs, we may say that this attitude is subjectively instrumental even if it does not serve the satisfaction of the need. But suppose we assume that all the premises in the process of attitude formation are of an unconscious character? In such case we cannot state even the subjectively instrumental character of attitude which are objectively non-instrumental, because we are unable to apply the criteria of internal consistency to such a process of attitude formation.

This may seem a bit of an academic question, because in empirical research even if attribute the relation of instrumentality to some attitudes with respect to inconscious needs, we usually have in mind such attitudes as are objectively instrumental to these needs, leading to their satisfaction. If the need is unconscious, and the attitude non-instrumental to it (in the objective sense of the term), why should we care about possible subjective instrumentality? But the question is by no means a simple one. If we admit the notion of intended but non-successful instrumentality for the phenomena and processes going on at the conscious level, why shouldn't we extend this to the unconscious motivational processes? The problem is, What would this really mean if we are unable to apply the criterion of internal consistency here?

Other problems related to the teleological style of explaining the origin of attitudes in terms of their functionality for needs are raised by the fact that a person always act under the pressure of *many needs and values* together, forming a more or less coherent system. Developing an attitude (or behaving in accord with a developed one) instrumental to one need often frustrates, sometimes severely, other needs or values. We might accept the apparent proposition that the *stronger need is always the winner*, or that each process of attitude formation involves an optimalization, conscious or not, of different incompatible tendencies. However, such a directive seems to involve the danger of tautological theorems, devoid of any predictive value, since it would only be possible ex post to tell which need was stronger. Besides, such a proposition would be at odds with what we know about the functioning of the human mind and its all too frequent failures in coping with motivational conflicts and in finding the best solutions for them. Attitudes developed as a result of ego-defensive mechanisms (Katz called them ego-defensive attitudes) are instrumental to the need of aggression and anxiety reduction, but they are rather essentially uninstrumental or even harmful to the whole set of adaptational needs of an individual. Here again the scientist may choose (unfortunately, rather arbitrarily) one of two solutions. He can say that for the given individual his 'ego-defensive needs' are more impotant than, for example, his survival needs; or he can say that the survival needs are more important for the given individual, but that he is unable to solve his motivational conflict in a way that would optimalize his hierarchy of needs. The second choice may be as justified as the first, because we know that some people are rather remote from the ideal of shaping their attitudes and behavior so as to make them optimally suitable for all their needs, objectives and value system as a whole, i.e. so as not to have some of their needs satisfied at the cost of a too extreme (even in terms of their own hierarchy of values) frustration of some others.

Thus even should we endorse the assumption of the teleological character of processes in the human mind and of the attitude formation among them, we must be aware of the various limitations of this general assumption, and consequently of the functionality principle, as the directive governing formulation of the laws specifying the relationship between attitudes and the needs or values that determine them.

4. DIFFERENT PATTERNS OF CONSCIOUS AND UNCONSCIOUS INSTRUMENTALITY

Up to now we have considered the problems of instrumentality of attitudes in general, looking at their possible functionality toward the psychological or social needs of individuals, and assuming that these needs may constitute the causes of attitudes on the basis of some teleological mechanism. But — as we will recall — attitude as defined above is a complex structure composed (potentially at least) of evaluative, cognitive and behavioral components. Also noted was the internal instrumentality of attitudes, meaning by that the 'domination' of some components over others. But the notion of interrelations between the components of attitude may also be posed with respect to their external instrumentality. We might ask which of these components is *more closely related causally to the given need* in the process of attitude formation, and which are *indirect effects* of the given need 'produced' directly by those components caused by the given need. If we come to the conclusion that there can be *various patterns of causal relations between the components of attitudes when regarded from the standpoint of their initial cause, i.e. the need*, we can distinguish various dynamic patterns in the process of attitude formation.

Let us consider first the most simple pattern, which I shall call the *completely rational pattern of attitude formation*. Suppose that as individuals we accept certain general values, certain *evaluative standards which may be applied to a certain object*. These values correspond in this case to our need, i.e. to our value system: we would like certain objects to be of the given kind and we would dislike them to be of other kinds. Now suppose that a certain object of this category comes into the area of our experience: we then learn something about its properties, *compare these properties with our evaluative standards*, come to a certain evaluation of the object, and eventually behave toward it in the way consistent with this knowledge and this evaluation.

While the foregoing example takes a set of social values as the starting point of the attitude-forming process, the situation looks exactly the same for 'more psychological' needs. Suppose we feel lonely; then we discover that, due to his or her personality characteristics, someone seems very likely to be able to reduce this loneliness; on the basis of this knowledge (and actual future tests of its validity) we develop a certain evaluation of this person and try to behave toward her or him in a certain way.

The structure of the attitude-forming process may also look different. Suppose our culture tells us that we should love our relatives, or that our relatives should be regarded as being close to us. On the basis of this general norm, we may develop a certain *positive evaluation* of our relatives and certain patterns of behaviors, *independently from our knowledge* about their real characteristics. Then, in striving to maintain consistency in our attitudes toward them (in order to reduce possible 'cognitive dissonance') we may be inclined to *perceive them and their behavior in a positive way* quite unrelated to how they actually behave. Even if the relatives are hardly lovable persons, we may feel a certain love for them and behave accordingly, even finally becoming convinced that 'in reality they are quite nice persons'.

A similar pattern may be initiated not by a normative social standard but by a more 'psychological' need. A child in order to satisfy this or her need for identification with mother may seek to be as close to her as possible and to perceive her in positive terms only. A man with strong aggressive tendencies may fix them on a certain person, and then 'adapt' the cognitive and evaluative components of his attitude to its behavioral component, which is directly and constantly related to the need. In all such cases more or less pronounced distortion of cognitions may be observed: they are 'instrumentally adapted' to the basic (evaluative or behavioral) component of the given attitude, which directly serves the satisfaction of the need.

What is interesting here is that from the standpoint of the need to which the given attitude is instrumentally related, the distortion of the cognitive process is not instrumental at all, either directly or indirectly. On the other hand, it is usually instrumental with respect to the other needs of the individual and especially to his other social values and norms. If we regard the tendency toward aggressive behavior in isolation, regardless of other needs and values, there is no logical necessity to justify this feeling by negative perception of the object of aggression. I may nicely kill an animal during a hunting expedition, feel a great release of aggressive tendencies, and at the same time not have to convince myself that this animal was 'bad' or threatening to me. This is so because our culture permits these kinds of aggressive behaviors. On the other hand, I would need substantial rationalization in order to justify an aggressive act against a human being, because there are norms in our culture which tell me not to do so. This means that *we have to look at the various patterns of deviation from strict rationality as well as the various 'defense mechanisme' operating in the process of attitude formation as the way the human mind deals, more or less successfully, with the multiple instrumentality of attitudes, with the task of satisfying by one attitude as many needs and values as possible*.

This process often takes the form of 'rationalization', for it is a process of functional adjustment of attitude toward different needs, working toward an eventual outcome which the individual can (unjustifiably) perceive as (what social scientists would objectively not characterize) a strictly rational attitude. We human beings change our perception of the object, or 'perceive' in our introspection only such needs and motives as would justify a given evaluation or course of action according to the strictly rational pattern of attitude formation, or in accordance with socially accepted values and norms. This seems to be the *tribute paid by the weaknesses of the human mind to the ideal of rationality in thinking and behavior*.

In most situations such as those noted above the individual may incorrectly perceive the causal-functional relations between the different components of this attitude, thus coming to the wrong conclusion about its character, but at least *he perceives the nature of the components of this attitude correctly*. In other words, he is at least partially *conscious of the character of his attitudes*. The situation becomes much more complicated in cases where the notion of attitude is applicable to unconscious states and predispositions. Not only may some emotions related to certain objects be 'repressed' to the unconscious area of our mind, but knowledge of certain features of the object of the attitude may also be repressed and, being unconscious, may strongly influence both our emotions and our behavioral predispositions toward this object. Finally, certain behavioral predispositions may be repressed to such a degree that the person 'does not know' why he behaves in the given way. Therefore the concept 'attitude' can denote 'multi-level' phenomena belonging to both the conscious and the unconscious areas of psychological reality. And here — in the subconscious part of the human mind — connections between phenomena whose existence the person is completely unaware of are ipso facto inaccessible to his introspection. Nevertheless the notion of the instrumentality of some of the components toward others, and the instrumentality of the whole attitude toward the needs it is supposed to satisfy, is not an empty one even in the area of the unconscious, *provided that this instrumentality is successful* and that we have the tools (e.g. clinical interviews, projective techniques) to reveal the existence of these components of inconscious attitude.

The problems relating to unconscious attitudes becomes even more complicated if we take into account the interaction of the two levels — the conscious and unconscious components of an attitude. It goes beyond the task of this chapter to discuss these problems in a more detailed manner,[17] since our goal here is to discuss some general methodological problems of 'functional thinking' in the social and behavioral sciences, taking the area of attitude studies as exemplification only.

5. THE PRINCIPLE OF FUNCTIONALITY AND THE PROBLEMS OF PREDICTION

We would expect that a theory explaining certain phenomena would also be able predict them. Two conditions must be met by a theory concerning the influence of needs and values on attitude formation to make it good not only for ex post explanation but also for predicting.

First, that the existence of needs which can influence attitude formation must be ascertained independently in the minds of the given persons from the existence of the respective attitudes. Such knowledge can be theoretical as for example in the lists, at one time frequently attempted by psychology, of all human needs and strivings; they provide deductive grounds for predicting that specified attitudes will appear in people in general or in certain individuals in some social situations. It can also be of a more empirical character, as when from the identified needs we can select such *observable indicators* that the presence of the relevant attitudes is not definitionally dependent on them. This refers, of course, to both conscious and unconscious needs, since we can assess the latter by, for example, certain projective techniques. Thus if we find a specific need in the given individual and assuming that it is the only psychological factor in the case, we could explain his attitude as well as predict that some still absent attitudes will eventually appear.

However, this kind of knowledge is not sufficient to predict the direction of attitude-formation processess since in real-life situations many needs are simultaneously operative, each of which may produce influences upon formation of various attitudes and which are likely to reduce the impact of each other. So it is necessary to be able to tell which of the competing needs or drives work more forcefully than others; in other words, it is necessary to know the hierarchy of human needs and strivings. Here, too, our knowledge can be theoretical. As an example of a theory ordering human needs into a hierarchical pattern we can point to Maslow's theory of the hierarchy of needs;[18] according to it, deprivational needs must be met first, and the fulfillment needs of development second. Without entering into the doubtful points of this theory but taking it for granted for the moment, we would predict that in people with acutely unsatisfied deprivational needs, those needs would predominate by shaping attitudes instrumental to them, whereas in people who are not severely deprived we might expect the shaping of attitudes functional toward the higher needs and strivings, i.e. toward various 'needs of development'.

A similar theoretical role can be played by certain models of *group cultures* in which values and recommended strivings are frequently ordered into quite coherent hierarchical systems. The knowledge of a group culture and of the hierarchy in its system of values can thus be significant evidence for predicting and explaining attitudes within the complex patterns of factors determining them.

It is still important, however, to obtain *observable indicators for the measurement of the relative strength of human needs and strivings*, so as to be able to predict concrete outcomes of attitude-formation processes. If we had such tools at our disposal, the principle of instrumentality of attitudes toward human needs and strivings would become testable in complex multivariable systems of needs and values, and would gain a great explanatory and predictive power.

It should also be kept in mind that all relationships of this kind are conditional, i.e. various kinds of modifiers of the role of needs in determining attitudes must be taken into account. Any specified need works in the context of other psychological predispositions of the individual, his other values and knowledge about the world, as well as in the context of his other already developed attitudes; it must be considered which of these are internal factors significant in the shaping of a specified attitude by a specified need, and which are indifferent. Other factors modifying the working of needs in attitude formation lie outside the range of the individual's needs, such as the social situations in which he participates, the groups to which he belongs, the objects and interactions within his experience. Thus the general pattern of a theorem defining the relationship between needs of a certain kind and the attitudes shaped by them can be presented as follows:

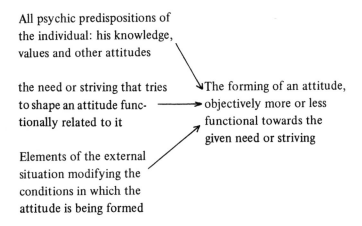

All psychic predispositions of
the individual: his knowledge,
values and other attitudes

the need or striving that tries
to shape an attitude functionally related to it

The forming of an attitude,
objectively more or less
functional towards the
given need or striving

Elements of the external
situation modifying the
conditions in which the
attitude is being formed

Assessment of the relative strength of a need as a potential factor in the attitude-formation process is a necessary but not a sufficient condition for predicting the *contents* of the incipient attitude. At most we are able to say that the attitude will be instrumental to this and not to other needs, or that the attitude will aim to satisfy several of the individual's most important needs in proportions corresponding to their relative strength. In order to predict the exact nature and content of attitudes, we will need a much more detailed theory of attitude formation.

One possible approach to the construction of such a theory is *strictly empirical* – we can try to find out in the course of empirical research what kinds of needs under what kinds of external and psychological conditions are likely to 'produce' attitudes of given 'shape'. Then we might discover, for example, what kinds of attitudes are likely to occur in a certain culture among people with a strong need for security, what kinds of political attitudes occur among those who — being very rich — are interested rather in the stabilization of wealth than in its further expansion, etc. But such propositions *will be of a rather low level of generality*, usually limited to one society or one specific social group. Hence construction of a theory relating attitudes to the psychological and social needs of people in a strictly empirical manner — 'from the bottom' — would be a very tiresome undertaking. Moreover, the theoretician is usually interested in constructing a theory before he makes his research (and testing it by the subsequent research) than in trying to derive a theory from his data. Thus we are faced with the importance of formulating rules for building a theory according to which the contents of attitudes might be predicted under the assumption that these attitudes will be instrumental to needs of a given type.

One rule seems fairly clear. Under the assumption that the attitudes will belong to the category characterized above as *strictly rational*, we can try to formulate predictions about their contents (i.e. about the subjects' evulation of the objects of attitudes, and their ensuing behavior). But — as we know — such predictions are valid only under the assumption that people behave completely rationally. Moreover such a theory, which would be useful for the purpose of prediction, requires some observable criteria of rationality of the studied persons.[19]

Unfortunately, a completely rational attitude-formation process is the exception rather than the rule in human thinking, feeling and behaving. I suggested above some of the more typical 'deviations' from this pattern; there are many others. Numerous socio-psychological theories have been put forth to deal with the phenomenon of limited human rationality — the theory of cognitive dissonance being typical. One interpretation of this their has it that people will tend to obtain an *emotionally balanced* picture of different aspects of an object, even if this costs them the truth of some of their partial judgments, rather than to seek a completely valid but 'emotionally dissonant' picture of the object. The predictive power of this theory, however, is fairly low; experiments in its support have at most been able to demonstrate that the reduction of dissonance *sometimes* really occurs.

Things are even worse with psychoanalysis, which due to the special structure of its propositions is able to 'explain' almost everything but to predict hardly anything. This is especially unfortunate for the theoretician of attitudes because we would like to find in the *theory of defence mechanisms* the source of predictive hypotheses for the attitude formation processes. But the theory of defense mechanisms is not constructed in the form of general (or probalistic) propositions which would specify that under such-and-such circumstances such-and such a defense mechanism myst (or is most likely) to occur. It is a classificational scheme of *possible* distorions of the rationality of human thinking and behaving, rather than the set of generalizations we usually call a theory. As long as his theory remains as it is, we may use it for ex post explanation of identified attitudes, but it has very little use in our attempts to construct a predictive theory.

For all its modifications and limitations, the postulate of the functionality of attitudes for human needs and values can plan *an important heuristic role* in the development of a future theory of attitudes. However, its utility does not cover all the theoretical problems concerned with attitudes, for theoretical propositions in which human values or needs is an independent variable are by no means the only type of propositions in a theory of attitudes. Here I would emphasize the importance of generalizations describing the relationships between certain *objective social factors* – especially factors referring to the position of the individual in the *social structure*, to the social roles performed by him, to factors related to his group belonginess – and the contents of attitudes formed under their influence. These factors may be considered within the 'functional approach' either as factors *shaping needs and values* and thus influencing attitudes in a relatively indirect way, or as *modifiers of relationships* between needs on the one side and attitudes on the other.

A social scientist is fully entitled, however, to look for relationships between social situations and human attitudes *without considering the mediating psychological mechanisms*. It is his right to consider what attitudes are shaped by a situation of social deprivation as opposed to a privileged position, what is characteristic of the attitudes of members of isolated groups as opposed to attitudes of those who have much interaction with other groups, etc. In the sociological and socio-psychological litereture there are numerous instances of the influence of certain social situations upon attitude formation, and these accounts ought to be fully used in constructing an integrated theory of attitudes.

But the influence of psychological factors ought to be set forth in such a theory, too. In other words, in constructing a theory of attitudes, all the achievements of sociology, psychology and social psychology ought to be made use of insofar as they uncover relationships between various social and psychological variables, irrespective of whether these relationships fit into the scheme of functional explanation or not.

When we look at the theories which either directly apply to the area of attitudes or might be indirectly used for the formulation of new hypotheses, we perceive that most are in rather poor shape. I mentioned earlier that they usually lack the specification of conditions that would transform them into generalizations or increase their predictive power. Another obvious lack in many of these theories is that they are *too general*, they do not specify which type of attitudes they are applicable to.

Some generalizations or theories aspire to explain all possible attitudes, whereas in fact they only explain *some* of them; i.e. they refer to attitudes toward specific types of objects. The theory of group interest primarily explains the formation of certain ideological attitudes. The theory of comparative reference groups explains the development of social feelings of deprivation and satisfaction, the shaping of levels of aspirations, etc. In order to increase the explanatory and prognostic utility of these theories, it will be necessary to specify their *substantive ranges*, i.e. to limit the class of attitude contents to which they are applicable.

6. SPONTANEOUSLY FORMED AND CULTURALLY IMPOSED ATTITUDES

For the sake of simplicity, I have been assuming that the attitudes which we attempt to explain have been shaped in *independent individual processes of attitude formation*, i.e. that they have *arisen spontaneously*, without imitation or borrowing of the ready patterns of attitudes from the surrounding societies. However, as we know, such an assumption is by no means always valid. Most of our attitudes are 'imprinted' into our minds by our social environment, by cultures of the groups in which we participate, by the powers of socialization. Obviously, in those cases too, our needs, both social and personal in character, may be quite important. When faced by a *choice* among various patterns of attitudes toward some object, we may be inclined in our more or less conscious decisions to choose one or the other set of attitudes depending upon the respective degrees to which the relevant patterns of attitudes seem to be instrumental to our signaficant needs. In a society with much class tension, a worker is apt to meet daily (in the mass media and elsewhere) various patterns of attitudes toward the capitalist system, and he accepts some while rejecting others, with a clear awareness of their functionality or dysfunctionality with respect to his own class interests. A man with 'authoritarian personality needs' is apt to choose from among all those that are suggested to him an ideology that will meet the needs rooted in his personality, even when his choice is likely to be much less conscious.

However, the fact that we find some *ready patterns of attitudes* in the cultures of groups familiar to us modifies in many ways the working of the functionality principle, even though our needs may play a significant role in our choices among those patterns. In the first place, since the patterns have already been socially modeled by some groups around us and usually in some way made socially objective (e.g. clearly and unambiguously verbalized), we accept them as ready-mades, whether or not they fully fit in with our individual needs. Sometimes we take over complex wholes of ideological patterns, internalizing them into our own cherished attitudes though only some part of them may be directly functional for our needs while others can be indifferent to or even incompatible with our actual needs and interests. A petty bourgeois might choose the fascist ideology, since it offered him an illusion of safety and thereby tended to reduce his anxiety in the face of the upheavals of a period of crisis. He would absorb this ideology in all its facets, though only some of its elements were directly instrumental to his needs while others such as its expansionist militarism, would be in the long run obviously dysfunctional. A man with a strong urge to religious experience might join a religious group and accept all its tenets, though many of them were a matter of indifference to him, and others even quite hard to swallow. It is rather difficult, considering the mechanisms of internalization and of social control, for a person to endorse only some selected patterns while rejecting others which do not fit in with his needs. The mechanisms of social control clearly favor ideological syndromes as integral wholes.

It can also be said that appropriating the ready patterns from the surrounding cultural environment can lead to more objectively instrumental attitudes than would ever be possible if attitudes were formed through individual spontaneous processes only, independent of any patterns imposed by our social environments. This tends to be the case especially in stable communities, where successive generations shape the patterns of attitudes, more and more adequately adjusted in their contents, to the *typical needs and values* of the members of the community. Individual processes of attitude learning of the trial-and-error type are complemented or even replaced by the cumulation of experiences of many generations. The theoretical model of individual attitude formation is in this case, where attitude has been "imprinted" by social environment, different from that of spontaneous processes controlled by individual values and knowledge only.

While socially 'imprinted' attitudes can be fully instrumental to one's individual needs, still the relations between the various contents of the given attitude and the needs of the individual is *not a causal but a spurious one* [20] Each attitude is independently 'imprinted' into the individual's mind in the course of his socialization. The pattern of relationships involved here can be presented as follows:

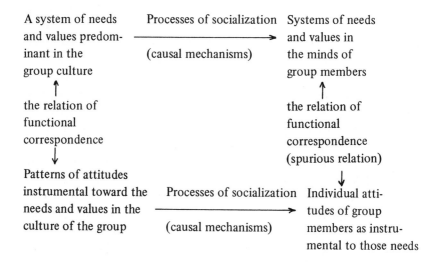

Sometimes an additional mechanism is at work. Individual needs and systems of general values can, under the influence of group values, be shaped, prior to the exposure to ready patterns of attitudes; the absorbing of such patterns is then facilitated by the fact that the already accepted values and internalized needs on one hand and the ready-made patterns on the other have a common social origin. The intervening mechanism is the instrumentality principle. There is thus a causal link between having certain needs/values and absorbing certain attitudes, though it is not the same as when attitudes are spontaneously developed only under the influence of values and needs. For here the relation between socially 'imprinted' needs/values and the individual's derived attitude is *partly causal and partly spurious*. But if at the same time other general values and patterns of attitudes are also absorbed, the relationship between them is wholly spurious and can be reduced to the fact that general values and patterns of attitudes coexist within a culture.

Finally, let us consider a situation where the attitudes absorbed during socialization are in their contents wholly uninstrumental to needs and values. Let us imagine a community in which the authorities decided one day to 'condition' the people positively to all 'things blue' and negatively to all 'things yellow'. Before that date both colors had been instrumentally indifferent for the community. There is no inherent need in human beings to see blue things rather than yellow ones. But now a complex system of pressures has been started; blue objects are ascribed emotionally positive labels, and yellow ones are linked with negative ideas; people are rewarded for painting anything blue and punished for the use of yellow. As a result, definite attitudes toward these colors and toward objects bearing them become imprinted in the society. Now, in my opinion, this kind of attitude formation would be outside the scope of the instrumentality principle as relating attitudes to human needs.

An adherent of the tradition of learning theory in psychology might object that in our hypothetical community a positive attitude toward blue objects has become highly functional for the very important need of securing as many rewards as possible, while antipathy toward yellow is instrumental in avoiding punishments. I fully agree with the objection. The point is, however, that in this case the colors in themselves are instrumentally indifferent for the needs and values of the people who react to them. Moreover, we shall assume that the other qualities of the yellow- or blue-painted objects are indifferent. The only factor that gives them emotional significance and makes them into objects of attitudes is the efficient working of the system of punishments and rewards related to the designated colors, the choice of which is quite arbitrary.

Certainly, such a state of lack of functional correspondence between attitudes and values cannot be stable; prolonged processes of 'color conditioning' would necessarily lead to corresponding changes in the system of the more general values of the group. It is highly probable, then, that in this hypothetical community blue would become a symbol of positive values and yellow of negative ones. Besides, various mechanisms distorting perception would probably be triggered off, causing the members of the community to 'see' many positive qualities in blue objects, and mainly negative qualities in yellow ones.

My intention here is to point of that, on the one hand, our attitudes to objects can result from the *inherent qualities* of such objects and the relations of these qualities to the satisfaction of needs already existing in us (other than the tendency to maximize rewards and minimize punishments in general) or that, on the other hand, our attitudes can be conditioned by *arbitrary relationships* between any of the qualities of objects and the consequences of our regarding or disregarding them. These two types of mechanism seem to be worthy of a theoretical distinction.

However, the fact that both types of mechanism are possible imposes still another limitation on the principle of functionality of attitudes. For – as we have seen in the above 'example' – in the long run it is the more general needs/values which are modified so as to become functional for the socially developed attitudes, for example by providing rationalizations for them. *The functionality principle is here reversed.* As the phenomenon of the adjustment of human needs and values to the requirements of the social environment and to the mechanisms of social control is quite universal, and people tend to adjust to and rationalize in general terms that which cannot be avoided, the principle of functionality of attitudes as a directive in theory building is, if not limited, then at least significantly modified.

The application of the principle of functionality at the level of whole social groups and societies, where the criteria of functionality are the 'needs' of groups or institutions, results in another class of problems. I distinguished above two kinds of instrumentalities: the objective and the subjective. In the subjective sense an attitude or behavior is instrumental if the individual *thinks or believes* it will lead to the satisfaction of a certain need he is aware of, or if the social scientists can derive this instrumental relation from the person's behavior or utterances. Such an attitude may be *more or less successful in satisfying the need to which it is causally* related. When it satisfied this need to a fairly high degree the attitude is treated as objectively instrumental. But in both cases, the terms needs values are assumed to refer to some *real phenomena existing in the conscious or unconscious areas of the human mind*. These more or less hypothetical, but postulated as 'really existing', needs provide here the criteria of subjective or objective instrumentality of attitudes.

The situation becomes much more complex when we try to discuss the instrumentality of attitudes and behaviors with respect to the needs of such social wholes as groups, institutions and societies. And — as we know — this is the problem area covered by so-called functional theories in contemporary sociology. It should be emphasized here that at the beginning of the 'functional school' — in the writings of B. Malinowski, for example — the ultimate reference points of the functionality of social institutions were, among others, the biological and psychological needs of individuals.[21] If we wish to follow Malinowski's tradition we could understand the term need at the societal level as well, having it refer to the conscious or unconscious needs of at least the majority of (if not all) members of a certain group, institution or society. Then the attitudes which are judged to be instrumental to these needs could be explained by the 'functionality principle' as discussed above under the additional assumption that the need of the majority are able to control the process of attitude formation at the societal level.

But — as we know — the notion of need (or its sociological correspondent: 'interest') is often used in another sense, which seems to assume that groups, institutions, organizations, or societies may have needs (interests) of this even independent from the needs and interests of a majority of its members. This again may have different meanings. It may refer to the needs or interests (understood here in the 'individualistic' sense) of a certain *minority* of members of the society or group who are in a position to control the 'social whole', its actions and 'behaviors' as well as the influence of their needs upon the processes of attitude formation, which they are able to impose upon the minds of the majority in such a way that they be functional to their own needs. In other cases — especially with respect to some purposeful institutions — the needs of the institution may be understood as determined by its formally defined tasks, which ultimately may refer either to the needs of those whom the institution is supposed to serve, or to the needs of those who manage it. Here again we may look at the degree of functionality these needs and the attitudes and behaviors of the members of the institution. But of course we should be aware of the exact meaning of our criterion of functionality in this case. We should also be aware that in all such cases the mechanisms of attitude formation are quite different than in the situations discussed previously in this chapter, for they constitute *mechanisms of social control* rather than mechanisms of spontaneous formation of attitudes under the influence of the needs of individuals. The discussion of methodological problems of this kind of theory goes beyond the task of this chapter.[22]

The term 'need' (or its sociological correspondent in some types of theories, namely the 'functional requirement') has quite another meaning when it refers to certain strivings and tendencies which can be *derived from the observation of actions of some collective bodies*, even if they do not correspond to any kind of psychological reality mentioned above. Thus we know that most organizations will have a *tendency to survive*, and will 'behave' in a way 'functional' to this tendency. We know that organizations of a certain kind will reveal a tendency toward expansion – whether they are business corporations or modern states. Assuming that the 'need to exist or to expand' can be assessed (derived) from the actions of the organization, we may later on try to see to what degree the values or attitudes of the individual members are instrumental with respect to this need. But here again we should be aware that the causal mechanisms involved are of quite a different kind than in the case of spontaneous formation of attitudes at the individual level.

With respect to these categories of situations we face an additional problem. In biology the idea of *persistence of individuals or species for better control of the surrounding environment is often used as the criterion of functionality* for certain properties of organisms. Suppose we were to apply the same criterion to societies, social groups, and institutions in a sense that involves not the needs of affected individuals but only the naintenance or disappearance of the societal artifact. It is obvious that attitudes which are functional for one kind of social structure may be dysfunctional for another. The attitudes expressed by Dante Alighieri in his 'Divine Comedy' were obviously dysfunctional for the existing social system of medieval Italy, made up of small – and constantly quarreling – republics. On the other hand, they were functional for his own vision of a unified Italy. Attitudes which are dysfunctional with respect to the capitalist socioeconomic system may be functional with respect to the socialist system, etc. Therefore when we conclude that something is functional with respect to the 'objective needs' of a certain social system or constitutes its 'functional requirement', the acceptance of this and not of some other kind of system as the critierion of functionality involves certain value judgments which can be accepted by some people, but are rejected by others. We have to *choose* from among a great number of possible criteria of functionality those which *correspond to our own value judgments*

Such normatively chosen criteria of functionality cannot be used for causal explanation of existing attitudes; it is needless to point out that only existing (not the desired and non-existing) phenomena can be used in hypotheses for explaining existing phenomena. Therefore, besides all the problems having to do with the normative choice of the criteria of functionality, this kind of 'functional assessment' cannot constitute a starting point for the formulation of an explanatory theory of attitudes.

NOTES

* Based on a paper presented at the Conference on the Theory of Attitudes at the Institute of Sociology of Warsaw University. The original paper titled 'Pojęcie postawy w teoriach i stosowanych badaniach spolecznych' (The Concept of Attitude in Theories and in Applied Social Research) was published, with other papers presented at this conference, in S. Nowak (ed.), *Teorie Postaw* (Theories of Attitudes), Warsaw, 1973.
[1] D. Katz, 'The Functional Approach to the Theory of Attitudes', *Public Opinion Quarterly*, 1960. Quoted from: M. Fishbein (ed.), *Readings in Attitude Theory and Measurement*, New York, 1967.
[2] D. Katz, *op. cit.*
[3] *Ibid.*
[4] *Ibid.*
[5] See W. I. Thomas and F. Znaniecki, *The Polish Peasant in Europe and America*, Boston, 1918 – 1920, vol. I, p. 27.
[6] See G. Allport, 'The Historical Background of Modern Social Psychology', in: G. Lindzey (ed.), *Handbook of Social Psychology*, vol. I. See also T. M. Osborn, 'The Emergence of Attitude Theory', in: A.S. Greenwald and T.C. Brooks (eds.), *Psychological Foundations of Attitudes*, New York, 1968.
[7] See D.T. Campbell, 'Social Attitudes and Other Acquired Behavioral Dispositions', in: S. Koch (ed.), *Psychology, the Study of a Science* vol. 6, 1963. See also L.W. Doob, 'The Behavior of Attitudes', in M. Fishbein (ed.), *Readings in Attitude Theory and Measurement*. See also L. Guttman, 'Attitude and Opinion Measurement', in: S. Stouffer *et al.*, *Studies in Social Psychology in World War II*, vol. IV, *Measurement and Prediction* Princeton, 1949 – 50.
[8] See Chapter I.

[9] See, e.g., E. Bogardus, *Fundamentals of Social Psychology*, New York, 1931. See also L. Thurstone, 'Attitudes Can Be Measured', in: M. Fishbein (ed.), *Readings in Attitude Theory and Measurement*.

[10] M.B. Smith, 'The Personal Setting of Public Opinion; a Study of Attitudes Toward Russia', in: M. Fishbein (ed.), *Readings in Attitude Theory and Measurement*

[11] See D. Krech, R.S. Crutchfield and E.L. Balachey, *The Individual in Society*, New York, 1962, p. 139. See also M. Rokeach, 'The Nature of Attitudes', in: *International Encyclopaedia of the Social Sciences*, vol. 1, pp. 444 – 457. It should be stressed here that for Rokeach attitude has a more complex structure than for, e.g., M.B. Smith, or for Krech, Crutchfield and Balachey. He defines attitude as a "relatively enduring organization of beliefs around an object or situation predisposing one to respond in some preferential manner", but later on he distinguishes cognitive, affective and behavioral components within each of the beliefs belonging to the given attitude.

[12] See D. Katz and A. Stotland, 'Preliminary Statement to a Theory of Attitudes', in: S. Koch (ed.), *Psychology, the Study of a Science* vol. 3, 1959.

[13] See L. Festinger, *Theory of Cognitive Dissonance,* Stanford, 1957.

[14] See M. Rokeach, *The Open and Closed Mind*, Basic Books, New York, 1960.

[15] See also D. Katz and A. Stotland, 'Preliminary Statement to a Theory of Attitudes', *ibid*.

[16] For an analysis of the dangers involved in functional thinking in the sciences, see E. Nagel, *The Structure of Science*, Chapter 12, New York, 1961.

[17] For an excellent discussion of the interaction of conscious and unconscious levels of attitude with conscious and unconscious needs, see Julia Sowa, 'Psychoanalityczna teoria postaw' (Psychoanalytic Theory of Attitudes), in: S. Nowak (ed.), *Teorie Postaw* (Theories of Attitudes), Warsaw, 1973.

[18] See A.H. Maslow, *Motivation and Personality*, New York, 1954.

[19] For an analysis of the problems involved in the assumption of rationality in social theories, see Chapter I.

[20] A relation is called spurious when neither is *x* the cause of *y* nor vice versa, but the two nevertheless tend to occur jointly or in a sequential way. See 'Causal Interpretation of Statistical Relationships in Social Research', in this volume, p. 165.

[21] See B. Malinowski, *Culture*, *Encyclopaedia of the Social Sciences*, New York, 1935, vol. 4.

[22] See R.K. Merton, 'Manifest and Latent Functions', in *Social Theory and Social Structure*, New York, 1968.

BIBLIOGRAPHY

Allport, G., 'The Historical Background of Modern Social Psychology', in: G. Linzey (ed.), *Handbook of Social Psychology*, vol. I.

Bogardus, E., *Fundamentals of Social Psychology*, New York, 1931.

Campbell, D.T., 'Social Attitudes and Other Acquired Behavioral Dispositions', in: S. Koch (ed.), *Psychology, the Study of a Science*, vol. 6, 1963.

Doob, L.W., 'The Behavior of Attitudes', in: M. Fishbein (ed.), *Readings in Attitude Theory and Measurement*, New York, 1967.

Festinger, L., *The Theory of Cognitive Dissonance*, New York.

Guttman, L., 'Attitude and Opinion Measurement', in: S. Stouffer *et al.*, (ed.) *Studies in Social Psychology in World War II*, vol. IV, Measurement and Predicition, *Princeton, 1949 – 50.*

Katz, D., *'The Functional Approach to the Theory of Attitudes', in: M. Fishbein (ed.), Readings in Attitude Theory and Measurement*, New York, 1967.

Katz, D., and A. Stotland, 'Preliminary Statement to a Theory of Attitudes', in: S. Koch (ed.), *Psychology, the Study of a Science*, vol. 3, 1959.

Krech, D., R.S. Crutchfield and E.L. Balachey, *The Individual in Society*, New York, 1962.

Malinowski, B., *Culture*, in: *Encyclopaedia of the Social Sciences*, vol. IV, New York, 1935.

Maslow, A.H., *Motivation and Personality*, New York, 1954.

Merton, R.K., 'Manifest and Latent Functions', in *Social Theory and Social Structure*, New York, 1968.

Nagel, E., *The Structure of Science*, New York, 1961.

Nowak, S., 'Causal Interpretation of Statistical Relationships in Social Research', in this volume, p. 165.

Nowak, S., 'Concept and Indicators in Humanistic Sociology, in this volume, p. 1.

Nowak, S., (ed.), *Teorie Postaw* (Theories of Attitudes), Warsaw, 1973.

Osborn, T.M., 'The Emergence of Attitude Theory', in: A.S. Greenwald and T.C. Brooks (eds.), *Psychological Foundations of Attitudes*, New York, 1968.

Smith, M.B., 'The Personal Setting of Public Opinion; a Study of Attitudes Toward Russia', in: M. Fishbein (ed.), *Readings in Attitude Theory and Measurement*, New York, 1967.

Sowa, J., 'Psychoanalityczna teoria postaw' (Psychoanalytic Theory of Attitudes), in: S. Nowak (ed.), *Teorie Postaw*, Warsaw, 1973.

Thomas, W.I., and F. Znaniecki, *The Polish Peasant in Europe and America*, vol. I, Boston, 1918 – 1920.

CHAPTER XI

THE LOGIC OF REDUCTIVE SYSTEMATIZATIONS OF SOCIAL AND BEHAVIORAL THEORIES*

1. THE 'PROBLEM OF REDUCTION' IN SOCIAL SCIENCES

One of the basic tasks of every theoretically oriented discipline is to discover and to test empirically various laws which explain the mechanisms governing the structure and dynamics of the phenomena studied by the given science. Its further task is to *systematize* these laws as consistently as possible into systems of propositions called *theories*. Theories may be classified according to various criteria: by the kind of phenomena they describe (substantive systematization), by the formal character of the variables and the relationships they involve (qualitative and quantitative theories), by their interest in either static or dynamic aspects of reality, etc. One possible approach to the classification of theories is to classify them on the basis of *relations between the propositions* which constitute the given theory. It seems especially worthwhile to distinguish two groups of theories from this point of view:

(I) Theories in which *some of the laws explain other laws*, i.e. in which some laws may be logically derived from the others.

(II) Theories which are systems of propositions 'put together' *in a way that does not permit such derivation*.

If theories of the first category might be called 'vertical theories' — since some of their propositions are 'more basic' than others which are derivable from them — theories in which no logical derivation is possible might be called 'horizontal'. A typical example of a 'horizontal' theory is one which tries to identify causes of the same dependent variable, to determine some of the causes as more direct than others, to look at the various causes from the standpoint of whether they are 'additive' or 'interactive', etc.[1] The theoretical results of a multivariate approach or of a 'path analysis' of a certain social process constitute the most typical examples of a horizontal theory.

In this chapter we are interested in 'vertical' theories, those in which some of the propositions are derivable from others. Here we have also a whole range of types of such theories. Any axiomatic theory constitutes a 'vertical' logical structure: the level of *postulates* of the theory constitutes the starting point for all future derivations, followed by one or several levels of *theorems* derived directly or indirectly (i.e., through other theorems) from these postulates.[2] But the axiomatic theory is not the only possible way of constructing a 'vertical theory', and the notion of vertical theories applies to many weaker forms of systematizations wherein we can trace the relation of derivability among the propositions.

Sometimes the relation of derivability connects laws within the same science or in the same area of research within a science. In other cases it goes 'across the boundary' of two or more sciences or at least of distinct research areas or of theories defined as distinct within the given sciences. For some peculiar reason we seem to believe that this latter is a special kind of explanation which deserves a special name. We say then that we have *reduced* the explained law or theory to the explaining one (or ones). As Ernest Nagel wrote in his classic study of 'reductions in sciences':

Reduction, in the sense in which the word is here employed, is the explanation of a theory or a set of experimental laws established in one area of inquiry, by a theory usually through not invariably formulated for some other domain.[3]

The main task of this chapter is to analyze certain problems of reduction — or as we say here, or reductive systematization of theories with respect to social sciences in general and to sociology in particular — and, to the degree that it is possible for the present writer, to clarify the basic misunderstandings around the 'problem of reduction' in sociology. I will try to limit my comments on reduction in general to the essential minimum, since the reader can find apply discussion in specialized studies written by philosophers and especially in the excellent analysis of this subject in Nagel's book cited above.

There is nevertheless one thing that should be made clear at the outset. When clarifying the problems of reduction in sociology and preparing for its use in theory building, one can declare 'reductionism' to be a kind of philosophical outlook, according to which all the sciences can be systematized into an all-embracing vertical theoretical structure with some of the most abstract laws of physics at the top, down through chemistry, biology, psychology, to sociology at the bottom of this pyramid of human knowledge. On the basis of such an outlook one can declare that reductions of all propositions from one level to the propositions of the higher level are a priori *possible* by the mere force of philosophical argument. It is not this kind of outlook that I would like to advocate in this chapter. But there is another attitude toward the *postulate of reduction of theories*. One can assume that any explanation of one theory by another, provided it is theoretically valid and fruitful, constitutes a step in the development of theoretical knowledge. Reductions, as understood here, are especially interesting types of explanations inasmuch as they 'connect' areas of studies previously unrelated; therefore *the reductions are worth trying*. This view would seem to be unrejectable provided we do not 'declare a priori' the basic reducibility of some groups of theories to some other ones, unless the former have been actually and successfully explained by the latter. And provided also that we understand that even if a theory has been 'reduced to another one', this does not make the reduced theory either invalid or *useless in practice*. It simply reveals that the reduced theory and the theory explaining it are interrelated in a way which, as we shall see below, may be profitable for understanding both of them.

2. TWO BASIC PATTERNS OF REDUCTION

When we look at the nature of theories related to each other by the rules of logical derivation, it seems we can distinguish two classes of situations.

(A) The first is when the two theories explain *phenomena of the same level of complexity* (or 'integration') and differ only in the *degree of their* generality. Thus, if we have, for example, a sociological theory which states that people in underpriviledged social positions tend to be prejudiced against minorities or are 'authoritarian', and then we explain this by a psychological theory which refers to the 'frustration-aggression mechanism', we have a situation of reduction according to the definition presented above — propositions of one science explaining the propositions of another science. But when we look at the 'objects' described by both theories, they turn out to be objects of 'the same level' — namely, human beings. The *extensions* of the concepts constituting the characteristics of the human beings dealt with by the two theories are by no means equivalent: the extensions of the concepts of the explaining (psychological) theory are *more general* than those of the explained (sociological) theory. We assume that all socially underprivileged people are frustrated, but not that all frustrated people are socially underprivileged. By reducing the 'sociological theory' to the psychological one in this particular case (and in all cases of the same logical class), we assume that the phenomena to which the less general theory refers *constitute a subclass* of the phenomena to which the explaining theory refers — when the phenomena dealt with by both theories refer to the same domain of objects: in this case, human beings. In other words, they are phenomena from the same 'level of integration'. I would call this the *reduction of a law (or of a theory) to a more general law of the same level.*[4]

In the social sciences phenomena 'from the same level' do not necessarily have to be human individuals. In a quite similar way we can explain propositions referring to a category of human groups or social systems by a more general proposition describing a broader class of phenomena of *the same level*. Thus, we can say, for instance, that a theory which has been tested in industrial sociology and describes the structural properties of industrial work teams may later on be explained by a more general theory about the functioning of 'tasks groups' in general, formulated independently from the former one and tested, for example, in laboratory experiments.

Here again, the phenomena dealt with by both theories are of the same level, namely the level of human groups, so that establishment of the relation between the extensions of their concepts is possible by a simple rule of logical subsumption. All 'industrial work teams' are 'task groups', but not vice versa.

(B) The second class of logical derivation of reductions is of a different kind. Here the explained theory and the explaining one refer to *phenomena of different 'levels of integration' of reality* where the relation between the levels can be characterized as that between a 'whole' and its 'parts'. This is the kind of reduction which has attracted most of the attention of philosophers. Thus for instance the classical example of this kind of reduction is the explanation of the laws of thermodynamics of gas (understood as a macroscopic 'whole') by Newtonian mechanics, describing the behavior of the particles of the gas understood as a statistical aggregate.[5] In sociology we meet this kind of reduction whenever we try to explain the regularities of behavior in the functioning of 'groups' or 'social systems' by propositions describing the behavior of 'human individuals'.[6]

The basic difference between these two kinds of reduction lies in the nature of the relations between the concepts referring to phenomena of the different levels. We cannot say — as we did in the case of reduction of the first kind — that the concepts of one level are more general than those of the second. Since no 'group' may be a 'human being' or vice versa these concepts of 'group' and 'human being' are simply *mutually exclusive*, in the same sense as the concept of 'cell' and 'organ' are mutually exclusive — no 'cell' is an 'organ' or vice versa. We must say, of course, that the group *is composed of its members*, and that the *organ is composed of its cells*, but this is quite another kind of relation than the relation between the extensions of two concepts denoting the phenomena from the same level.

What is interesting is that most of the 'philosophical anxieties' aroused by the 'problem of reduction' and most of the misunderstandings around it seem to be related to the second kind of reductions only, i.e. reduction of 'wholes' to the 'parts'. And many overall 'objections' to reduction in general seem to be addressed to this kind of reduction only.[7]

As we shall see below, most of the logical problems of reduction seem to be common for both these kinds of reductions, since they are simply *general features of logical derivation of some general propositions from others*. On the other hand, there are some features of reduction of 'wholes to parts' which have to be discussed separately, and these problems seem to take especially interesting form in the case of the social sciences.

3. THE REDUCTION OF LAWS TO MORE GENERAL LAWS

Let us look now at the first category of reductions, i.e., those in which the laws or theories refer to phenomena of 'the same level' (either the level of individual human beings or that of social groups, aggregates or social systems) but which differ in the degree of their generality. For the sake of simplicity (and also because the degree of logical consistency of social or behavioral theories seldom is high enough to permit *derivation of a whole complex theory from another theory*) I shall illustrate the reasoning here with reductions of singular generalizations rather than with reduction of whole theories.[8]

The basic rule of such reduction is quite simple. If a law applies to a general class of phenomena, it applies necessarily to all subclasses of this more general class. If we assume that the laws of learning theory as presented by B.F. Skinner apply to *all animals*, they have to apply also to that subclass of animals we call human beings.[9] Therefore if we find regularities of human behavior which have been established by observation of social interactions, and then later on we come to the conclusion that these constitute a subcategory or more general laws of learning (for both humans and animals in general), we may explain the laws of social interaction by demonstrating their derivability from the laws of learning theory. The same may be done – as I mentioned above – for the laws of any level, provided that both the explaining and the explained laws are from the same level.

But in order to make such a derivation we have to demonstrate that the concepts in the antecedent and in the consequent of the two laws or theories have such extensional relations as necessary for the proper derivation. Therefore we need certain additional assumptions connecting the concepts of these two propositions which, as E. Nagel stated, will create the 'conditions of derivability', namely certain 'rules of reductive correspondence'. Such rules will tell us, for example, that 'social approval... is a generalized reinforcer',[10] permitting us to connect reductionally the propositions about social interaction with the propositions of learning theory, in which the term 'reinforcer' is one of the basic terms. Or we may have another construction written according to the rules of Hempel's deductive-nomological explanation in which the role of the rule of reductive correspondence becomes visible:

Explanans:
(1) *More general law:* Every privileged person tends to be oriented toward the preservation of the status quo.
(2) *The rule of reductive correspondence:* Very rich people are privileged.

Explanandum:
Less general law: Very rich people tend to be oriented toward the preservation of the status quo.

I do not want to discuss the truth of any of these propositions or by the same token the validity of the whole reductive construction. What I would like to stress at the moment is the *connective role* of the rule of reductive correspondence. The logic of derivation in this (simple) case corresponds to a simple syllogism and may be symbolically presented as follows:

More general law: $A(x) \to B(x)$
Rule of reductive
   correspondence: $A_1(x) \to A(x)$

Less general law: $A_1(x) \to B(x)$

Let us look now at the logical character of the rule of reductive correspondence. As we know, these rules may be either empirical or analytical,[11] i.e. their acceptance may be based either on a strictly *definitional relation* between the meanings of the concepts (a relation already established in the given language or introduced as a new terminological convention) or on the establishment of certain empirical relations. But if we refer this problem to the definition of reduction presented above, we can see that one of these possibilities seems to be excluded: we assume that the reductions are explanations of laws established in *one domain of research by some other laws established independently in another domain of study.*

This means that at least some concepts of both laws, the explaining one and the explained one, have meanings independent of one another; this applies to both their theoretical and their empirical meanings. Their theoretical meanings were defined by means of various concepts, specific to a given discipline, while their 'operational', i.e. empirical, meanings were defined by means of various indicators and measurement techniques, specific to the discipline or research area in question. Hence *the establishing of the proper extensional relation between the terms which occur in the explained laws must be empirical in nature;* it consists in stating that the properties or events which are denoted by the terms of the derived law constitute, due to certain empirical relationships, the subclass of the class of phenomena denoted by terms which occur in the explaining law. *Hence in the case of a reduction sensu stricto the rules of reductive correspondence must have the nature of empirical propositions*, i.e. propositions subject at least in principle to observational verification.

Let us look at the rule of reductive correspondence from the above example. In order to see whether it is an empirical proposition or a definitional relation, we should ask whether it is logically thinkable that in some special conditions very rich people would not be privileged, or even be underprivileged. The answer is clear and affirmative therefore our rule of correspondence is an empirical statement. There may be situations where very rich people will not be privileged, or may even be 'underprivileged'. But this has also additional consequences for the explained theory because in such conditions rich will not be interested in preserving the status quo. To the contrary, they would be interested in such form of social change as would restore their privileges.

This means that our explained law is too general, because it does not specify the conditions under which the rich will tend to preserve the status quo. Therefore it should be reformulated in a way that will reveal its conditional character. *Very simple 'reduction' to a rather obvious psychological principle reveals at once the limits of applicability of the less general law.* This is one of the most important 'practical gains' from reductive systematizations.

The empirical relations described in the rules of reductive correspondence may be of fairly diverse types provided that they specify the relation which is here sufficient for a given reductive explanation. They may consist in the statement that, for instance, the phenomena referred to by the antecedent of the explained law include, among its properties, that property which is referred to by the antecedent of the explaining law. For instance:

> *Explaining law:*
> In every human collectivity with a strongly stratified structure there are phenomena of 'perspectivist perception' – i.e., the same phenomena are perceived differently by people holding different positions in the structure of a given community.
> *Empirical rule of reductive correspondence:*
> Armies are collectivities with strongly stratified structures.
>
> *Explained Law:*
> In armies we observe phenomena of perspectivist perception.[12]

The rule of reductive correspondence is here an empirical statement which says that *among their properties* those collectivities which are called 'armies' have the property of being strongly stratified. It is empirical because we could imagine armies (e.g. certain revolutionary armies and people's militias) whose structures would not be differentiated vertically; hence, inner stratification is not analytically linked with the meaning of the term 'army'. On closer examination of this example we notice that if the explained regularity is to take place, then *out of all the properties which armies have, only that property is essential for the given explained regularity which is equivalent to the property described in the antecedent of the explaining law.* All other properties are inessential for the occurrence of the phenomenon described in the consequent of the explained law, since *on the strength of the explaining law* stratification of an organization or of a social system is itself a sufficient condition of what is termed 'perspectivist perception'. It will be said, therefore, that *the reduction of a law to a more general one by indication that the phenomena denoted by the antecedent of the explained law include both properties denoted by the antecedent of the explaining law and some other properties, specific only to the phenomena referred to in the explained law, is equivalent to dismissing those other specific properties as inessential for the occurrence of the consequent of the explained law. For that consequent to occur, only that property is essential which corresponds to the antecedent of the explaining law.*

In other cases the rule of reductive correspondence may be a causal proposition stating that the phenomena denoted by the antecedent of the explained law are the cause of the phenomena denoted by the antecedent of the explaining law. The antecedent (independent variable) of the explaining law then becomes the mediating variable of the explained relationship. This means that the independent variable of the explained law (A_1) evokes the effect B by causing the occurrence of the phenomenon A, to which the more general law refers. Of course, for the explained law to follow from the explaining law the rule of reductive correspondence must guarantee the inclusion $A_1 \subset A$ (or $A_1 \rightarrow A$), so that A_1 is a sufficient condition for A.

The following simple reduction illustrates this:

> *Explaining law:* Dissatisfaction with a political system breeds the tendency to change the system.
> *Empirical rule of reductive correspondence:* Economic deprivation results in dissatisfaction with the political system.
>
> *Explained law:* Economic deprivation breeds the tendency to change the political system.

I think this example represents a whole category of explanations that are frequently met in situations where we try to explain a 'social' regularity by a mediating psychological process independently of whether this psychological process is described by a 'psychological' or a 'sociological' theory. Thus we assume that there is a whole range of possible social antecedents of a certain psychological sequence which lead to the same consequence as described by a psychological regularity. In our example, there may be a number of alternative causes of 'dissatisfaction with the political system' (economic deprivation being one of them) and once any of them is able to produce this dissatisfaction, it indirectly produces the tendency to change it. Therefore the psychological mediating mechanism is in this case more general than any of the relations between the social causes of dissatisfaction *and their impact upon the tendency to change the social system.*

In the cases analyzed above the rule of reductive correspondence links the denotations of the terms A_1 and A by the relation of inclusion, or by a one-way implication where A_1 and A stand for certain qualitatively defined classes of phenomena. But analogical explanation may also be used when both laws, the explained and the explaining, involve functional relations between quantitative variables. Such is the case, for instance, of the following reduction:

> *Explaining law:* The greater the external threat to a group, the stronger the integration of that group.
> *Rule of reductive correspondence:* The greater the difference between the values of a minority group and those of the surrounding community, the greater the external threat to that group.
>
> *Explained law:* The greater the difference between the values of a minority group and those of the surrounding community, the stronger the integration of that group.

In all the cases discussed above the rule of reductive correspondence is an empirical proposition, which directly or indirectly follows from empirical data.

As we can see, the explaining and the explained generalizations do not necessarily have to belong to different sciences; it is enough that they have been established independently of each other. They may both be 'strictly sociological' propositions as in the case of the propositions in the last example where they referred to groups. They may both be 'psychological' ones; the most famous example of an attempt toward reduction within psychology is that to reduce the laws of psychoanalysis to the laws of learning theory in the book by Dollard and Miller, *Personality and Psycho-*

therapy.[13] A well known example of the reduction of 'sociological' regularities to 'psychological' ones is the explanation of the ideological syndrome
constituting 'fascist ideology' by certain underlying psychological
mechanisms described in terms of psychoanalytic theory.[14]

4. REDUCTIONS AND PSEUDO-REDUCTIONS

We have already seen that the typical reductions relate two theories previously formulated and tested independently of each other, in consequence
of which the rules of reductive correspondence must be the propositions of
an empirical character. Let us look now at situations where the rule involves
definitional conventions. Two kinds of such situations can be distinguished.

In situations of the first kind laws of a lower, less general level are
derived from more general laws by defining certain less general concepts
according to the rule of genus proximum et differentia specifica. *Then, of
course, whatever is true for the whole class has by definition to be true for
any of its subclasses in whatever way this subclass should be defined.* If we
have a general theory that applies to all frustrating situations, it has also to
apply to situations of frustration of any particular kind. Thus we can say,
for example, 'that people frustrated by lack of social advancement will react
aggressively'. If we have a theory which states certain regularities of the
functioning of all 'class societies', it has by definition to apply also to 'class
societies of the feudal type', just as a theory that describes the 'behavior' of
all iron bars in a magnetic field applies also to iron bars that have been
painted pink(in case anyone should wish to create a theory on the
magnetism of pink iron bars).

We could construct any number of 'lower level theories' by the rules
exemplified above, and they would have to be true to the same degree as the
more general theory is true. If we do not do that in the sciences, one reason
is that such lower level theories would not increase our knowledge in any
meaningful way. Moreover, they would not satisfy one of the postulates of a
good scientific law or theory: that it cover adequately a whole class of
phenomena to which it would validly apply, and not only a part of this
class.

Situations of the second kind are more interesting, and in a way even
fruitful for the development of theories. For in this case the concepts of the

more general theory are definitionally dependent upon the concepts of lower level generalizations in such a way that these *less general concepts constitute partial definitions of the more general ones.* If these less general concepts are of a strictly observational character, they may at the same time constitute partial operational definitions of the more general concept which except for this set of partial operationalizations *does not have any meaning in itself.* This is not an artificial situation, it actually occurred at a certain period of development of behavioristically oriented learning theory. For some learning theorists operationalizations such as 'food' or 'electric shock' constituted partial operational definitions of concepts of the theory which was supposed to explain why food or electric shock reinforce consecutive behavioral patterns, namely the concepts of 'reward' and 'punishment'. If we look more closely at such situations, it becomes obvious that explanation of the lower level generalizations (e.g. 'food reinforces consecutive behavior') is not possible in such a structure.[15] Therefore by the same token the lower level generalizations cannot be reduced to a theory which in itself does not have any meaning independent from them.

To say this does not mean that formulations of this kind are not important in the development of science. Take again the same example: There were many different alternative operationalizations of the term 'reward' in learning theory. The fact that all were applied to the same term suggested that they might have something in common which was responsible for their functioning as 'reinforcers' in the process of learning. But what was it that they were supposed to have in common? When we say, for instance, that the nature of 'reward' lies in the 'release of tension', this applies to a quite different category of behavior, either social or non-social, than when we say that the nature of reward lies in the 'increase of happiness'. If, on the other hand, we say that the nature of reinforcement lies in its being associated with the stimulation of certain centers of the brain,[16] we have quite another theory which may or may not be empirically equivalent to any of the others. And each of them may apply to different regularities, at the lower level.

In any case, the use of the terms of learning theory in a way which made their meanings completely dependent upon a set of partial operationalizations *indicated rather the need for formulation of a more general learning theory, than the existence of such a theory.*

To summarize these comments, I would propose that the 'vertical theoretical structures' in which the connections between concepts from different levels are of strictly definitional character be called *psuedo-reductions*. But it is not always the case that those explanations in which terms occurring in both laws are linked by definitional relations in the structure of a certain theory deserve the somewhat derogatory name of 'pseudo-reductions'. For it can happen that subsequently adopted *definitions of terms which occur in a given theory will be based on previously found empirical relationships* between denotations of terms occurring in laws of different levels. Consider the following class of situations. There are two laws, previously formulated and verified in two different fields of research. Further research reveals empirical relations between the denotations of the terms occurring in the two laws, enabling the formulation of empirical rules of reductive correspondence. This operation makes it possible to reduce the less general law to a more general one. In the structure of the theory involved, the terms that occur in both laws obviously continue to be independent of one another.

But in the next stage the researcher may decide to redefine the terms which occur in the explained law by *linking them definitionally* with the terms that occur in the explaining law. *In this definitional operation he takes into account the previously found relations between the denotations of terms occurring in laws of different levels, so that the definitions he suggests reflect objective relations between phenomena at the two levels.* In this way he transforms the empirical correspondence of the given reduction rules into definitional propositions, but these propositions are not arbitrary terminological conventions, *they are based on previous factual findings*.

Such a redefining of the meanings of the terms which occur in the explained law complies with the requirement that those elements of the meaning of a scientific term which are important for some theoretical reasons be above all taken into consideration. *Defining the terms which occur in lower-level laws in terms that occur in higher-level laws is necessary if we want to formulate a given theory as, for example, an axiomatic theory.* In such a axiomatic theory the terms which occur in the explaining law (or at least some of such terms) function as primitive terms when the terms that occur in the explained law are defined by reference to the terms of explaining laws.

When we look at such an axiomatic theory *after it has been constructed* we have the impression that the terms from all its levels are connected by or strictly definitional rules of correspondence; therefore we might suspect that the whole reductional structure constitutes an example of pseudo-reduction. But when we look at the problem more 'historically' (i.e. when we ask why these definitions have been constructed in this particular way) we understand that the scientists were by no means free when establishing these conventional relations but were bound by the empirical relation discovered earlier. Therefore this kind of theoretical structure 'deserves' to be called 'genuine reduction' and not a pseudo-reduction.

5. PROBLEMS OF INCOMPLETE REDUCTIONS

Can we say that in the examples discussed above we were able to reduce completely one law to a more general law? Actually we had among the premises from which the less general law would be derived two propositions. One was the 'explaining law' while the other, equally important in the reductive structure, was the rule of reductive correspondence. Only in cases where the concepts of a less general law belong by explicit definition to the denotations of the more general concepts of the explaining law can one say that the more general law alone can completely explain the less general one. But such 'reductions' (or as I called them 'pseudo-reductions') are seldom theoretically interesting.

However, the incompleteness of reductive explanations may also be of a different, more serious nature. The reductions discussed above were extremely simple, such that we could apply a simple syllogism for the control of correctness of the derivation. We assumed that the two laws differ in the generality of their antecedents but that they have *either identical or at least equivalent consequent.* We face much more serious problems when we do not make the second assumption, i.e. in the situation in which there is a relation of logical inclusions *both between the antecedents and between the consequents* of the explained law and the explaining one. Suppose that we would like to have the following type of derivation:

Explaining law:	$A(x) \rightarrow B(x)$
Cor. Rule 1	$A_1(x) \rightarrow A(x)$
Cor. Rule 2	$B_1(x) \rightarrow B(x)$
'Explained' law:	$A_1(x) \rightarrow B_1(x)$

It is obvious that this would be an incorrect logical derivation, because the explained law cannot be derived from the set of premises presented above. Unfortunately, this is not a rare situation; we meet it quite often, for example, in the cases which are believed to be the *explanation of a sociological regularity of a psychological mechanism*. Let us return again to the example where we explain the well-known regularity that (at least in some societies) people in economically underprivileged groups are likely to be racists by the frustration-aggression theory. We have here two laws or law-like statements which are then 'connected' by two correspondence rules. One of these, as stated above, treats economic deprivation as a subcategory of the frustrating situations, the other treats racist attitudes as a subcategory of aggressive predispositions. But from these two premises (plus the frustration-aggression theory) the explained sociological regularity cannot be logically derived. What we can derive from the premises is the following:

(1) Assuming only the first correspondence rule (which states the relation of inclusion between economic deprivation and frustration) we can conclude that economic deprivation *must lead to aggressive tendencies of some form.*

(2) Assuming that racism is a form of aggression (among many other possible forms of agression) we can then conclude that economic deprivation *may lead to racism, as well as to any other form of aggression.*

Thus we see that in this class of situations we are not able to make a complete reduction of one generalization to another generalization if the *consequents of these two propositions are not empirically equivalent*. To put it another way, on the basis of the above premises we are not able to explain the necessity of the occurrence of racism in the situation of economic deprivation, we are only able to explain the necessity of occurrence of *one of the features of this consequent — either racism or one of the many other possible consequences of economic deprivation, aggressiveness of people in economically frustrating situations*. Our explanation of the derived relationship is here only *partial, incomplete*; we are able to explain one of the components (or features) of the consequent of the less general regularity, but not the explained phenomenon in its entire specificity (or complexity).

If we wish to find a complete explanation of 'racism' in economically underprivileged groups, then our explanans has to be more complex. Besides the explaining law and the two rules of reductive correspondence noted above, we need a premise which states that under certain specified conditions the more general regularity will lead precisely to those consequences which are described as the consequent of the less general law. Let us call such a premise *the specifying rule*. The specifying rule may have a variety of forms. Let me mention some of them. First, we may say that in any circumstances if the antecedent of the more general law (A) takes the value (A_1); then in all possible conditions the consequent (B) of the more general law will take the value (B_1). Our explanation looks then as follows:

More general law:	$A \to B$
Cor. rule 1	$A_1 \to A$
Cor. rule 2	$B_1 \to B$
Specifying rule	$A_1 \to (B=B_1)$
Less general law:	$A_1 \to B_1$

To give a simple example of this kind of reasoning: Suppose the more general law says that 'people who are dissatisfied with their situation try to change it' (which seems to be as true as it is banal). Suppose also that the explaining law says 'people who have an underprivileged economic position in the society will try to change the economic structure of this society'. The first 'law' seems to 'somehow explain' the second one, because we intuitively accept the correspondence rule which rells us that economic deprivation is a subcategory of unsatisfactory conditions, and change of the economic structure is a subcategory of kinds of behavior aimed toward elimination of the sources of dissatisfaction. Nevertheless, in order to make such an explanation (reduction) complete, we have to assume that whenever the sources of dissatisfaction lie in people's economic deprivation, the action toward elimination of the sources of dissatisfaction will be oriented toward change of the economic structure and not toward other ways of improving one's economic situation. Such a specifying rule is, of course, an empirical proposition and does not necessarily have to be a priori true. It could happen that people will rather seek improvement of their situation in hard work or in striving toward their own individual promotion up the social ladder. But the specifying rule is necessary for our reduction, which then can legitimately relate the explained regularity to the more general class of phenomena and constitutes an essential part of the reductive theory.

Our specifying rule may also have the form of a *conditional proposition* which states that generally under the additional conditions C, if $A = A_1$ then $B = B_1$. The specifying rule for the above example would be that in societies where there are practically no chances for individual improvement of the economic position of an underprivileged person – i.e. if the economically underprivileged situation invariably or inevitably leads to dissatisfaction – the attempts to improve one's position will take the form of attempts to change the economic structure of the whole society.

Another kind of specifying rules are the propositions which state that in the case of *one specific society, group or culture, for reasons we are unable to explain at the moment, the more general phenomenons described by the consequent of the explaining law takes the particular form which is described as the consequent of the explained law.* Then the specifying rule has the form not of a universal but of a *historical proposition*, stating that in the concrete society, group or culture H the more general phenomenon B must have the form B_1. But then, by necessity, we have to treat the explained regularity not as a universal law but as a 'historical generalization' valid for the society H only. Our reductive theory looks then as follows:

More general law:	$A \rightarrow B$
Cor. Rule 1	$A_1 \rightarrow A$
Cor. Rule 2	$B_1 \rightarrow B$
Specifying rule:	$H \rightarrow (B = B_1)$

The explained historical generalization:	$(H \text{ and } A_1) \rightarrow B_1$

Here again we can refer to the many examples of explanation of racism by the frustration-aggression theory, in which the necessary specifying rule is the assumption that in the given society racism is the only (or at least the most common) outlet for the aggressive tendencies. *Such reductions are valid for the specified societies only.* And of course if we come to the conclusion that the specifying rule is not empirically valid even for a particular society H, then we have the situation of an incomplete reduction, in which we van say we have proved that, according to a more general theory, in the given society A_1 *may* (but does not necessarily have to) lead to B_1.

6. EXPLAINING A LAW BY A NUMBER
OF ADDITIVE LAWS

The cases considered so far consisted in explaining one law by another, more general, law, while the rule of reductive correspondence in that type of inference played auxiliary functions by connecting the denotations of terms occurring in both laws by the inclusion relation. But in many cases the reduction of the explained law requires considering, in the explanans, many laws working simultaneously and *additively contributing* to the mechanism of the explained regularity. Distinction has to be made between two kinds of cases: those in which the explaining laws differ from one another by their independent variables but have one and *the same dependent variable*, and those in which the explaining laws differ from one another by *both independent and dependent variables*.

Suppose we wish to explain reductively the generalization of the theory of social mobility, stating that the children of peasants very rarely acquire the social status of professionals. We know there are a certain number of much more general propositions which in the consequent have the variable: the chance of entering the group of professionals.

(1) The higher the income in the child's family, the greater his or her chances of entering the group of professionals.

(2) The more the parents emphasize the child's education, the greater chances he or she has of entering the group of professionals.

(3) The higher the evaluation of education in the child's family, the greater his or her chances of entering the group of professionals.

(4) The more numerous the personal contacts of the child's family with professionals, the greater his or her chances of entering the group of professionals.

(5) The less the (non-professional) family urges the son to take up his father's occupation, the greater the son's chances of entering the group of professionals, etc.

We state at the same time that in the presant class the independent variables occurring in all these functional relationships have very low values. Now, *on the condition that the consequences of those relationships act additively*, it follows therefrom that children from the peasant class have very small chances of entering the group of professionals as a result of upward mobility processes. By referring to the same laws, but make the assumption that the independent variables occurring in the applicable laws have very high values, we explain the law which states that children of professions have good chances of remaining in that group.

The logic of this reduction (corresponding of course to Hempel's scheme for deductive nomological explanation of laws) looks as follows:

> *Explaining laws:*
> $B = f(A_1)$; $B = f(A_2)$; ... $B = f(A_n)$.
> *Rule of correspondence:*
> $C = A'_1$ and A'_2 ... and A'_n where A'_1 ... A'_n refer to specific values of the variables A_1 ... A_n; C stands for the whole syndrome equivalent to the antecedent variable of the explained relationship.

> *Explained law:* $C \rightarrow [B = f(A'_1) + f(A'_2) + ... f(A'_n)]$

In other cases the additive character of the set of the set of laws looks different, because *each of the explaining laws explains another aspect of the explained regularity*. Suppose we would like to explain that groups of a special kind (C) have certain value systems (V). We might say that these groups are characterized by a set of characteristics $C_1, C_2 ... C_n$, when the value system characteristic for them is composed of values $V_1, V_2 ... V_n$. If we find later on in our more general sociopsychological theory a set of laws each of which relates a specific kind of value to a specific condition according to the scheme

$$(C_1 \rightarrow V_1); (C_2 \rightarrow V_2); ... (C_n \rightarrow V_n),$$

then we can reductively explain the relationship $C \rightarrow V$. But we have of course included in our reduction the correspondence rule $C = (C_1$ and C_2 ... and C_n) and also another rule $V = (V_1$ and V_2 ... and V_n). In such case the *set of explaining laws, under the assumption that their antecedents are interrelated into a syndrome C, permits us to explain a whole complex set of characteristics composing the syndrome V.*

In order to make such an explanation possible we have to make two additional assumptions:

(a) that all other characteristics of the syndrome C (other than the properties C_1 ... C_n) are *irrelevant for any of the explaining regularities*, i.e., they are not able to 'cancel' it or to modify its strength or direction.

(b) that the antecedent of each of the explaining regularities is not able to cancel, modify or change the strength of any of the other relationships used for the reductive explanation of our regularity.

To put it another way, in the case of reduction of a law to a whole set of more general laws, we have to assume that there is *no interaction* between any of these relations described by the explaining laws and any other characteristics of the less general phenomenon *C*, either those which occur in any of the other explaining laws or those which have been omitted in the scheme of our explanation. This means we have to assume that all our explaining regularities work in the given set of conditions additively, i.e. in the same way as they work 'in isolation', as they were described in the more general theory. In other words, to use a term which will be discussed in following sections of this chapter, *we have to assume that there is no 'emergence' of some basically new regularities as the result of the joint action of several explaining laws.*

7. REDUCTION OF STATISTICAL LAWS

What has been discussed so far applies only to general laws or at least to 'historical' but exception-less regularities, because only for such cases can we apply the rules of syllogistic reasoning or some other logical rules of derivation. But, as we know, there are very few (if any) propositions in the social sciences which are really general, and it is much more realistic to assume that practically all the regularities we discover in our research, or the laws that we have in our theories, are *statistical propositions* which may have different forms:

(a) They may state certain *probabilistic relations* between a more or less complex antecedent and the subsequent of the given law, stating that the probability of the occurrence of *B* when a certain *A* has occurred is equal to a certain *p*.

(b) They may describe the strength of *correlations* between *A* and *B* in as unambiguous numerical terms as possible:

(c) They may have the character of propositions stating that there is — on a general scale — a *positive relationship* between one and the other variable so that the occurrence of the event denoted by the antecedent of our law somehow increases the probability of the occurrence of the event denoted by its consequent, but we are unable to say how strong (in general terms) this relationship is, since its strength may vary in different conditions.

(I omit here for the sake of simplicity the various kinds of propositions which may refer to the *statistical relations between the quantitative variables.*)

Actually when we look at the propositions in sociological and psychological theories, we find that on the basis of survey or experimental findings of a correlational character the theoreticians usually devise statements in which there are formulations like: '*A* tends to be followed (accompanied) by *B*', which means that they assume there is a relation of the third kind mentioned above, i.e. that there will be a positive relation between *A* and *B*, but possibly differing in strength under differing conditions.

If we face the problem of explanation of such a law by another law, i.e., its reduction to another, this reduction may be one of two different kinds (assuming we have decided to reduce it to a proposition of the same level):

(a) We may try to explain such a law by a certain *general causal law*. We assume then that a statistical regularity (of either probabilistic or correlational character) is the effect of action of a certain general, exception-less conditional regularity, and we do not know, or we do not observe or control, the conditions under which our relationship is really general. The logic of such explanations is discussed in another chapter of this volume.[17]

(b) We may try to reduce our statistical law to *another statistical law which applies to a more general class of phenomena.* It is this type of situation that will be discussed here.

Let us assume that we have the following situation:

We know a law which states that there is a specified probability that A will be followed by B: $P(B/A) = p$.

We also know that the phenomena A_1 constitute a subclass of phenomena of the type $A : A_1 \rightarrow A$ (Rule of reductive correspondence).

It is obvious that in this case we *cannot derive* the statement that $P(B/A_1) = p$ without additional assumptions.

Even of we have a proposition of a much weaker type stating only that there is a positive (numerically unspecified) relation between A and B, (when the term A is more general than A_1), we cannot say in advance that there will be a positive relationship between A_1 and B. In order to make such a derivation we would have to formulate some additional assumptions, which are by no means always true.

As we will recall, in the case of simple reduction of a law to a more general law we had to assume that the other properties which codefine the concept A_1 as a subclass of phenomena A are irrelevant for the occurrence of the relationship $A \to B$ so that the relationship $A_1 \to B$ may occur. Our assumption in the case of statistical explanation must be an analogous character: We must assume that the other properties which make the phenomena A_1 only a subclass of the class of phenomena A do not basically change the explaining statistical relationship P (B/A) = p. Under this assumption we may derive from the statistical relationship between A and B the explained relationship between A_1 and B.

This assumption may have either a *stronger* or a *weaker* form. Let us assume that we know a statistical law which says there is a specified probability p that A will be followed by B. If we would like to derive *the same probabilistic* value p for the relation between a less general class of phenomena A_1 (subclass of A) and the same dependent variable B, we must assume that whatever makes the phenomena A_1 different from A has not impact upon the probability of occurrence of B when A has occurred. In other words, we must assume that all properties of the phenomena A_1 other than the property A are statistically independent from B (at least as long as we are observing the sequence of events $A \to B$). Suppose that the phenomena A_1 being a subcategory of the phenomena A possess also the property C so that $A_1 = (A$ and $C)$. We have to assume then that C is independent from B within the sequence of events $A \to B$. Our derivation looks than as follows:

$$P\,(B/A) = p$$
$$A_1 = (A \text{ and } C)$$
$$P\,(B/AC) = P\,(B/A\overline{C})$$
$$\overline{}$$
$$P\,(B/A_1) = p.$$

In such case we can reduce the proposition $P(B/A_1) = p$ to the proposition p (B/A) = p, because we can derive the explained law with its specified probabilistic value from the specified probabilistic value of the explaining law.

In most cases the explanation of a statistical law by another statistical law looks different from the above derivation, either because the probabilistic values of the two laws are different, or because one or both laws are formulated in terms of numerically unspecified statistical tendencies (or positive relationships) only. In such case it is sufficient to make much weaker assumptions: We have to assume that the *properties which are characteristic for the phenomena of the explained level do not change the direction of the explained relationship as compared to the explaining one*, although they may change the probabilistic value of B/A for the relationship $A_1 \to B$ or the strength of their correlation. This means that we accept the possibility, for example (to return to the above illustration), that the variable C (or some of its correlates) may be the *additive cause of B for A_1* thus increasing our decreasing $P(B/A_1)$ as compared with the relationship $P(B/A)$. In any case *we exclude the possibility that C may be the interacting cause* of the relationship $A \to B$, which could 'cancel' or change the direction of this relationship.

It may also happen, as in the case of general laws, that we reduce the relationship $A_1 \to B$ to several additive statistical relationships, for instance, to $P(B/A)$ and $P(B/C)$. The logic of such reductions is discussed in another chapter of this volume.[18]

Finally, we may have the situation in which (at least in theoretical formulations) both laws refer to numerically unspecified statistical tendencies. The explained law and the explaining one both state that their antecedents 'tend' to produce B. In such case the derivation of a numerical relation is not possible, but we can still explain one law by the other assuming that the properties which are characteristic of the phenomena of the less general (explained) level do not interact with the antecedent of the explaining law.

In cases where we do not derive the probabilistic value of the explained statistical law from the given value of probability in the expleining law, but only the positive relationship between A_n and B_n then of course one of our assumptions can be weaker; we should then assume only that the direction of the statistical relationship between B and A_1 is the same as between B and A.

The same kind of reasoning can be extended to the reduction of a statistical regularity to several additive statistical laws: In each such reduction we have to add the specifying rule stating that the explaining laws under the specific conditions of the explained laws lead to the same subcategory B_1 of the dependent variable of the explained law.

Since most of the examples presented in Sectiond 3 to 6 of this chapter illustrating reductions of a law to another, more general one refer (given the reality of contemporary social theory and of contemporary social research) to statistical regularities only, the reader should look at them again and reinterpret them in the context of the foregoing remarks in this section.

8. THE 'OBJECTIVE EXISTENCE' OF COLLECTIVITIES AND THE DEFINITIONAL DEPENDENCE OF THEIR CHARACTERISTICS UPON THE CHARACTERISTICS OF THEIR MEMBERS

Up to now the discussion has centered on problems of reduction in which both the explaining and the explained laws refer to phenomena of the same level of the studied reality, and differ only in their degree of generality. Let us now consider a second class of reductions, in which the two theories we would like to relate to each other in a deductive manner refer to *different levels* of integration social reality. Such multi-level relations between phenomena are often characterized by the terms 'whole' and 'parts', the latter constituting the former.[19] As I noted earlier, the notion of different degrees of generality does not apply to concepts from different levels (or propositions in which these concepts occur), because the concepts of 'part' and 'whole' are simply mutually exclusive – no 'part' is the given 'whole' or vice versa. Therefore let me say that in the case of the explaining level (the level of the 'parts of the given whole'), I shall use the term 'more elementary laws', or for the given reduction simply 'elementary laws', while for the level of explained phenomena, occurring at the level of the 'whole', I shall use the term 'macro-laws'.

Here I should make one terminological remark in order to avoid future misunderstandings. In speaking of 'macro-laws', I have in mind not only those propositions which describe regularities at the level of large societies or human collectivities. A given regularity is of the 'macro' type only with respect to certain more elementary regularities in the context of which the macro-law is being analyzed. As rightly stated by C.P. Wolf in his introduction to *Macrosociology:*

Relative to cultures or communities, small groups are microscopic phenomena-but not when compared to their constituent units, usually individuals or persons.[20]

As we know, most philosophical controversies around reduction in sociology refer to this class of situations. One of the first questions that have to be asked in sociology is whether the phenomena described by the macro-laws 'do exist'. Here we face at once one of the most visible differences between social and (at least some) natural sciences. In biology no reasonable person doubts that the body exists in no less a degree than its organs, or that the organs exist as well as the cells from which they are composed. On the other hand, many philosophers of the social sciences express their doubts as to whether units of the higher level of social reality 'really do exist', and whether only the individual human being exists, and whether it makes sense at all to say that the properties of the higher level units exist. The dispute about the 'existence' of higher-level units constitutes one of the dividing lines between the school of *methodological individualists* and those who call themselves *holists*.

In order to avoid the metaphysical problems involved in the meaning of the term 'existence' itself, let me popose that we understand it in the following way: *We can say that a certain object or phenomenon exists when we know what we mean by the term referring to it and how we recognize its referents and we find these referents in the empirical reality.* Now assume that we agree that the units of the lower social level (human individuals) 'really do exist'. Then the question arises, what do we mean when we say that certain higher-level social units (e.g., groups) exist too?

First, we may mean that we are able to observe their existence and 'behavior' in a direct way, and therefore we make observational judgments about them. In my opinion it is to be assumed that the term 'observable' is a quantitative variable: objects and phenomena may be more or less directly observable. A small group of people can be observed in its actions as directly as a single person, whereas assessments of events at the level of nations or social classes are as a rule inferences from longer or shorter series of more detailed observations. But this is not a situation unique to the social sciences. If we took the term 'to observe' in its literal sense, we would say that before the time of artificial satellites the existence of earth and all its global characteristics were of inferential nature. This 'conclusions' was the result of integration of a great number of singular observations covering the earth's smaller or larger parts only.

The question of observability may be posed in a more specific way. Suppose we agree that we are able to observe a certain object as a whole (a stone, a group in action). Can we observe it independently from the observation of its composite units at a given lower level? In some sciences the answer is obviously affirmative, and by changing the conditions of observation and the observational techniques we can 'switch' from one to another level of observation. Thus, instead of using our eyes only, we can use the microscope and observe the macro- and micro-levels of biological phenomena independently from each other. We can independently observe certain macro-properties of a piece of metal and the separate crystals of which it is composed. The same may be true, although to a lesser degree, for some social units. A crowd seen from an airplane may look like a large blotch when we are unable to see its separate members but later on we may see them from a lower attitude. In other cases, when we observe certain social units, whether our attention concentrates upon the 'whole' or upon its 'composite parts' (the members) is a matter of the focus of our attention, in the same way as when watching a machine we may concentrate our attention either on the machinery as a whole, or on its parts and the interrelations between them. But the fact that we are able to concentrate upon the parts does not mean that the machine exists 'to a lesser degree'. The fact that when we observe a task group we see a number of workers does not imply that task groups do not exist in this observational sense, just as galaxies exist even when some of them are observed only as white spots rather than, like others, as configurations of stars.

But the question of the observability of units of a certain level does not solve the problem: *What kinds of properties of these units are we interested in?* If we see a crowd from an airplane, we can see only its physical properties (like size, shape, color), but for us, *as social scientists*, these properties are, at the most, of secondary importance. What we would like to know would be such things as: Why are those people gathered together at the same place? What are the reasons that brought them there? What do they intend to do? Are they only an amorphic mass, or is there an internal social structure within the crowd? In other words, we ask specific kinds of question in terms which have *at least an indirect reference to certain characteristics and behaviors of the human individuals composing these units.* This is true both for situations in which we are able to observe a unit without observing directly its 'composite parts' (a crowd from an airplane) and when we observe the unit jointly with the behavior and characteristics of its members (the task group). Thus the observability and existence of collectivities does not change the fact that we are primarily interested in such their properties, the meanings of which are somehow related to the behavior and characteristics of their members.

Now, when the size of a certain 'object' is too large to be accessible to a global observation, we may also and rightly so insist upon its existence, even if our judgments are of an inferential nature. Thus we may say that certain social units of a higher level exist because we define them in such- and such a way (in terms of certain characteristics and relations between their composite parts) and find these relations in empirical reality.

The observational or inferential character of phenomena from a given level and the dependence of the characteristics of higher level units on the characteristics of individual members are to a considerable degree determined by the *natural observational perspective of human beings*. Let me give an illustrative example. As mentioned above, the phenomena of the level of whole biological organisms are accessible to direct human observation. Suppose now that there exists a micro-organism the size of a single cell of our body which, for some fantastic reason, possesses the observational and intellectual capacities of a human scientist. From *'his' cognitive perspective* the natural observational level would be the level of single cells; being an intelligent scientist (and being able to 'travel' freely within the human organism), this 'micro-scientist' would come to the conclusion that certain configurations of the cell are distinct enough and sufficiently separated spatially from other cells to be called higher level units (organs). He would also study the interactions between these higher level units and the functions of one configuration for another one. As the result of long studies our 'micro-scientist' would even arrive at the notion of the organism as a whole and describe the patterns of its 'behavior'. But in doing so, he would always have to *define* all higher level phenomena in terms of those units which were accessible to his direct observation and to make judgments about them in an inferential way, by observation of those units observable from his level. *Our natural observational perspective and the kinds of questions we are interested in seem to constitute for us the ultimate reference both for our judgment about reality and for the specification of meanings* of the concepts we use in our studies. This does not of course invalidate the theoretical usefulness of the concepts denoting phenomena from the inferred level, but it makes them definitionally and observationally (operationally) dependent upon the concepts denoting phenomena from the level accessible to our direct observation.

In the case of larger social units, inaccessible to direct observation, the definitional dependence of properties of higher levels upon the characteristics of lower level units is not a matter of choice of problems, as in the case where we might be interested in the shape or color of a crowd seen from an airplane; but we think that such problems are not interesting for social scientists. In the case of larger social units their properties must be defined in terms of properties of their members because *we do not have any direct observational data from the higher level which could give any independent meaning to the terms for this level.* As we know, in sociology many 'collective properties', especially the properties of human communities and aggregates, are simply *defined* in terms of properties of their component parts, (i.e. singular persons). They may be *absolute* properties of these component parts, as in the case of those properties of human collectivities which are described by various statistical parameters, where given mathematical formulae define the meanings of some collective terms. P.F. Lazarsfeld and H. Menzel[21] proposed to call such properties of human collectivities their *analytical properties*, and the collectivities distinguished by such characteristics are usually called (aggregates'. Other collective properties are defined in terms of relations between the parts of the given 'whole'. Such is the nature, for instance, of the various *structural properties* – like e.g. the structure of small groups defined in terms of the ratio of reciprocated choices in a group to the total of all possible choices. We usually call such collectivities 'systems'. Both the analytical and structural properties of collectivities are of course definitionally dependent upon the absolute and relational characteristics of their members.

The question arises whether *all* properties of human collectivities are in their meanings related to absolute or relational properties *of the members of the given collectivity.* There is definitely one category of properties for which this is not true; these arethe concepts referring to *relational properties of the collectivities themselves.* If we wish to specify the meaning of the concept 'power elite', then we have to take into account not only certain patterns of behaviors (and predispositions to these behaviors) among the members of the elite, but also the properties and predispositions of those who obey the power elite. If we say that a certain group has the highest average intelligence among a number of groups, we necessarily refer not only to the absolute measures of the highest group members on the intelligence test but also to the absolute properties of members of all other groups.

But Lazarsfeld and Menzel distinguished from among the other properties of human collectivities their 'global characteristics' as well, given the following examples:

American Indian tribes have been characterized by the frequency with which the themes of 'achievement motive' make their appearance in their folk tales. Societies have been classified as to the presence of money as a medium of exchange, of a written language, etc... Army companies may be characterized by the cleanliness of their mess equipment.[22]

When we look more closely at these 'global properties' we see two aspects of their meanings. First, we find the reference to some *generalized patterns of behavior of all or some usually unspecified group members*. Thus we may say that at least some group members must be in the habit of keeping the things clean since their mess equipment is clean, and we may treat in the same way the occurrence of the 'achievement theme' in the folk takes of a given society. The existence of money as a means of exchange can be treated as a very complex pattern of interactions between the group members. There is nevertheless one thing that seems to make these kinds of generalized (absolute and relational) properties of group members different from the typical 'structural' or 'analytical' properties distinguished by Lazarsfeld and Menzel. We infer the existence of the given behavioral pattern not by consecutive observations of all group members, one after another, but — as James Coleman stresses[23] — in a more 'wholistic' or 'impressionistic' manner so that we are unable to say which of the group members keep the mess equipment clean, which are responsible for the invention or repetition of the folk tales in the given group, etc., although we have no doubt that *some* individuals must behave in the given way inasmuch as we observe the given phenomena 'at the group level'. These individuals may remain unspecified even with respect to strictly conceptual aspects of the collective properties. Suppose we say that 'a certain value exists in a certain society when at least more than half of its members accept it'. Then, of course, the problem of which group members accept it and which don't is, from the standpoint of this 'analytical' group property, irrelevant; the only thing that matters is that there be a sufficient number of those who accept this value. *The exchangeability of particular individuals in 'creating' a given property at the group level* is another factor contributing to the sense that these properties are to some degree definitionally independent from the properties of concrete individuals.

Group properties assessed in this 'impressionistic manner' have been the subject of a thorough analysis by James Coleman, who states:

Our assumption is that the observation on the group as a whole will capture something usually missed in observations of discrete individuals apart from their relationship to others in the group. Though these observations are often unsystematic and impressionistic, they are ordinarily insightful in capturing aspects of group processes.[24]

I agree completely with this statement. I agree also when Coleman says that we need concepts denoting such group properties in our theories. In order to justify this, it

is enough to say that social scientists *do* make generalizations about social units as such, that these generalizations are useful at the present state of affairs, and that they would be more useful if there were a welldeveloped methodology for making them.[25]

We can also say that theories which refer to phenomena of the group level, understood in a more or less vague way, may be verified independently from their strict 'redefinition' in the language of properties of individuals. When we look at the history of social theories we can find quite a few theories *which were formulated and* (at least partly) *verified* on the basis, for example, of *historical data*, which constitute a typical kind of evidence for this kind of theories — and there is no reason to stop this kind of verification studies now. But, as May Brodbeck writes:

In principle, of course, for whatever comfort that may be, all such concepts must be definable in terms of individual behavior. In practice, however, we frequently cannot do this. Are we then prepared to say that in whatever context these terms occur they are wholly ambiguous as to meaning? I hardly think so. The course of science is not always as smooth as the logical analysts would like. And it seems to me that there are cases in which the best we can do is to point out the distinctions and the difficulties. The most that we can ask of the social scientist whose subject matter requires him to use such 'open' concepts is that he keep the principle of methodological individualism firmly in mind as a devoutly to be wished-for consummation, an ideal to be approximated as nearly as possible.[26]

One can also add that if at a certain stage of understanding of a macro-theory this ideal cannot be realized to a high degree, then such a theory cannot be reduced to a behavioral theory. This does not mean, of course, that it cannot be developed and tested at its own level (and by the indicators applying to it in their more or less 'global' meanings.) In some cases the methodological ideal of reductive definitions of macroprocesses is being approximated quite closely even outside the area of reductive thinking, and many scientists undertake the reinterpretation of such macro-concepts in terms of properties of individuals and the relations between them only in order to give the corresponding concepts *more precise meanings*.

A good example of the closer analysis of such a collective concept bringing the reference closer and closer to the level of human individuals is the analysis of the concept of 'integration' by W.S. Landecker. He wrote:

> In order to distinguish among the different ways in which a group may be integrated, it is necessary to assume the existence of different types of group elements. A typology of integration can be developed on the premise that for sociological purposes the smallest units of group life are cultural standards on the one hand, and persons and their behaviors, on the other. If one uses this premise as a criterion of the types of integration, three varieties suggest themselves: integration among cultural standards, integration between the cultural standards and the behavior of persons, and integration among persons.[27]

Then, if we ask what is meant by 'cultural standards', it turns out that it means certain value orientations or certain attitudes of the group members. Thus, in a fairly clear way, the vague notion of 'integration' of the group has been definitionally referred to the properties and behaviors of the group members.

In some cases this kind of reference may seem questionable, especially if we take into account those characteristics which constitute certain propensities[28] of the groups or institutions, and which can only be derived from their 'actions' and 'behaviors'. The group may 'tend to survive' and to behave accordingly, even if none of its members (at least consciously) wants it to survive. This would seem to be a typical case of what is called an 'emergent' property, and one which deserves sociologists' attention. On the other hand, a group may be 'well integrated', even if none of its members is aware of or concerned with its integration.

A sociologist may define any concept at the group level if such is required by his study. But when it comes to a closer specification of what he means by the finding that the group tends to survive or that it is integrated, this turns out to be impossible without his specifying what kinds of behavior or at least some of its members he has in mind when speaking about the group's integration or its tendency to survive, even if he has been able to observe the behavior of the group as a unit.

The second aspect of the meaning of 'global properties' — as distinct from the generalized patterns of group members' behavior — lies in their reference to certain *material objects, which are either correlates of group attitudes or the results of corresponding behaviors.* By this I mean things as 'coins', 'clean mess equipment', 'books with folk tales of a given kind', etc. With respect to these objects we face two alternatives. We may treat them as composite elements of the group itself, premising that the social reality is composed both of human beings and of certain material objects essentially related to the members of the groups, just as we say the factory is composed of workers and machines, the army of soldiers, uniforms and guns, etc. Then, when analyzing the property of the given group, we have to treat these material objects as elements of the given part of social reality in much the same way we treat the human beings who compose the group. Then of course the 'global properties' referred to by Lazarsfeld and Menzel become either analytical or structural ones.

We may also say that these material objects are external with respect to the group itself; they constitute the *environmental* context of its functioning by reason of the fact that they are (spatially or otherwise) realted to it. Then the fact of possession of certain objects by the group is *a relational property of the group* itself or of its members.

It should be kept in mind, too, that the properties of individuals constitute, in some cases, a *direct* definitional reference for the concepts of the 'collective' level but, in other cases, only a *very indirect one.* If we want to specify the meaning of a structural property of a collectivity having a very complex structure of a multi-level kind (e.g. a large organization), then of course, we have first to define it in terms of the units of the nearest lower level of integration, and in such a way as will reveal the relations between them. Thus the natural 'next level unit' in the anatomy of the human body would be the level of the separate organs, which later on we might describe in terms of the cells composing each organ (or we might find that for our given purpose the description in terms of organis is sufficient). But should we wish to characterize the structural properties of modern states, the description of these properties directly in terms of human individuals would be as useless as it would be technically impossible; we have to specify the nearest 'natural level' of integration of such a structure as a modern state, and then reveal, step by step, the nature of the relations between the units at each particular level which constitute the next higher level unit. In some cases we can say that for our purpose the relations at a given level (e.g., in the case of 'state', at the level of certain administrative regions or sectors of an organization) are sufficient, and we do not need to go 'down' to the level of constituent individuals. But when we begin to think *what we mean* by certain characteristics of the units of this level, then more or less indirect reference to certain characteristics or behaviors of the individuals comes immediately to our minds. The multi level character of a given structure should not obscure the ultimate definitional dependence of most characteristics of social collectivities — at all levels — upon the properties and behaviors of individuals.

Suppose now we agree that at least in the social sciences the characteristics of collectivities are definitionally dependent *directly or indirectly* upon the characteristics of the human beings composing the collectivity, and possibly the characteristics of members of other groups, if the collective property is itself of the relational type. Does this mean that these characteristics are *definitionally completely reducible* to the characteristics of their composite elements?

This is another aspect of the dispute between 'methodological individualists' and 'wholists' (besides the question whether group 'really exist'), which was presented by May Brodbeck in the following way:

The denial that there are such definable group properties or, such superentities is the view usually known as *methodological individualism*. It is contradictory to metaphysical *holism*. It is called 'holism' because its proponents generally maintain that there are no so-called wholes, group entities, which have undefinable properties of their own.[29]

I would be inclined to answer this question in the following way. When we define the meaning of a collective property (or at least think about it in a more or less impressionistic manner), we necessarily *refer* to certain characteristics of behavior of all or some of its members. But we also have in mind that these members are elements of a certain collection, structure, or configuration. The notions of *structure, collection,* or *configuration belong definitively to the same collective level as the property of the collectivity itself*, and they seem also to involve the meaning of collective property over and above the properties of individuals. In some cases this is clearly visible in the definition of property itself. When we use, e.g., the notion of 'mean income' of a certain group, we have in mind the kind of value we obtain by adding the incomes of each of the members of this group taken separately and dividing the sum by the *number* of group members. The notion of number 'belongs' definitely to the collective level, although, the *mean income depends* on the incomes of individuals. The same seems to be true for more structural characteristics. Suppose we wish to define the structure of a group having one 'sociometric star' and no relations between the other group members but only those between each member and the star. This notion depends of course on the individual sociometric relations between the group members, but the notion of configuration which 'emerges' from them belongs to the collective level.

The only problem is that those sciences where there are no independent observable properties at the higher levels of configuration, these *configurations, etc. are of a strictly formal character*. They have to be formal precisely because we are unable to observe any phenomena at the global level which could give them any empirical meaning independent from the empirical properties of individuals. Being strictly formal (e.g. analytical, structural) properties, they can apply to all possible aggregates and systems which satisfy their requirements from a strictly formal point of view; the structure of a television set and the structure of an organization as described by a graph may be quite similar, as from a strictly 'structural' point of view two persons as well as two stars may be thought of as 'interacting dyads'. *As a result of the 'interaction' of these formal characteristics of a configuration of elements with the empirical properties of the elements involved in it, we obtain the proper meaning of the collective properties in cases where the level of the collectivity is not subject to independent observation*. The definitional dependence of collective properties upon individual ones is therefore only partial, but it is a very essential part because it *gives to the collective notions a definite empirical sense*. Once we see *how the kinds of elementary properties and their relations to each other within the given configuration determine this meaning*, we can later on of course perform any kind of logical or mathematical operations on them and define any new concept for the collective level. We can speak of group actions or group propensities; we can characterize a group's structure in qualitative and qualitative terms; etc. While doing this, *we operate all the time at the collective level*. I would like to stress that such reference to the properties of individuals in the interpretation of collective terms does not mean that we are *replacing* collective properties with individual ones, but we are rather *using* the individual properties in order to define in an empirically meaningful way the collective phenomena.

9. GENERAL PATTERN OF INTER-LEVEL REDUCTIONS AND THE CHARACTER OF CORRESPONDENCE RULES IN REDUCTIVE SOCIAL THEORIES

When we try to relate a macro-law to some elementary laws in a reductive theory, the main problem is to find (or formulate) rules of reductive correspondence connecting the phenomena or processes from the two levels which make our derivations possible. In some sciences, as I have stressed several times, such rules may be of a clearly empirical character, as for instance when we relate the 'hardness' of a piece of metal to its specific crystalline structure or the behavior of a certain organ to the pattern of functioning of its cells, etc. How does this problem look in the social sciences?

From what I said above one could derive the conclusions that if most of the sociologically interesting properties of human collectivities seem to some extent to be definitionally dependent upon the characteristics of the individuals belonging to the given collectivity, these rules must be definitional. Actually they *may* be definitional statements, but *they may also be empirical ones*, and I think that the empirical rules of reductive correspondence will be more frequent than definitional ones in future reductive theories in the social sciences. But *these empirical relations will be of a special kind, namely they will connect the variables from the same, elementary level*. Let me explain this.

Suppose we have a macro-theory in sociology which relates two variables with the level of collectivities. The meaning and the indicators of these variables may stress their *strict definitional and operational dependence* on the properties of individuals, as structural or analytical properties of the collectivity, or they may be understood in a more or less vague way and assessed in a more 'impressionistic' manner. In the latter case, before we undertake any reductive explanations of the corresponding macro-law, we have to specify the meanings of their concepts in such a way that the dependence of the corresponding macro-properties upon the properties and behaviors of individuals becomes as clear as possible. In either case it may happen that the properties of individuals (which we use for the specification of meanings of the properties of collectivities to which the macro-theory refers) *will at the same time be involved in an elementary theory which we have to use in our reductions*. Then, of course, our correspondence rules will be strictly definitional. But it may also happen that the properties of individuals which we have to use in our specifications of meanings of collective properties will not belong to any elementary theory at all – or at least that the theory in which they occur is not suitable for the explanation of this macro-theory we would like to explain. If at the same time we find another elementary theory which could be used for the reduction of our macro-theory, the relation between the indicated concepts belonging to the specification of meanings of collective properties and the individual concepts figuring in the elementary theory has to be an empirical one. To give a most simple example: Suppose we specify the meaning of certain processes at the level of a small group in terms of a certain configuration of roles, and that we interpret each of social roles in this group as a certain configuration of the expectations of other members of the group directed toward the role-holder and the social sanctions they have at their disposal. Suppose now that we would like to explain this mechanism by the *S-R* theory of behavior. Then we have to find the correspondence between the 'role-behaviors' and 'reinforced behaviors' (in the way discussed above,) and the relations between these *concepts at the elementary level have to be empirical onces*.

In the case where the elementary (explaining) regularity itself refers to a lower but still collective level (e.g. to the level of the functioning of small groups), then the situation is even more complex. It may happen that one set of properties of individuals will be used for the specification of meanings of the properties of higher level collectivities, and quite another set of properties of *the same individuals* as the specifications of meanings of the concepts to which the elementary law applies. Before we relate reductively our macro-theory to the explaining theory we have to establish the empirical correspondence of these two different sets of properties of individuals in a way that will make our reduction both theoretically consistent and empirically meaningful. For reasons of simplicity I shall not discuss this category of reductions here, limiting my attention at the moment only to reductions in which the explaining theory applies to the level of the members of the given collectivity. Let us look a little closer at the logical structure of our reasoning in such reductions.

Suppose we have a generalization which refers to the collective level of social phenomena and which states that when certain 'social wholes' *W* have the property *A* they also have concurrently, or as a consequence also the property B, which may be stated symbolically as $A(W){\rightarrow}B(W)$. Now we would like to reduce this generalization to a certain regularity (described by a generalization, law or theory) applying to the level of elements e (in our case, the individuals belonging to the collectivities). We have to assume first that these individuals are the 'composing parts' of their wholes, which (according to our arguments presented above) constitute certain *configurations* of the individuals with their absolute and relational properties and which also have some formal properties 'of their own' configurational level as, e.g., proportions of individuals of different kinds, the number of individuals involved, the overall pattern of totality of relations between these individuals, etc. Let us denote these configurations by C and, of course,

$$C(e) = W.$$

The exact nature of this configuration $C(e)$ does not have to be specified in our reductive theory as long as we more or less clearly know what it means and especially how we can recognize the wholes W and the individuals e belonging to them, and provided (as will be discussed later) the configuration does not interact with the elementary regularities of our reductive theory. What we have to specify in as precise a manner as possible are those (additional) properties of our 'whole' to which our macro-theory refers. Thus, for example, when we find a certain regularity at the level of 'nation-states' and try to explain it by certain micro-regularities of individual behavior we do not have to specify what we mean by such a complex structure (configuration of elements e) as that we call 'state'. On the other hand, we must clearly specify the character of the configuration of properties of individuals corresponding to those properties of states which our macro-regularity refers to. The exact *reductive meaning* of our whole may remain unspecified as long as we know how to recognize it among other wholes and how to recognize its composing elements.

Our macro-law was denoted above in the most simple way by $A(W) \rightarrow B(W)$, and we are interested in reductive explanation of the relationship between the collective properties A and B of the whole W. This implies that we have to specify the exact reductive meaning of the terms $A(W)$ and $B(W)$ at the level of our elements e. Suppose now we come to the conclusion that the whole W $[=C(e)]$ has *additionally* the property A when the elements e of this whole have the property a in a configuration which can be denoted by $C_1 a(e)$. Then our (difinitional) rule of reductive correspondence may be written as follows:

$$[C(e) \cdot C_1 a(e)] = A(W).$$

We can similarly come to the conclusion that when the elements e have the property B in a way that can be described by the configuration $C_2 b(e)$, then our whole W has the property B, the rule being written as:

$$[C(e) \cdot C_2 b(e)] = B(W).$$

We might call the formula $C_1 \, a \, (e)$ *reductive definition of macro-property A*, and the formula $C_2 b(e)$ *reductive definition of macro-property B*.

Since we are not interested in the functioning of the whole as such in all its complexity but only in our regularity occurring in it, namely $A(W) \rightarrow B(W)$, we can treat the formula $C_1 a(e) \rightarrow C_2 b(e)$ *as the elementary correspondent of our macro-regularity*. This means that whenever in the whole W $[=C(e)]$ the sequence $C_1 a(e) \rightarrow C_2 b(e)$ occurs we have at the macro-level the regularity $A(W) \rightarrow B(W)$. The relation between our elementary correspondent of macro-regularity and the macro-regularity itself is (in the social sciences) of an analytical character.

Now — as we remember — we may have two different kinds of situations. In situations of the first kind we are lucky, because there is an elementary theory or an elementary law in which the properties $a(e)$ and $b(e)$ also occur in such a way that they permit us to explain the *necessary character of the sequence $C_1 a(e) \rightarrow C_2 b(e)$*. Let us denote this elementary law by $l(a,b,e)$ (what might e.g. correspond to the formula: $a(e) \rightarrow b(e)$). Then the way to explain the sequence constituting the elementary correspondent of the macro-law is to *derive $C_2 b(e)$ from the elementary law $l(a, b, e)$ and the statement that the properties $a(e)$ occur in the configuration $C_1 a(e)$*. The logical structure according to the pattern of deductive-nomological explanations looks as follows:

> *Explanans:* $l(a,b,e)$
> $\dfrac{\qquad C_1 a(e). \qquad}{}$
> *Explanandum:* $C_2 b(e).$

Thus, for example, if we assume that $l(a,b,e)$ has the character of a general proposition stating that 'a is always followed by b' and $C_1 a(e)$ means that '*all* the elements e of the given whole are a', this of course implies or explains $C_2 b(e)$, which means here that 'all e's are also b'. If (under the same general law) $C_1a(e)$ means that a majority (or a certain proportion) of e's in this population have the property a, then this logically implies that the same proportion of e's must be also b, etc. More complex patterns of derivations and the special problems they pose will be discussed below, the the above scheme presents the general pattern of *explanation of one configuration of elementary properties by another configuration under the assumption that these two sets of elementary properties are related to each other in the way described by our elementary theory*.

Our whole reductive structure might then be schematically presented as follows:

Explaining law:	$1(a,b,e)$
	$[1\,(a,b,e) \cdot C_1a(e)] \rightarrow C_2b(e)$
Cor. Rule 1	$[C(e) \cdot C_1a(e)] = A(W)$
Cor. Rule 2	$[C(e) \cdot C_2b(e)] = B(W)$

Macro-law	$A(W) \rightarrow B(W)$

It should be stressed here that in the above reasoning *we do not explain the existence or the character of the configuration $C_1a(e)$*, nor do we explain the existence or the shape of the whole $W=C(e)$. *They constitute in this kind of theory the ultimate premises for the explanation of the specific processes or regularities occurring in them and we simply take their existence for granted*. This does not mean of course that the nature of the configuration is in itself not explainable by other theories or approaches. Thus we might explain, e.g. the existence of the configuration by a more *historical approach*, demonstrating the nature of the historical process and different factors operating in it, due to which this social configuration occurred in the given reality. We might also try to explain it in terms of '*functional theory*', demonstrating its instrumentality for some other coexisting aspects of social reality, etc. *Reductive theories are by no means the only theories in any science*, and phenomena which in some of their aspects cannot be explained in a reductive manner may be succesfully explained, when regarded from the standpoint of another theory or approach.

We have considered so far the kinds of reductions in which the terms specifying at the elementary level the meanings of macro-phenomena are at the same time the terms of a proper macro-theory. When this is not the case, the structure of our reasoning (and of the reductive theory itself) is more complex. We still have to specify the elementary correspondent of the macro-law in the analytical way as described above, but the properties $a(e)$ and $b(e)$ do not belong in this case to a theory necessary for our reductive purposes. Therefore the proposition which states that they are related to each other in a way that permits us to derive configuration C_2 from the configuration C_1 cannot be called a law of science. Let us say that we have here an *elementary regularity, the methodological character and degree of generality of which will not be specified* more closely at the moment, and denote it by $r(e,b,e)$. We assume of course that if this regularity is valid at least within the whole W it permits us to make — within the limits of this whole — derivations analogous to the derivation presented for the class of situations, namely:

$$[r(a,b,e) \cdot C_1 a(e)] \rightarrow C_2 b(e).$$

But since the proposition $r(a,b,e)$ does not constitute a general scientific law, our next step is to find an elementary law from the same level from which, with the use of proper *correspondence rules of empirical character*, the regularity $r(e,b,e)$ could be derived. Suppose there is in our behavioral theory a law stating that some other properties, namely a' and b', are related to each other in the way described by the law $l(a',b',e)$. Suppose additionally that the properties a, and b, of our elementary law are empirically related to the properties a and b in such a way that we may formulate correspondence rules permitting explanation of $r(a,b,e)$ by $l(a',b',e)$. These correspondence rules will be denoted by $[a(e)CR_1 a'(e)]$ and $[b(e)CR_2 b'(e)]$. Then the law $l(a',b',e')$ becomes the elementary law of our reductive theory, permitting us to explain the behavioral sequence $r(a,b,e)$ occurring in reductive theory. The scheme of explanation of $r(a,b,e)$ looks as follows:

$$l(a',b'e)$$
$$a(e)CR_1 a'(e)$$
$$b(e)CR_2 b'(e)$$
$$\overline{r(a,b,e)}$$

Thus, if we assume that the elementary law $l(a'b',e)$ has the character of the general relation of the kind $a'(e) \rightarrow b'(e)$ and the correspondence rules will have the nature of relations $a \rightarrow a'$ and $b = b'$, then of course we may derive from these premises the generalization $a(e) \rightarrow b(e)$ in the way discussed in Section 3 of this chapter. To say it in a more general way, *we have to explain the behavioral regularity (occurring in our 'whole' and constituting an essential element in our reductive reasoning) which in itself does not constitute a law or theory by deriving it from another (usually more general) theory from the same level.* It is needless to stress that all that was said at the beginning of this chapter with respect to this kind of explanation is valid for these situations too. This is my main reason for discussing earlier in such detailed manner these kinds of (non-controversial) reductions of one theory to another, more general one.

Once we have explained the elementary regularity $r(a,e,b)$ by a proper elementary law $l(a',b',e)$, and this regularity $r(a,e,b)$ permits us to explain the occurrence of $C_2 b(e)$ by deriving it from $C_1 a(e)$, and finally these two configurations $C_1 a(e)$ and $C_2 b(e)$ constitute definitional correspondents of the collective properties A and B involved in our macro-law, our reduction has been completed. Its complete structure may now be presented as follows:

Explanans

(1) The explaining law of the
     micro-theory (elementary law) $l(a,b,e)$

(2) The application of the elementary
     law for the explanation of the elementary
     regularity r occurring in the whole W
     with the use of empirical correspondence
     rules CR
     $$[l(a,b,e) \cdot (a'(e)\, CR_1 a(e) \cdot (b'(e)\, CR_2\ b(e))] \rightarrow r(a,b,e)$$

(3) The application of elementary
     regularity r for the derivation
     of C_2 from C_1 $[r(a,b,e) \cdot C_1 a(e)] \rightarrow C_2 b(e)$

(4) Correspondence rule defining
     the collective property A
     as configuration of a $[C(e) \cdot C_1 a(e)] = A(W)$

(5) Correspondence rule defining
     the collective property B as
     configuration of b $[C(e) \cdot C_2 b(e)] = B(W)$

Explananandum; the reduced macro-law $A(W) \rightarrow B(W)$.

In order to make our reductive theory more simple, we may now, on the basis of correspondence rules $(a'CRa)$ and $(b'CRb)$, *'redefine' our macroproperties A(W) and B(W) in terms of the concepts of the explaining theory and of the configurations in which the elementary properties denoted by them have to occur.* Then of course the 'redefinition' of our macro-concepts in terms of properties a' and b' permits us to present our reductive theory in a more simple way, as presented above. Nevertheless we should remember that the basic premises for our definitional decisions were the empirical correspondences of the properties involved in the elementary law to those elementary properties which previously specified for us the meanings of our macro-properties.

Let us look now at the formula which above was called 'the application of the elementary regularity or for the derivation of C_2 from C_1' and denoted by $[r(a,b,e) \cdot C_1 a(e)] \rightarrow C_2 b(e)$. It seems that this formula corresponds to what is usually called 'theoretical models of macro-phenomena' where between the elements composing these macro-phenomena these models refer to certain relations and the regularities of behavior of these composite parts. In order to construct such a model we have to assume that (1) the elements of the given whole 'behave' in the way $r(a,b,e)$ and *at the same time* (2) that they are coexisting and arranged in a way which corresponds to the configuration $C_1 a(e)$. This permits us to derive the (logically or mathematically) necessary character of the occurrence of the configuration $C_2 b(e)$ of the given model. If the configuration $C_1 a(e)$ is very complex and our model involves many different regularities of the behavior of the elements, we often *simulate* the functioning of the given model *in a computer*; after feeding both the whole configuration C_1 and the elementary regularities r of functioning of the elements e into the computers memory, we let it run and see whether we really obtain as the result of these transformations the configuration $C_2 b(e)$, the occurrence of which we observe in the 'modeled process' as the *effect* (outcome) of our model. If we do, we can conclude that our model adequately *represents* a certain macro-process or macro-regularity in terms of some micro-mechanisms. If we do not, we conclude that some of the assumptions of our model must be wrong.

The question is, how should these models (provided they are adequate) be regarded from the standpoint of the problems discussed in this paper? The models are internally consistent (logically or mathematically) formal theories, and this is true independently of whether there exists any social reality at all that satisfies their assumptions. If they turn out to be empirically valid, they may be treated -- as I said above -- as adequate formulations of certain macro-mechanisms in terms of the behavior of and interrelations between their 'component parts', or as reductive correspondents of macro-regularities.

A further question arises: Since these models involve in their formulations reference to the properties and regularities of behavior of particular elements (e.g. individuals), *can they also be called reductive theories of the social sciences*? The answer depends of course upon the character of the elementary regularities $r(a,b,e)$ occurring in them. If we think that these regularities in themselves constitute unconditional laws (general or statistical) of social behavior, the validity of which does not depend upon certain other properties of the social contexts $C(e)$ in which our models work or upon the nature of the configuration $C_1 a(e)$ in which the antecedents of the given model occur, *then our model constitutes at the same time a reductive theory of the given macro-process* (of course with additional, definitional correspondence rules relating given configurations to proper macro-phenomena). If we do not think the regularities of behavior of individuals figuring in our model are actually unconditional, then our model does not constitute yet a reductive theory. But *even in this case it constitutes an essential step toward future reduction, being the elementary correspondent of the given macro-law.*

If we take into account the number of models of social processes and regularities which are now formulated (especially in the area of mathematical sociology) with the reference to the pattens of behaviors of interacting individuals, or any other 'parts' of 'social wholes', we can say that *the program of construction of reductive theories is more advanced in our discipline than one would have suspected*. Of course, the particular models do not necessarily have to be formulated with regularities in terms of the behavior of individuals and the relations between them; they may refer as well to smaller units of a more comprehensive aggregate or system, if we decide that this is the proper level to which our macro-theory should be reduced, because we know some laws of functioning of these smaller units. I have simply concentrated my attention in this paper on reductive theories referring to the level of individuals, but as we will recall these do not exhaust all possible reductions in sociology.

10. THE PROBLEM OF 'EMERGENCE'

Let us consider now the most controversial philosophical problem of the reduction of 'wholes' to parts, usually called the problem of 'emergence'. The philosophers who are opposed to reduction stress that when we pass from one to a higher level of the studied reality new properties 'emerge' which cannot be predicted on the basis of knowledge of the elementary properties of the lower level, and that the same is true for the laws and regularities governing relationships between the phenomena of the higher level, i.e. they are basically underivable from the laws of the lower level. The postulate of 'emergence' may take a stronger or a weaker form. In its stronger form it would apply to *all* properties and regularities; in the weaker it would apply only to *some*, merely classifying the laws and properties from any level into two categories, those which can be reduced to the lower level phenomena and laws about them and those which cannot.

First, let us look at the problem of the *emergence of properties*. It is obvious that in those sciences where the phenomena are accessible to independent observation at the two levels, the 'emergence' of the undefinable property is an empirical fact; since the concepts from any level have their independent empirical reference they are therefore (or at least may be) undefinable one by another. The propositions relating the data of our experience from the two levels are basically empirical ones. Also, the question whether we will be able to find relations between the phenomena from the two levels which will connect them in the necessary way is an open empirical one – in some cases we are able to say that a certain configuration at a lower level corresponds to a certain macro-property at the higher level, in others we simply cannot. Physicists are quite aware of the nature and characteristics of particles of hydrogen and oxygen and the configuration necessary for the existence of water, but on the other hand they don't know what lower level configuration (if any) would correspond to such elementary particles as protons or electrons.

Once we have discovered such correspondence in empirical studies, we may of course later on redefine our macro-concepts in such a way as to take these empirical inter-level relations into account, but then our definitional conventions will have to conform to the empirical findings.

In sociology, due to our natural human perspective in the observation of social phenomena, the situation seems to be different. All the 'emergent' properties of human collectivities are definitionally dependent upon certain configurations of absolute and relational characteristics of human individuals which constitute them or upon any other kinds of properties of higher level collectivities which can be further on analytically related to the properties of individuals. If we cannot specify the 'human components' of these properties this usually means that we do not understand them clearly enough. Since I discussed this problem in Section 8 of this chapter there is no need to develop it again here.

The problem of 'emergence' seems to be much more serious with respect to *laws*[30]. In the opinion of some philosophers of science, at least some laws regarding wholes cannot be reduced to (i.e. be explained by) laws governing component parts of those wholes, since, as they claim, at every new level of complexity there 'emerge' regularities which are essentially new and which thus cannot be deduced from laws governing component parts of the wholes involved. For instance, from de laws on hydrogen and oxygen, as we now know them, we cannot deduce the laws on water as the chemical compound of the two. From the laws on human behavior under laboratory conditions, we cannot deduce laws on human behavior in a crowd or laws about man's behavior when he plays a certain role in a certain organization, etc. These laws are essentially new, and *every level of integration is marked by the fact that new laws may emerge in it.* Emergence would thus be a natural limit to a reductive systematization of human knowledge. Let us examine this issue in greater detail.

First, it is obvious that definitional reducibility constitutes a necessary but not a sufficient condition for theoretical reduction. Even if we know what properties of elements are involved into the reductive correspondents of given macromechanism, we still need elementary laws in our reductive theory. Many methodologists have emphasized that having at one's disposal not just any, but definite, elementary laws is a condition for explaining a global law.[31] Our theoretical knowledge about 'parts' is never complete, and it very often happens that *at a given level of development of science we do not know all the elementary laws which are indispensable to explain a given global law*, but later stages bring discoveries of such laws, thus making a reduction of the global law in question possible. As many philosophers have pointed out, laws which are 'emergent' at certain stages of development of science often cease to be 'emergent' and become 'reducible' at a later stage. In the case of sociology this would mean that we are often able to point out the elementary correspondents of a certain 'global regularity' – i.e. to say how people behave when a certain macro-process is going on – but we are unable to specify any general or statistical laws that could be used for the description of these behavioral sequences, and in consequence for the explanation of the given macro-law.

There are two possible reasons for our being unable to explain a certain macro-law by the kind of elementary law (or set of such laws) that could figure in a consistent reductive theory. First, it may be that the elementary (e.g. individual) *terms* which specify for us the empirical aspects of the macro-terms *are not suitable for the formulation of a corresponding elementary theory*, and at the same time we are unable to find any empirical correspondence of these terms to the terms of another theory that would be useful for such purpose. In situations of the second kind, which we shall consider here in more detail, the *elementary terms are potentially useful* for the construction of the explaining elementary theory, but the propositions in which they occur in our elementary regularity – denoted in the previous section by $r(a,b,e)$ – are theoretically incomplete. They do not take into account the *interaction* of the variable a of the antecedent of this regularity with some other properties of the elements of the whole W; or its interaction with the configurational characteristics C_1 of the way a occurs in the given whole; or finally, the interaction with other features of the whole W denoted above by $C(e)$. Let us look first at the interaction of the last category.

In discussing the general scheme of reduction, I stressed that the configuration $C(e)$ corresponding to our whole W does not have to be specified in reductive terms; we may take it more or less for granted – otherwise no reductive theory of complex multi-level systems would be practically possible. Here I should make it clear that we may take $C(e)$ for granted under one assumption only. Suppose that the elementary law our theory has the character of a general proposition $a(e) \rightarrow b(e)$. In order to take the nature of $C(e)$ 'for granted' we have to assume that the elements e, if they have the property a, *will always have (or tend to have) the property b, independently of whether they occur within the configuration C(e) (i.e. within the whole W) or 'outside' of it*. The logical form of this assumption may be presented as follows:

$$[a(e) \rightarrow b(e)] \rightarrow ([C(e) \cdot a(e) \rightarrow b(e)]).$$

Let us call this the *assumption of irrelevance of the surrounding context of the whole W for the elementary law a(e) → b(e)*. To illustrate we might say (if we think this is generally true) that frustrated people will tend to react aggressively, independently of whether they are alone (or facing the source of aggression only) or are members of any social group and acting within this group, just as we assume that stones will have the same acceleration in the field of the earth's gravity whether they are falling 'alone' or there are numbers of them falling together.

This postulate of irrelevance of context of the given whole seems to be the simple logical conclusion of the elementary law. On the other hand, we know that it does not necessarily have to be true. If it is not true, then the only possible conclusion is that *the elementary law is in its general, unconditional formulation false*; it applies to the elements only when they 'behave' in isolation, but not when they belong to the whole described by the configuration $C(e)$. Our law is then *too general*; it should be formulated in such a way as to make clear that it does not apply to the conditions $C(e)$. Therefore if the postulate of irrelevance of context is not true, the proper formation of the elementary law should look as follows:

$$[a(e) \cdot \overline{C}(e)] \rightarrow b(e).$$

And then, of course, we will know that we are not allowed to use this law for configurations $C(e)$, and by the same token neither for the explanation of the global laws applying to the wholes W. Otherwise we would be entitled to derive its conclusions of this law for all possible e's including those whoch form configurations $C(e)$. This means that in the simplest type of inter-level reduction pressented above *we assume that the context described by the configuration of elements C(e) does not have an interactive effect being involved in the action of the elementary law a(e) → b(e)*.

In order to say that the elementary relation $r(a,b,e)$, having for example the form of a general law $a(e) \rightarrow b(e)$, may be called an unconditional law of science, it is not enough to assume that the total context $C(e)$ does not have a canceling effect upon it. We also have to assume that it is *irrelevant also in the positive sense*. To say that the overall context of the whole $[W = C(e)]$ is irrelevant for the given elementary regularity means that the elements e behave in the same way both 'within' this whole as 'outside' it. But this has another consequence too: It implies that in whatever context of the phenomena of the higher level the configuration $C_1 a(e)$ occurs, it will necessarily lead — *due to the action of this elementary law alone* — to the configuration of $C_2 b(e)$.

If this cannot be assumed, it means that some unspecified properties of elements belonging to the totality of properties of elements characterizing our 'whole' or some strictly configurational aspects of this whole are also involved in our elementary regularity, which therefore is *incomplete in its theoretical formulation*. In order to transform our elementary relation into a really unconditional (general or statistical) law of science we have to discover these properties of our whole which *interact* with the elementary antecedents a of our relation in producing b. Then of course we will have *another* elementary law than the one we had previously. And its validity will now be theoretically unconditional even if these additional modifiers of the relation $a \rightarrow b$ occurred empirically only within the given 'whole'. Theoretical generality of a law is not identical with empirical occurrence of the sequences described by this law 'practically everywhere'. It only means that our law is theoretically complete, nothing more.

Let us now look at another, less obvious class of assumptions of reductive systematizations which, similar to the postulate of irrelevance of context of the whole W, do not necessarily have to be true. When stating that the elementary law $l(a,b,e)$ permits us to explain the necessary character of the occurrence of the configuration $C_2 b(e)$ when the configuration $C_1 a(e)$ has occurred, we assume *that this law constitutes sufficient theoretical knowledge for this purpose*. In other words, we assume that all the other rules and premises necessary for the derivation $C_1 \rightarrow C_2$ are of a *strictly formal* character. They may be rules of logic or mathematical theorems, they may belong to another strictly formal theory like cybernetics or the theory of games (in their strictly mathematical formulations), or to such deductive constructions as, e.g. 'general system theory'. In some cases, when our configuration refers to certain spatial arrangements of the elements and spatial relations between them, we may have to use some tools of geometry or topology. All these formal rules permit us to derive (to predict or explain) the *overall outcome of the behaviors of individuals, assuming that these individuals behave according to the law $l(a,b,e)$ and that the starting point for their behavior is adequately described by the configuration $C_1 a(e)$*. The pattern we have to derive is of course the configuration $C_2 b(e)$.

If the rules of reductive correspondence between the properties of collectivities and their individualistic counterparts are simple (as in the case of 'analytical properties' of collectivities) and the elementary law permits a simple derivation of C_2 from C_1, then the relation between the macro-law and the elementary law is clearly visible. The more complex (the more structural) these rules are and the more complex is the set of formal rules necessary for this derivation, the more our macro-law seems to be 'emergent' at the level of the collectivity as compared to the level of its individual members and the more these analytical relations seem to 'interact' with empirical ones. When we use the propositions of mathematical game theory or the complex theorem of formal system theory in explaining the dynamics of a dyad in the situation of a simple game or in explaining the transformation of a complex system, we have the feeling that these configurational mechanisms interact with the behaviors of the individuals in a way that makes the notion of emergence quite clear. In a sense these processes are 'emergent', but their emergence is of a strictly analytical character, and from the strictly logical point of view the most complex formal mechanisms do not differ from the most simple derivations. They 'emerge' in the same way the relation between the two arithmetical means at the collectivity level emerges from the relations between the characteristics of individuals and from the assumption that the distributions of given characteristics of individuals change in time. As long as these formal rules refer to patterns in which the properties of individuals are distributed and we apply some laws of elementary social behavior as the *only empirical premises* in that transformation of one configuration into another and, the emergence is of an analytical and not an empirical character.

But now we have to consider the situations in which the *emergence is of an empirical character*. By this, I mean a situation in which the configuration of the kind $C_1a(e)$ is *not irrelevant* to the behaviors of the elements themselves. It may happen that the elements e behave differently when they are interrelated in the way described by $C_1a(e)$, whatever the nature of the configuration might be. Thus it may happen e.g. that the number of members of the given group is not irrelevant for the way these members behave under certain conditions. The structure of the group as a whole will co-determine the behavioral patterns of its members in a way impossible outside a group having this structure. Then, of course, knowledge of the laws of elementary behavior of the elements in isolation is not sufficient for the prediction of the outcome C_2 from the preceding configuration C_1. This is what might be called the *empirical emergence* of new elementary regularities which are now co-determined by the context (in a more narrow sense than discussed above and denoted by $C(e) = W$) of the given pattern in which the antecedents of the given elementary regularity are arranged in the given whole W. Some more typical sociological categories of this kind of emergence will be discussed below.

Finally, the variables involved in the interaction with the antecedents of our elementary regularity may be certain characteristics of the elements (individuals) themselves which we did not take into account in our analysis, but which must necessarily characterize them (at least in the wholes W) if this regularity 'works'.

Thus we see that the variables involved in possible interactions may consist of either 'individualistic' antecedents of the elementary laws themselves, or the structural or compositional aspects of $C(e)$ or $C\, a(e)$, *wrongly assumed to be irrelevant for the action of the given elementary laws*.

Laws referring to the interaction between the characteristics of individuals and their context that produces the individuals' behaviors are sometimes called 'composition laws' or 'composition rules'. To quote May Brodbeck again:

In addition to elementary laws telling how the individual acts in the presence of one or a few others we must also have composition laws stating what happens under certain conditions as the number of people he is with increases. The latter of course state how he behaves in a group.[32]

The terminology is of course irrelevant, but I think these propositions should simply be treated as 'normal' laws of behavioral science because they do not differ from other laws which have also to take the possible interaction of some factors into account. Moreover, in our search of reduction *we do not add them to the laws of behavior of individuals 'in isolation'; we replace the laws referring to the behavior of individuals 'in isolation' by laws which describe their behavior in different social contexts*. Only from the latter can we derive our macro-theory, if it is actually reducible to them.

The fact that certain elements of the context $C(e)$ or of the configuration $C_1 a(e)$ 'cancel' the action of some elementary laws or produce in interaction with other antecedents [i.e. $a(e)$] some new laws, and that some of these modifying factors may occur only within certain social configurations, (groups or systems) does not mean that such laws are not *theoretically general*; they are, as long as we are able to specify these factors in our theories. To employ an analogy — the process of atomic fission may occur only within a special (and very 'artificial') set of conditions, but this does not mean that the theory denoting this process is not *theoretically general*. The analogy, however has special implications for the social and behavioral sciences. *When formulating laws about human behavior, we ought to take into consideration the possible modifying role of the various situational and contextual factors which may account for the fact that people respond differently to 'the same' stimulus conditions* and modify the regularites of social behavior. It is only such knowledge of 'elementary forms of social behavior' that will make it possible to carry out with increasing efficiency the program of reducing laws on complex social aggregates, groups and social systems to laws on human behavior, in as many different sets of conditions as possible, thus contributing to an ever improving theoretical knowledge of man and society.

11. LAWS OF BEHAVIOR AND LAWS OF SOCIAL INTERACTION IN REDUCTIVE SOCIAL THEORIES

Let us return now to more simple situations in which — at least as we supposed when building our reductive theory — unknown characteristics do not interact with the known ones, i.e. where we know the proper elementary laws. It seems that for the purpose of sociological reduction (as well as reductions in other sciences) we can classify our elementary laws into two categories: those in which the antecedents and consequents occur *in the same elements* and those in which the *antecedent refers to the properties or behaviors of* one element (e_1), while the consequent refers to the properties or behavioral reactions of another element (e_2). *Let me give an example of the first category.*[33]

> *Elementary law*:
> People tend to accept opinions which justify the privileges they possess:

$$a(e) \rightarrow b(e)$$

Rule of red. cor. 1
A privileged social class exists when there is a group of people different from others with respect to their social position $C(e)$ and this difference consists of their being more privileged than the others: $C_1 a(e)$:

$$[C(e) \cdot C_1 a(e)] = A(W).$$

Rule of red. cor. 2
A social class tends to accept opinions justifying their privileges when a certain number of people differ from the others with respect to their social positions $C(e)$ and at least the majority of them tend to accept these opinions: $C_2 b(e)$:

$$[C(e) \cdot C_2 b(e)] = B(W)$$

Macro-law:
Privileged social classes tend to accept opinions justifying their privileges.

$$[A(W) \to B(W)].$$

Here the postulate of the irrelevance of configuration means that we assume it does not matter whether the privileges people possess are of an individual or a class character: people will in either case tend to accept the available justifications.

In this example (as well as in the logical pattern of reduction proposed in the previous section) we took into account only one special category of reductions, namely those in which elementary law operates at the level of *each of the elements separately*, the relations connecting them – even if they are relevant for the existence of that 'whole' and its properties – are not relevant for the action of the elementary regularity in question, and the explaining law has as its antecedent and its consequent the absolute *properties of the same elements*. If we were to concentrate on reductions in sociology where the reductive level is that of human individuals, we would say that the elementary laws in such reductions are of an *intrapersonal* character; i.e. we would explain the global law reducing it to certain laws of behavior of members of the given aggregate or system, which do not refer to interactions of different people. However, as we know many regularities of social behavior are of an *interpersonal* character: the characteristic a of one of the elements of the category (e_1) (or its given behavior) may cause (produce, influence) the occurrence of another characteristic (behavior) of element (e_2) of another category. Our regularity may then be described by the formula: $a(e_1) \rightarrow b(e_2)$. But it is usually the case that such interactions (not between laws, as understood above, but between the elements of the given 'whole') are possible only under certain additional conditions. In order to influence someone's attitude by a verbal message I must be in the kind of relation with the receiver of my message that permits me to 'transfer' it to him. In order to modify someone's behavior by a pattern of 'rewards' and 'punishments' I must be in a position to observe this behavior and to 'administer' my rewards accordingly, etc. Let me define the kind of relations between the elements e_1 and e_2 which are necessary for the given law of interaction by $e_1 Re_2$. Our elementary law will now have the following form: $[(e_1 Re_2) \cdot a(e_1)] \rightarrow b(e_2)$.

It should be kept in mind that the relations $e_1 Re_2$ between the elements e_1 and e_2 are here not identical with the total configurations of the absolute and relational characteristics of the elements $[C_1 e$ and $C_2 e]$ to which our reductive definitions of macro-properties refer; $(e_1 Re_2)$ refers to the *nearest social context* of the individuals which is *necessary for the operation of the law describing certain interactions between individuals at the elementary level*, and therefore it should be stated separately in our reductive theory.

Besides this elementary law we need in our explanans the rules of reductive correspondence between the properties of the whole and the properties of its elements. But now the global properties will be reduced to the properties of *different categories of elements*.[34] We may have to state, for example, that the property $A(W)$ is reducible to the property a of the elements of the class e_1 from our elementary law, while the property $B(W)$ is reducible to the property b of the elements of e_2 type. And, of course, the rules of reductive correspondence will include additionally the totality of properties of the elements e_1 and e_2 and the relations between them $C_1(e_1)$ and $C_2(e_2)$ necessary for the existence of the global properties A and B of the given whole, although they are also, as noted above, irrelevant for the functioning of our interpersonal law.

Finally, let us consider situations in which the elementary law is of a *mixed character*, i.e. it describes certain behaviors of the elements e_1 which are caused both by certain of their antecedent characteristics $a(e_1)$ and by certain characteristics (behaviors) of the other elements, e.g. $d(e_2)$. Its structure may then be written as follows:

$$[(e_1 R e_2) \cdot a(e_1) \cdot d(e_2)] \rightarrow b(e_1).$$

Needless to say, most of the 'behavioral regularities' we will use as explanatory laws in sociological reductions will belong to the last category. We assume that people behave in a given way $b(e_1)$ *both* because they have certain characteristics $a(e_1)$ *and* because some other persons (e_2) who are related to them in the way $(e_1 R e_2)$ possess the characteristics $d(e_2)$ or behave in this way $d(e_2)$.

Here again it should be stressed that we assume that the totality of the characteristics (both relational and absolute) of the elements e, other than those specified by the antecedent of our elementary law (which are necessary for the existence of the whole and its corresponding characteristics $A(W)$ and $B(W)$ are irrelevant for the action of the explaining law at the elementary level.

If we cannot assume this, it means that *we do not know the elementary law* of social interaction which is necessary for the explanation of the given global law; therefore we cannot have a theoretical reduction of the global law to a law (or a set of laws) of more elementary character.

12. STATISTICAL RELATIONSHIPS IN MULTI-LEVEL REDUCTIONS

Up to now we have assumed that we are dealing with *general* relationships at both level — let us now look at the different, quite simple situation in which the relationships of one or both levels are of a statistical character. The most simple case might be stated as follows:

(1) The elementary law has the character of a *general proposition* stating that $a(e) \rightarrow b(e)$ (a is always followed by b);

(2) The rules of reductive correspondence identify the global properties $A(W)$ and $B(W)$ as quantitative variables, *being the functions of the frequency* (F) of the elementary properties $a(e)$ and $b(e)$ in the given configuration $C(e)$ according to the formulas:

$$A(W) = f\,[F(a(e) \cdot C(e)]\quad \text{and}\quad B(W) = f\,[F(b(e) \cdot C(e)].$$

In such case the *macro-law will have the nature of a general, monotonic functional relation* between two *quantitative* 'macro-variables' A and B and can be presented as $B(W) = f[A(W)]$. Thus we may say that if the elementary laws refer to the relations between the dichotomous attributes, but the frequency of occurrence of the atributes at the elementary level is related by the rules of reductive correspondence to some quantitative variables at the global level, the corresponding global law will have the character of a functional relationship between these global properties.

We have a very similar pattern of reduction when the underlying elementary law has the character of a *probabilistic relation* of the shape $P[b(e)/a(e)] = p$ (and the rules of reductive correspondence are as stated above). The only difference here is that the value of the dependent variable *at the global level* will now be the function of the frequency of the independent variable at the elementary level multiplied by p (b/a) at the elementary level, which means in practice that the functional relation at the global level will be weaker than in the case where the underlying elementary relation is a general one. The increase of frequency of $a(e)$ in the given population will have here a lesser impact upon the increase of frequency of $b(e)$, and the increase of the intensity of the global variable $A(W)$ will have lesser impact consequently upon the increase of the value of the dependent global variable B(W), than in the case of a general elementary law.

The situation looks much the same if at the elementary level we can assume only a positive statistical relation between a and b, without the specification (in general terms) of the strength of this relationship.

This seems to be an important category of situations in sociology, because in 'macrosociology' *many 'global properties' are identified as the functions of frequencies of some properties of individuals or smaller social units composing them,* which leads to interesting implications for 'macrosociological theory'. As we know, *the laws of social behavior of individuals* (whether formulated in 'sociological' or in 'psychological' language) *are — as a rule — statistical ones*. But this does not mean that *the laws describing the regularities at the level of the whole social systems* or *aggregates* must also be statistical in their nature, even if they are 'reducible' to some statistical laws of human behavior. There *may be quite 'nice' functional (monotonic) relations*, stating, for example, that the greater the 'mobility rate' in the given system, the greater the 'criminality rate' in it, even if we know that not every mobile person is a criminal.

It was assumed above that the elementary laws (whether general or statistical) are of an intrapersonal character, i.e. their antecedents and consequents are the properties and/or behavior of a single individual. The situation is exactly the same for probabilistic relations when the elementary law refers to an 'interpersonal' relation $a(e_i) \to b(e_j)$ (or a similar relationship of statistical character) and the corresponding global properties $A(W)$ and $B(W)$ are identified as the functions of frequencies of $a(e_i)$ and $b(e_j)$, then of course there *may be* a linear functional relationship between $A(W)$ and $B(W)$ at the macro-level.

I would like to stress strongly the words 'may be' in the last sentence. Even if our elementary relation $a(e) \to b(e)$ is complete (unconditional) in its general or statistical form, and the global variables $A(W)$ and $B(W)$ are rightly identified as the functions of frequencies of the corresponding elementary variables in the different compared groups or societies, there will be no relationship at the global level if the compared societies do not differ on relative frequencies (or other statistical parameters) of the variable $a(e)$ and consequently $b(e)$. In order to assess a relationship between the collectivity level variables we must be able to state their different values; this means that the distributions of the elementary variables must be different in different groups and societies. When this is not the case the compared societies will be more or less similar on the macro-variables A and B and no relation between these variables can be observed.[35]

We discussed above the situations where the elementary relation $a(e) \to b(e)$ is complete, stating that this implied a functional relation between macro-variables A and B provided that the compared groups differ on the frequency of the elementary events $a(e)$. But it may also happen that our elementary regularity $a(e) \to b(e)$ is incomplete, i.e. it is *conditional* with respect to some other variable $c(e)$, and at the same time the compared societies *differ in an essential way on the occurrence of c(e)*. Suppose that the complete (but unknown to us) formula for an unconditional (general or statistical) elementary law should be $[a(e) \cdot c(e)] \to b(e)$, but at the same time in some societies the property $c(e)$ is a fairly common characteristic of its members, while in others it is absent (or almost absent). Then, of course, at the global level we will discover only a statistical relationship between the quantitative variables $A(W)$ and $B(W)$, and the smaller the proportion of the societies with $c(e)$ as the common characteristic of its members the weaker the statistical correlation between $A(W)$ and $B(W)$ at the global level. A similar, but even more striking, effect will be observed if for instance $c(e)$ should have an effect such that a will produce b when non-c determines that a will necessarily lead to non-b at the elementary level, because then in societies of the $c(e)$ type [let us denote them by $C(W)$], $A(W)$ will be monotonically positively related to $B(W)$, while in societies of the non-$C(W)$ type these two variables will be in a negative monotonic relation.

At the same time, since we are unable to distinguish these two kinds of societies and observe only the overall relation between the macrovariables A and B for all of them taken together, there may be no relationship between them at all. It does not matter, of course, whether the property $c(e)$ *is of an absolute or a relational character as long as due to different patterns of its* occurrence in different societies, it has a modifying impact upon the final outcome of the elementary processes on a mass scale, and as a result, upon relations between the global variables, related to the frequency of these elementary events.

It may happen that the elementary property $b(e)$ will be the function of *several additive relationships* – let us say that $a_1(e) \rightarrow b(e)e$, $a_2(e) \rightarrow b(e)$. Suppose now that we identify our global variables in such a way that $A(W)$ is the function of frequency of $a_1(e)$ when $B(W)$ – of $b(e)$. Now, if with respect to the pattern of distribution of the other variable $a_2(e)$ the compared societies are similar (and the statistical relationships between a_1 and a_2 are also rather the same in the different societies), then there will be a monotonic relation between $A(W)$ and $B(W)$ at the global level even if we fail to take into account the other additive mechanism. On the other hand, if this is not true, any of the other possible outcomes at the global level may happen. Suppose that the greater the frequency of a_1 the smaller the frequency of a_2 in the given society (with a_1 and a_2 being statistically independent). Then there will be no relationship at all between the variables $A(W)$ and $B(W)$ at the global level. But if the frequencies of a_1 and a_2 are positively related to each other at the societal level, then the linear relation between A and B will be stronger than in the case of the single action of one elementary law only. I discuss these problems more thoroughly in another chapter of this volume.[36]

In any event, we may say in general that the action of other variables, both those which have an additive and those which have an interactive impact upon the dependent variable at the elementary level, may lead to wrong predictions of the global relations between the variables identified as the functions of the frequencies of the elementary events or characteristics.

In sociology we often face the reverse situation; we find a relationship between the properties at the global level (as measured, e.g., by the statistical parameters describing the frequencies of certain attributes at the elementary level) and later on we try to determine a corresponding relationship at the elementary level, i.e. we try to 'derive' the explaining elementary regularity from the global one. This situation poses certain special problems. The first problem we face here is that even if we correctly find out how the variables are related to each other at the elementary level, this does not mean that we have found an explaining elementary law. The latter would be the case only if we were able to specify all the variables at the elementary level which were necessary for the given elementary (general or statistical but not limited to one society only) relation. Otherwise *we have only the elementary correspondent of the given global regularity*. But even the identification of this elementary correspondent may pose certain problems. As has been stated by many writers, such conclusions may be erronenous, and we may commit here the 'ecological fallacy'. This fallacy may be one of two different kinds: (a) First we might conclude, from the strength of the relationship at the global level, that the corresponding relationship at the elementary level is *equally strong*. As we can see from the argumentation above, this does not have to be true, because even a fairly weak probabilistic relation may produce a perfect linear relationship (to say nothing of a strong statistical one) at the global level.

(b) Moreover, we might commit another kind of 'ecological fallacy' by concluding that there must be an analogous (underlying) relationship at the elementary level of an intrapersonal nature, when in reality this is not the case. Here we may distinguish two subcategories of such situations.

(b_1) In the first type of situation the corresponding underlying regularity will be of an interpersonal character, i.e. it will have the form $a(e_i) \rightarrow b(e_j)$. It is obvious that in this case we will have *a negative correlation between the properties a and b when they are assessed for the same e elements*.

(b_2) In any other kind of situation there will be a perfect independence between the characteristics (a) and (b) at the level of elements of the given population, while at the same time we will observe a functional relation between their frequencies (or the empirical correspondents of their properties) at the global level.

I do not want to discuss all the strictly mathematical aspects of 'ecological fallacy' since they have been sufficiently analyzed already by other authors.[37] What I would like to stress is that *when we try to reach conclusions about relationships at the individual level from observation of the relationships between the statistical parameters of the population level or vice versa, we are – knowingly or unknowingly –* making reductions of macro-regularities to micro-regularities in predicting macro-regularities from knowledge of elementary regularities. Whether these micro-regularities 'deserve' the name of explaining laws or are only elementary correspondents of the global regularities depends only on the degree to which we are able to specify which conditions of the surrounding social context are relevant for the given elementary relation and which are not.

If this is the case, we can say as I said about mathematical models of social processes *that contemporary sociology is much more advanced in the area of reduction than one might expect*. It is most interesting that powerful premises for further reductive systematizations have been built up by those sociologists who, at least by their declarations, seem to be most opposed to reductions – by 'macro-sociologists' who study the relationships between statistical parameters characterizing whole social aggregates and who are very much aware of the dangers of incorrect reductive inferences or, as they prefer to say, of 'ecological fallacies' of various kinds.

13. SOME SPECIAL PROBLEMS OF 'EMERGENCE' IN SOCIOLOGY

In discussing problems of the emergence of certain regularities at the macro-level with respect to the *known* elementary regularities of social behavior, I stressed the fact that such emergence involves the interaction of some additional unknown factors which occur in the population. Once we discover them and include them in our regularity the elementary law becomes unconditional. At the same time I distinguished three kinds of such situations: those in which these new factors may be interpreted as the characteristics (absolute or relational) of the elements themselves, those in which they belong to the configuration of the properties of the antecedents, and finally those which belong to the other, unspecified characteristics of our 'whole'. In the last two cases, although our elementary regularity becomes (when we discover this interaction) unconditional in its theoretical formulation, the occurrence of sequences to which it refers is by definition limited to the configurations of the given kind (or to the wholes of the given kind). We might say that in such a case we observe the *interaction of variables from a higher level with the 'individualistic'* antecedents of our elementary regularity, and the explaining law is of a multi-level character[38].

Let us look more closely at the problems of *levels* of social theories. First we may say that the level of certain theories is clearly specified by the nature of the concepts with the use of which the given theory has been formulated. If all the concepts of a certain law apply, according to their meanings, only to the characteristics of individuals, then this is obvious *one-level theory* which deals with regularities at the level of individuals. If all the concepts of the given theory are characteristics of collectivities, then this theory obviously applies to regularities at the collective level, independently of whether or not we are able to reduce the theory later on to regularities governing the behavior of individuals. It may happen that a certain concept can meaningfully apply (if its level has not been additionally specified) to different levels of social reality, but the theory in which the concept occurs is true only for some but not all levels to which this concept might validly apply, or to collectivities of one but not another kind.

Thus, the notion of 'wealth' applies both to individual persons and to corporations or states, and even to administrative regions within the given state, some of which may be 'richer' than others. But the proposition stating that 'wealth is positively related to power' is true for individuals, corporations or states, and not for administrative regions. This means that the *level of the given generalization and the type of collectivities from the given level to which it validly applies have to be specified* as clearly as possible, if we want our generalization to be empirically valid.

As I mentioned above, some theoretical propositions in sociology are valid for different levels (as, e.g., the proposition stating that the integration of the given group increases with any perceived external threat to this group). Others are valid only for one specific level and for collectivities of a certain kind. Some propositions in social theories include variables from *more than one level* simultaneously and these multi-level generalizations are especially interesting for the problems of reduction in sociology.

When discussing the emergence of new (unexplainable or unpredictable) regularities as the result of interaction of some unknown or unaccounted for variables, I stressed that one category of such factors may be the configurational aspects of the way in which the characteristics a of the elements e occur. But, according to what has been said in the discussion of the concepts referring to characteristics of collectivities, these configurational aspects of $a(e)$ belong clearly to the same level as the collectivity itself. Therefore even if we later on discover their interaction with the characteristics of individuals, and are thus able to build a reductive theory in which the laws of the macro-level are related deductively to proper theoretically complete laws about elements, the fact remains that the laws about elements involve at least some properties from the strictly collective level itself by reference to the whole configurations of these elements as necessary for explaining the behaviors of individuals. *The elementary laws of such reductive theories are multi-level propositions.*

I think that one should distinguish the cases of emergence which refer to simple underivability of the macro-law from the micro-law due to our inability to formulate the micro-law properly, from those situations in which the micro-law has been properly formulated and we see that the behavior of individuals is co-determined by the nature of the context and the configuration in which they occur within the given collectivity. Those macro-laws which can be explained only by some multi-level propositions involving also some properties from the collective level can be called 'emergent' in the strict sense of the term.

The configurational characteristics involved in the laws of human behavior may be quite diverse. The most simple ones refer to the *number* of persons in the group to which the given individual belongs. Georg Simmel has demonstrated how basically the laws of human interaction change with the increase of the size of the group in which the interaction takes place. The *number of group members* is in such regularities of human interaction the configurational characteristic of the collectivity. Another kind of finding is that the tendency to 'helping behavior' turns out to be strongly influenced by the number of persons present in the situation when help is needed — the greater the number, the smaller the tendency to help on the part of each particular member of the group.[39] Another example of findings in which the number of persons is one of the independent variables is the result of the famous experiment by Ash on the relationship between conformity in perception and external pressure toward perceiving a certain object in a given way.

Many other, more complex characteristics of collectivities as a whole have been found to be relevant as modifiers of the behavioral regularities of individuals. We often use to describe such situations the term 'contextual effect', and the characteristics of collectivities involved in regularities in the behavior of individuals are usually called the 'contextual characteristics' of these individuals or referred to as the 'contextual variables' characterizing them.

The notion of contextual variable[40] deserves some comment. In the previous discussion of the characteristics of collectivities I stressed their definitional dependence on the characteristics of the individuals composing them. In the case of contextual variables, we have a definitional dependence which seems to go in the opposite direction. In general we attribute a contextual property to an individual when we say that he is a 'member of the collectivity of a given kind'; it becomes obvious that it refers to a structural or analytic property of this collectivity which has to be interpreted as discussed above. When we look at the typical 'operationalizations' of such contextual properties, we can also see that the indicators of contextual variables are such characteristics as 'proportion of religious persons in the group of which the given person is the member', 'degree of integration of the group to which he belongs', etc. Therefore, in order to give a clear meaning to the contextual property, we first have to specify the nature of the property of the group itself in one of the ways discussed above. This higher-level property, even when it is a relational property of the individual, represents the mechanism of interaction between the characteristics of the individual and the configuration to which he belongs, and has a reciprocal role in determining the individual's behavior.

Let us look first at the action of contextual variables understood as analytical properties of collectivities and described by the various statistical parameters characterizing them. In order to consider a most simple illustration of such mechanisms let us assume that the probability of occurrence of $b(e)$ when $a(e)$ has occurred depends not only upon the occurrence of $a(e)$ or some other individual characteristics of e, which we might eventually discover, but also upon the *relative frequency of a(e)* in the given population of e's. Let us also assume that the higher the relative frequency of $a(e)$ in the given population the higher the probability of b when a has occurred. Or formally,

$$P(b/a) = f\,[P(a)].$$

In such case we will have a typical *interaction effect of the statistical composition of the population*; the probability of occurrence of *b* in our elementary law is dependent not only upon the occurrence of the antecedent *a(e)* but also upon the composition of the given population with respect to the frequency of *a(e)*. There are many interesting sociological examples of this kind of dependence of the probabilistic relations upon the statistical composition of the population.[41] This being so, the *relationship between the variables at the global level will be much stronger than what one could derive by observing the frequencies of a(e) in each of the populations separately* and predicting on the basis of this the corresponding frequency of *b(e)*, and correspondingly the value of *B(W)*.

But it may also turn out that the increase of frequency of *a(e)* will influence negatively the probability of *(b/a)*. Thus for instance one could say that 'other things being equal' the greater the number of candidates aspiring to a certain position in society, the smaller the probability that any one of them will get this position. In such case there will be no relation between the ratio of candidates to the given positions and the ratio of persons occupying them at the global level, even if in each of the societies taken separately the relationship between aspiration and getting the position may be quite visible (although the greater the number of candidates, the weaker this relation). We might call this the *negative impact of statistical composition* upon the elementary relation, or the negative contextual effect.

The overall configuration of relational properties constituting a structural property of collectivities may also be involved in interaction with the behavioral regularities of individuals. As examples: people dissatisfied with their social system will behave differently depending on whether the power structure of the system is weak or strong; the incidence of suicide, while undoubtedly related to certain personality characteristics of individuals, will nevertheless (if we accept Durkheim's theory) be different in strongly integrated groups from that in groups whose integration is low; the satisfaction group members derive from a cooperative enterprise depends additionally upon the pattern of their interaction within the group; etc.

The regularities referring to the interaction of individuals within a certain system of relations and the dependence of the regularities of their behavior upon the characteristics of the system itself deserve special attention in sociological theory, and overlooking this may lead to an 'individualistic

fallacy', as Erwin Scheuch proposed to call it. He wrote that the 'individualistic fallacy' is, for instance

> to count the percentage of authoritarian persons or to ascertain the proportion of
> individuals who some closer to a particular notion of 'democratic, civic culture' in their
> opinions, and to take this as an index of the degree to which the society is democratic.
> In such studies as international citizenship survey, what is ignored is that one may have
> a democratic system with few 'democratic' personalities and various types of
> authoritarian systems with high percentages of democratic personalities. Democracy is
> the form of a political system and a political system is obviously not just the aggregate
> of the individuals comprising it.[42]

It might be noted here, however, that if by the 'political culture' of a given
country we understand the 'modal political values of its citizens', then of
course this is a typical 'analytical property' of an aggregate and from this
standpoint these 'individualistic' findings would be correct. On the other
hand, if we are interested in how a political system functions, then of course
the modal values of citizens may have a greater or a lesser impact upon this
functioning, depending upon the kind of structure within which the given
political process and its regularities occur, and *our reductive theory should
take the relations composing this structure into account.* These relations will
play two different roles in our reductive theory. First, we must take into
account those relations which co-determine the occurrence and 'shape' of
regularities of behavior of the elements of the system, i.e. those which
interact with the characteristics of individuals (or smaller units within the
system) and without which an adequate explanation of prediction of the
behaviors of elements of the system is not possible. To give a simple
example: Suppose we would like to explain in a reductive manner the
regularities of functioning of a certain formal group. As we know, the
actions of members of such groups are to a high degree co-determined by
the role requirements 'sent' to them by other members or specified by
certain codified rules describing the group's structure. These role requirements determine behavior somewhat as the shape of a maze determines the
behavior of the experimental rat. If we want to explain why a certain action
occurred at all, we can forget about the exact contents of role requirements
and concentrate only upon the motivational factors of individuals. If we
want to explain their behaviors in their more complex characteristics, the
role requirements will have to be also taken into account. When we want the
ultimate explanation, we have to take into account as well the impact of the

immediate social milieu on the individuals, i.e. the reactions of the other members of the group to these role behaviors. The formula presented here can be specified as proposed above:

$$[(e_1 Re_2) \cdot a(e_1) \cdot c(e_2)] \to b(e_1).$$

Once we have explained in this way the repetitious behavior of all the members of the group, (or all units of a certain system) we then face the problem of *integrating the regularities into an overall pattern*, which in our symbolics is described by $C_1 a(e)$, with the expectation that the overall effect of this will be $C_2 b(e)$. The transition from C_1 *to* C_2 will be made possible by the elementary regularities working at the level of behaviors of the elements, and the interactions between them. The overall *theoretical model* of such a process will be much more similar to the diagram of a television set with a great number of 'feedbacks' between its elements, than to a simple formula in which the behaviors of the elements have been aggregated into an analytical structure of the collectivity as a whole and then related to each other by a simple law of behavior. But this does not mean that such complex structural processes are not explainable in a reductive way, as we do not doubt that the functioning of a TV set as a whole is explainable in a reductive way by reference to the laws of electricity and electronics, under the assumption that these laws operate under a complex set of conditions described by the diagram of a TV set.

The only (but essential) thing which differentiates the functioning of the parts of a TV set from the functioning of the 'human parts' of a social system is that besides their 'individual feedbacks' with other components of the system, human beings may also be influenced in their behavior by their perceptions of the system as a whole, and these perceptions may have an impact upon the way they behave in the system, independently from the fact that the overall pattern of the system integrates their individual behaviors into the 'functioning whole'. When this is the case, we have to use certain collective (structural) characteristics of the system at the level of explaining theory as modifiers of the regularities of individual behaviors. As we know, these collective characteristics can be interpreted as 'contextual variables' of individuals, but this is rather a verbal than a theoretical modification of our theory.

Nevertheless, when we discover the importance of contextual characteristics of any kind for the proper formulation of an elementary law, this does not mean that at least some of them cannot be later on interpreted as strictly individual characteristics of the group members. I mentioned above the dependence of the tendency to 'helping behavior' on the number of persons present at the moment the help is needed. But in more detailed experimental studies, the experimenters discovered that it was not so much the real number of persons that mattered here as the conviction of the person who was supposed to help that a certain number of other persons were also present. It was enough to create by proper experimental instruction[43] the conviction that other persons were also listening to the call for help (even if the 'presence' of others was a simple experimental simulation) in order to obtain the same linear negative correlation between the 'number of people present' and the tendency to help. The effects of the actual presence of others and the effects of the conviction of the subject that others were also present were the same. Therefore we may say that in this particular case the subjects' conviction about the situation was the 'real' variable which interacted with the tendency to help.

One reason the contextual variables are so important in the regularities of social behavior is that people are able to *observe* more or less directly the structure and functioning of certain collectivities, and are able to make more or less valid inferences about the structure and functioning of some larger ones. Their judgments as to the characteristics of various social wholes may modify the ways they behave in them. In this way, by mediating the cognitive processes of individuals and all the motivational and affective correlates of these processes, the most complex features of human collectivities may be involved as modifiers of regularities of human behavior.

But to the degree that these collective properties mediate 'through the heads' of individuals, we can construct a theory that will concentrate on a closer causal link in these processes, i.e. on the psychological states, attitudes, opinions and perceptions of individuals. Then we can transform a contextual variable into a real (absolute) characteristic individual what makes from it a 'contextual variable with humanistic coefficient'.[44] This does not mean of course that we say that these properties of collectivities can create different behaviors in individuals, which from the standpoint of a particular contextual varible are identical. We have then to assume that even within a given collectivity its properties are perceived or evaluated differently by different individuals.

But this interpretation of contextual variables by strictly individualistic mediating psychological processes permits us to explain only one type of mechanism of their impact upon the behavior of individuals. Behavior is in its course and consequences determined not only by our motives and the immediate stimuli which cause it, but also to a lesser or greater extent by the objective features of the natural and social environment in which it occurs. Our attitudes (including those referring to the characteristics of groups and societies) may sometimes explain *why we decided to undertake a certain action*, but its course and consequences will be co-determined by the nature of the social reality itself, and especially by the narrower or broader context of this behavior. A public demonstration of dissatisfied citizens may be co-determined by the belief that the power structure of the state is weak, but the demonstrators will learn they were wrong as soon as they face the wall of police across the street.

These two kinds of mechanisms may of course work together. Let me give another example from sociological research. The contextual variable 'the average school success of students of a given class in high school' was related to the success of particular individuals. Students having the same individual characteristics but differing on this contextual variable also differed in their school successes. We can assume that this was partly due to the fact that the perception of the average success in the given class constituted an additional motive to 'keep up with the Joneses' among those students who otherwise would have had a much lower level of school aspirations. But we must also assume the possibility of a real interaction of students with their colleagues in class, and that interaction with a number of intelligent and knowledgeable students helps each member of the class to develop his potentialities to the maximum. In this case our contextual variable remains *really contextual* in reductive theory; it cannot be validly transformed into strictly individual characteristics without losing from our theory the fact of interaction between the individual and his environment.

14. POSTULATES OF REDUCTIONISM AND THE RELATIONS BETWEEN SOCIOLOGY AND PSYCHOLOGY

For some strange reason it is fairly commonly believed that in sociology 'reduction' always means the explanation of sociological laws by the laws of psychology. For some writers the program of reductionism means the 'elimination of sociology'[45] altogether. Others practice reduction in such a way that it may create the impression that it should always be understood as reduction of sociological propositions to laws of psychological theories.

It is quite understandable that such an approach to the program or practice of reductionism elicits little enthusiasm among those who, being sociologists, still believe they have something to do in the area of study of man's social behavior and the functioning of social systems. They often over-react in the other direction, denying either the possibility or the utility of reductive systematization in the social sciences. It seems that in this case both sides are wrong.

I have mentioned several times that reduction of the laws of one theory does not mean that the explained laws become useless and that the theory or science to which they belong should be 'liquidated'. Even when we know in principle how to explain the laws of thermodynamics by more elementary laws we still use them, because they describe the relations between the parameters of the volume of gas on the macro-level in a much more simple and economic way than could be done on the level of singular particles of gas. When we succeed in reducing the macro-laws of macro-genetics to laws of molecular biology, we still will be using Mendelian laws at the level of explaining the heritability of singular traits of organisms. The fact that these macro-laws have been successfully reduced will help us in better understanding them or to propose certain their modifications, but it will not invalidate them. Nor will it make them useless at the level where their use is most fruitful. The same is true for sociology, or any other behavioral or social science. *For both theoretical and practical reasons we need in our sciences laws at all levels of generality and elementarity.*

Besides, it is by no means true that in our disciplines reduction always implies explanation of sociological propositions by psychological ones. Even without entering into a long (and in my opinion hopeless) discussion of the problem of differences and boundaries between psychology and sociology, I could cite many examples of full or at least partial reductions which connect two laws or two theories belonging to the same discipline – either sociology or psychology.

Even if one accepts that psychological propositions about human behavior are usually more general than sociological ones, when we vertically order the different propositions on human behavior in a reductive theoretical structure in which less statements are derived from more general ones, we still have a continuum of propositions of different levels of generality rather than a dichotomous structure. The problem of defining the exact field of our disciplines or the exact location of the borderline between different disciplines studying human behavior is not really very important. It is much less important, in any case, than the implementation of a program of reductive systematization of propositions of different levels of generality, which would provide a better understanding of the mechanisms of regularity of lower levels of generality when we explain them as special cases of more general laws, and a better empirical verification of the more general laws when we find that these are confirmed by their specific less general conclusions.

Another type of reductive systematization of propositions is, we will recall, the explanation of laws about social groups and systems in terms of some more elementary laws referring to the elements of which these social 'groups' wholes are composed. Here again we may say that the reduction of some laws about the functioning of human groups and aggregates may proceed toward any lower level of social phenomena, provided we find there a theory which is useful for our reductive purposes, and not necessarily a psychological theory. In methodological terms, the contributions of small group theory to the theory of large organizations are at least partial reductions of the regularities of large human aggregates by propositions about the functioning of the 'parts' of these aggregates, but 'parts' in these cases are not individual human beings but smaller human groups. Some 'functional analyses' of interactions between different parts of social 'wholes' (especially if these parts are composed of different persons) are not so different from the realization of postulates of reduction of laws about social wholes to more elementary social laws, and could probably be 'translated' into reductionist language. Cybernetic analysis of systems in terms of properties of elements and the connections between them may also be seen as a special kind of reductive explanation of the functioning and dynamics of social wholes, if the functional elements are distinguished in a way that corresponds to our meaning of 'parts' of a 'whole'.

Now what if we would like to reduce a macro-theory to a theory of individual human behavior? A necessary condition here is that we are able to describe the macro-regularity of the macro-process in terms of individual human behavior, i.e. to find at least its elementary correspondent for that level. It may turn out that we come to the conclusion that the regularity of behavior figuring in the elementary correspondent of the global law constitutes a *sociological law* itself, i.e. we find that certain conditions defined in sociological but 'individualistic' terms produce on a fairly general scale a certain behavioral sequence of interest to a sociologist.

Then our next step *may be* to explain this sociological regularity (applying to the social behavior of individuals on a mass scale) by a *more general* psychological law or theory. But we may also stop at the level of a 'strictly sociological' regularity of behavior in our reductions.

Whatever the character of the explaining regularity, whether we call it a 'sociological' or a 'psychological' law of human behavior, we should remember that the elementary law is only a part of the whole reductive structure: it interacts with all configurations of the absolute and relational characteristics of all the elements, and it may also interact with the collective properties of the whole group or system involved – as contextual variables – in the elementary regularity itself. It is obvious that *it is the task of the sociologist to study this overall pattern of relations and that he is the one who is competent to study them.* To build a reductive theory of the social process of a sociological macro-regularity is to build a complex theoretical model *one of the elements of which is the laws of behavior of the elements.* From simple behavioral laws nothing at all can be deduced for laws of the macro-type in the social sciences.

Nevertheless, reductions of sociological propositions to laws of psychology, if they are successful, can be especially valuable because the laws of psychology are usually much better formulated and much more thoroughly tested in experimental conditions than most of the laws of social behavior we can find in textbooks of sociology. This means that the effective reduction of sociological propositions to psychological ones usually leads either to improving the formulation of the sociological propositions by showing that some essential factors of their validity are missing in their formulation, or to increasing our confidence in the validity and generality of these sociological laws.

Even if we succeed in explaining some 'sociological' regularities by some 'psychological' ones, we should be aware of the possibility of *causal relations* going in the opposite direction, too. If a psychological mechanism is used for the explanation of a 'sociological' relationship because it constitutes an intervening process, or in other words a mediating link in a causal chain, this means that certain 'sociological' processes or phenomena are the causes of the mediating psychological process. *They are 'responsible' for its occurrence on a mass scale.* To say that we can reduce some mechanisms of the formation of certain ideological attitudes to the frustration – agression theory does not imply that certain sociological phenomena – e.g. an extremely rigid class structure or poverty – are not responsible for the frustration of the underprivileged in the given society.

We might even go one step further. I think we have to accept that the general regularities of the psychological and behavioral functioning of man constituting the subject matter of psychology are the result of interaction of two kinds of uniformities:

(a) uniformities of human anatomy and physiology, and especially the uniformities of structure and functioning of the human brain, i.e., the mechanisms comprising the 'hardware' of man's functioning;

(b) uniformities of social conditions, such as certain of the most general features of cultures and certain of the most general features of man's social experience, which imprint certain 'software' into his thinking and functioning – to continue this computer analogy.

The weight of the factors belonging to either of these classes, in a given particular psychological mechanism, or for the explanation of any particular psychological theory, is, of course, an open empirical question. But whatever the solution to this problem for any given psychological theory, many of such theories will be at least partly explained by reference to certain uniformities in man's social experience.

This does not mean that, even without explaining these theories in this or some other way, we cannot use them for a reductive explanation of another sociological theory.

To say it in more general terms: *When explaining sociological mechanisms by psychological ones, we should always be aware that the society as a whole or certain of its smaller parts co-determine human personalities of the given types and create the conditions in which the given personality types function in the given way. The order of reductive explanation of theories and the order of socio-psychological causation do not always have to go in the same direction. Psychology is equally (although in different ways) as dependent upon sociology as is sociology upon psychology. The main problem is to see clearly the different natures of their mutual dependencies.*

NOTES

* Basically modified version of the paper presented at the Conference on Methodological Problems of Sociological Theories organized by the Institute of Sociology of Warsaw University in 1970. Most of the papers from this conference, including the first version of the paper, were published in S. Nowak (ed.), *Metodologiczne problemy teorii socjologicnych (Methodological Problems of Sociological Theories)*, Warsaw 1971, and in their English translation in *The Polish Sociological Bulletin* (Nr. 2, 1970), and in a special issue of *Quality and Quantity* (Nr. 2, 1972).

[1] For a more extended analysis of some typical patterns of 'horizontal' systematization see Chapter V. See also H. Blalock, *Causal Inference in Non-Experimental Research*, Chapel Hill, 1961.

[2] For an example of axiomatized theory, see H. Zetterberg, 'On Axiomatic Theories in Sociology', in: P.F. Lazarsfeld and M. Rosenberg (eds.), *The Language of Social Research*, New York, 1950. See also: H. Hochberg, 'Axiomatic Systems, Formalization and Scientific Theories', in: L. Gross (ed.), *Symposium on Sociological Theories*, New York, 1959.

[3] E. Nagel, *The Structure of Science*, New York, 1961, p. 338.

[4] Many examples of reductions of this kind can be found in A. Malewski, *O Zastosowaniach teorii zachowania (On the Applications of the Theory of Behavior)*, Warsaw, 1965, although the author did not use the term reduction at all and refers to the above procedure as explanation. For Malewski the explanatory theory is *S-R* learning theory.

[5] See the detailed analysis of this particular reduction in: E. Nagel, *The Structure of Science*, Ch. XI, 'Reduction of Theories'.

[6] For typical examples of reductions of small group mechanisms see G. Homans, *Social Behavior, Its Elementary Forms*, New York, 196 . See also John W. Thibaut and H.H. Kelley, *The Social Psychology of Groups*, New York, 1959.

[7] The different approaches to reduction and different philosophical controversies and reduction in social sciences can be found in the papers reprinted by M. Brodbeck in *Readings in the Philosophy of the Social Sciences*, New York, 1968, Part IV, 'Social Facts, Social Law and Reduction'.

[8] Some philosophers insist that in order to reduce one theory to another one we have to deal with two completely consistent, axiomatized theories. (See the discussion of these views in K. Schaffner, 'Approaches to Reduction', *Philosophy of Science*, Vol. 34, 1967.) In my opinion such a postulate is much too strong for the social sciences, where we rarely deal with consistent, axiomatized theories and usually have more or less unrelated *propositions* applying to a certain area of social or behavioral phenomena. As we shall see, even for particular singular generalizations, it is not easy to find a complete reduction to another generalization, and usually we have to be satisfied with partial reduction only. Moreover, it often happens that in order to explain a certain generalization we have to use more than one explaining theory. It would, of course, be better if the contemporary social theories were sufficiently consistent (e.g., axiomatized) that at least some of them could be derived from other axiomatized theories, but I do not think we should wait that long with our attempt to explain at least some of the propositions of contemporary sociology by reducing them to other more elementary or more general propositions.

[9] See G. Homans, *Social Behavior, Its Elementary Forms*

[10] Ibid., p. 89.

[11] See E. Nagel, 'Reduction of Theories', in: *The Structure of Science*, Ch. XI.

[12] This is a modified example from Hans Speier, 'The American Soldier and the Sociology of Military Organizations', in: R.K. Merton and P.F. Lazarsfeld (eds.), *The American Soldier, Continuities in Social Research*, Glencoe, 1950.

[13] J. Dollard and N.E. Miller, *Personality and Psychotherapy*, New York, 1950.

[14] See Adorno and others, *The Authoritarian Personality*, New York, 1950.

[15] See P. Meehl, 'On the Tautology of the Law of Effect', *Psychological Bulletin*, 1950.

[16] See J. Olds, 'Physiological Mechanism of Reward', *Nebraska Symposium on Motivation*, 1965.

[17] Chapter V.

[18] For a more detailed analysis of this assumption see Chapter VI.

[19] See: E. Nagel, op. cit.

[20] C.P. Wolf, 'Foreword' to: J. Coleman, A. Etzioni, and J. Porter, *Macrosociology, Research and Theory*, Boston, 1970, p. xii.

[21] P.F. Lazarsfeld and H. Menzel, 'On the Relations between the Individual and Collective Properties', in: P.F. Lazarsfeld, A. Pasanella, and M. Rosenberg, *Continuities in the Language of Social Research*. New York, 1972.

[22] *Ibid.*, p. 228.

[23] J. Coleman, 'Properties of Collectivities', in: J. Coleman, A. Etzioni, and J. Porter, *Macrosociology, Research and Theory*, Boston, 1970.

[24] *Ibid.*, p. 10.

[25] *Ibid.*, p. 6-7.

[26] May Brodbeck, 'Methodological Individualism, Definition and Reduction', in: M. Brodbeck (ed.), *Readings in the Philosophy of the Social Sciences*, New York, 1968, p. 283.

[27] W.S. Landecker, 'Types of Integration and Their Measurement', in: P.F. Lazarsfeld and M. Rosenberg (eds.), *The Language of Social Research*, Glencoe, 1955, p. 20.

[28] See J. Coleman, *op.cit.*

[29] *Op.cit.*, p. 283.

[30] See May Brodbeck, *op.cit.*

[31] See E. Nagel, 'Reduction of Theories', *op.cit.*; see also M. Brodbeck, 'Methodological Individualism, Definition and Reduction', *op.cit.*

[32] May Brodbeck, *op.cit., p. 298.*

[33] The example is taken from a paper by Andrzej Malewski: 'Tresc empiryczna teorii materializmu historycznego' (The Empirical Content of the Theory of Historical Materialism), which was published in *Mysl Filozoficzna*, Warsaw, 1958.

[34] Many examples of such a situation were given by Herbert Menzel in his comments on Robinson's 'Ecological Correlation and the Behavior of Individuals', *American Sociological Review*, 15 (1950), p. 674.

[35] See W.S. Robinson, 'Ecological Correlation and the Behavior of Individuals', *American Sociological Review,* 15 (1950).

[36] See the last section of Chapter V.

[37] See H.R. Ahlers, 'A Typology of Ecological Fallacies', in M. Dogan and S. Rokkan (eds.), *Quantitative Analysis in the Social Sciences*, Cambridge, 1969.

[38] See A. Edel, 'The Concept of Levels in Social Theory', in: L. Gross (ed.), *Symposium on Sociological Theory*, New York, 1969.

[39] See B. Latane and J.M. Darley, 'Social Determinants of Bystander Intervention in Emergencies', in: J. Maculay and L. Berkowitz, (eds.), *Altruism and Helping Behavior*, New York, 1970.

[40] See P.F. Lazarsfeld and H. Menzel, 'On the Relations between the Individual and Collective Properties', *op. cit.*

[41] For a detailed discussion of the category of contextual variables understood as aggregate characteristics of collectivities, see a special symposium on the role of the 'climate of opinion variable': D. Sills, Moderator; J. Coleman, Commentator, *Public Opinion Quarterly*, 1961. See also the many examples of studies in which the impact of this kind of variable has been demonstrated, reprinted in P.F. Lazarsfeld, A. Pasanella and M. Rosenberg, *Continuities in the Language of Social Research*, New York, 1972, Section III B, 'Weight of Collective Characteristics' Section III C, 'Contextual Analysis'.

[42] E. Scheuch, 'Cross-National Comparisons Using Aggregate Data: Some Substantive and Methodological Problems', in: R. Merrit and S. Rokkan (eds.) *Comparing Nations*, New Haven, 1966.

[43] See B. Latane and J.M. Darley, *op. cit.*

[44] For the analysis of concepts defined with 'humanistic coefficient' see 'Concepts and Indicators in Humanistic Sociology', in this volume, p. 1.

[45] A typical example of such a viewpoint, according to which the explanation of one theory by another is equivalent to the elimination of the explained theory, may be found in the paper by Kummel and Opp, 'Sociology without Sociology', *Inquiry, 11,* 1968. On the other hand, H.F. Spinner, in criticizing the paper by Hummel and Opp, seems to go too far in his rejection of reductivism, stating that reduction constitutes an obstacle in the development of science. I would be inclined to say that successful reductions constitutes one essential (but by no means the only) component of the process of development of sciences. For the presentation of Spinner's views see H.F. Spinner, 'Science without Reduction, A Criticism of Reductionism with Special Reference to Hummel and Opp's "Sociology without Sociology" ', *Inquiry* 16, 1973.

BIBLIOGRAPHY

Adorno, *et al., The Authoritarian Personality*, New York, 1950.

Ahlers, H.R., 'A Typology of Ecological Fallacies' in: M. Dogan and S. Rokkan (eds.), *Quantitative Analysis in the Social Sciences*, Cambridge, 1969.

Blalock, H., *Causal Inference in Non-Experimental Research*, Chapel Hill, 1961.

Brodbeck, M., 'Methodological Individualism. Definition and Reduction', in: M. Brodbeck (ed.), *Readings in the Philosophy of the Social Sciences*, New York, 1968, p. 283.

Coleman, J., A. Etzioni, and J. Porter, *Macrosociology, Research and Theory*, Boston, 1970.

Dollard, J., and M. Miller, *Personality and Psychotherapy*, New York, 1950.

Edel, A., 'The Concept of Levels in Social Theory', in: L. Gross (ed.), *Symposium on Sociological Theory*, New York, 1969.

Hochberg, H., 'Axiomatic Systems, Formalization and Scientific Theories', in: L. Gross (ed.), *Symposium on Sociological Theories*, New York, 1959.

Homans, G., *Social Behavior, Its Elementary Forms*, New York, 1961.

Kummel, and Opp, *Sociology without Sociology Inquiry*, vol. 11, 1968.

Landecker, W.S., 'Types of Integration and Their Measurement', in: P.F. Lazarsfeld and M. Rosenberg (eds.), *The Language of Social Research*, Glencoe, 1955.

Latane, B., and J.M. Darley, 'Social Determinants of Bystander Intervention in Emergencies', in: J. Maculay and L. Berkowitz, (eds.), *Altruism and Helping Behavior*, New York, 1970.

Lazarsfeld P.F., and H. Menzel, 'On the Relations between the Individual and Collective Properties', in: P.F. Lazarsfeld, A. Pasanella and M. Rosenberg (eds.), *Continuities in the Language of Social Research*.

Malewski, A., *O Zastosowaniach teorii zachowania* (On the Applications of the Theory of Behavior), Warsaw, 1965.

Malewski, A., 'Tresc empiryczna teorii materializmu historycznego' (The Empirical Content of the Theory of Historical Materialism), which was published in *Mysl Filozoficzna*, Warsaw, 1958.

Meeh, P., 'On the Tautology of the Law of Effect', *Psychological Bulletin*, 1950.

Menzel, H., 'Comments on Robinson's "Ecological Correlation and the Behavior of Individuals" ', *American Sociological Review,* 15, 1950.

Nagel, E., *The Structure of Science*, New York, 1961.

Nowak, S., 'Causal Interpretation of Statistical Relationships in Social Research', in this volume, p. 165.

Nowak, S., 'Inductive Inconsistencies and the Problems of Probabilistic Predictions', in this volume, p. 228.

Olds, J., 'Physiological Mechanism of Reward', *Nebrasca: Symposium on Motivation* . 1965.

Robinson, W.S., 'Ecological Correlation and the Behavior of Individuals', *American Sociological Review* 15, 1950.

Schaffner, K., 'Approaches to Reduction', *Philosophy of Science* 34, 1967.

Scheuch, E., 'Cross-National Comparison Using Aggregate Data: Some Substantive and Methodological Problems', in: R. Merrit and S. Rokkan, (eds.) *Comparing Nations*, New Haven, 1966.

Sils, D., J. Coleman, *et al.*, Symposium on the "Climate of Opinion Variable", *Public Opinion Quarterly*, 1961.

Speier, H., 'The American Soldier and the Sociology of Military Organizations', in: R.K. Merton and, P.F. Lazarsfeld (eds.), *The American Soldier, Continuities in Social Research*, Glencoe, 1950.

Spinner, H.F., 'Science without Reduction, A Criticism of Reductionism with Special Reference to Hummel and Opp's "Sociology without Sociology" ', *Inquiry,* 16, 1973.

Thibault, J.W., and H. Kelley, *The Social Psychology of Groups*, New York, 1959.

Zetterberg, H., 'Axiomatic Theories in Sociology', in: P.F. Lazarsfeld, and M. Rosenberg (eds.), *The Language of Social Research*.

CHAPTER XII

VALUES AND KNOWLEDGE IN
THE THEORY OF EDUCATION:*
A PARADIGM FOR AN APPLIED SOCIAL SCIENCE

1. THEORY OR THEORIES OF EDUCATION

The scope of interest of the theoreticians of education is (at least some-times) sufficiently wide that it embraces the whole of social and psycholo-gical factors upon which depends the final cast of man in society with his knowledge about the world, with his personality traits and his system of values, of norms and evaluations, his behavioral predispositions, and some-times also his physical aspect. This puts the theory of education in a situation which is advantageous and disadvantageous at the same time: advantageous because there is no threat of lack of problems or excessive specialistic narrowness in taking them up, disadvantageous because the wider the the richer the problems the more difficult it is to introduce some theoretical-methodological order and to systematize directives and propo-sitions. Some of these difficulties I want to discuss in this chapter.

Let us start with the problem stated in the title. The subject of our interest is the theory of education. However, the term 'theory' is under-stood in many different ways. Sometimes one has in mind the meaning in which theory is opposed to 'practice' or at least is differentiated from it. Educational practice would then be a set of generalized methods applied with greater or lesser success in 'shaping' people's personalities, teaching them necessary knowledge, and transmitting to them a system of norms and values. The theory of education would thus be simple a scientific reflection on these educational practices.

Recently there seems to be an increasingly greater recognition given to a somewhat different and narrower understanding of the term 'theory' in the area of various more mature sciences. According to this understanding, theory is a relatively uniform and internally logically connected system of propositions somehow related to each other (e.g. implied logically one by another) and explaining a certain defined area or sphere of phenomena. Every scientific discipline contains, as a rule, many such theories sometimes completely independent of each other, or inrelated to each other.

It is well known that there is no 'theory of health' in medicine but that there are many *partial theories* explaining special aspect of 'proper' functioning of the organism and deviations from this functioning. There is no uniform theory of society, but there are many different theories explaining different categories of social phenomena.[1] This does not mean that there is no theoretical knowledge about human organism, no theoretical sociology, theoretical psychology, or finally no theoretical science of education.

If, however, we agree on an understanding of the term 'theory' which, more and more generally, is accepted in the various sciences, then we can draw the obvious conclusion that there is and for a long time will be no all-inclusive theory of education. This does not mean that the science which is interested in educational processes is not a theoretical science, that is, one which formulates general laws and uses them in its predictions. I think that in the theory of education, along with numerous purely practical observations stemming from the experience of the great masters of education, there are many elements of a genuinely theoretical character taken from psychology, sociology, or social psychology, as well as those elements which are the result of theoretical generalizations by the investigators of educational institutions and processes. I don't think, however, that one can speak here of any single all-inclusive theory. I think that what we have are many partial theories. Some investigate the results and consequences of specified types of educational systems and processes; others search for factors which determine the shaping of definite types of personalities, or factors which assure the efficacy of processes of learning; while still others investigate the role of personality factors in educational processes, etc. As in any other developing empirical science, this is far from being an all-embracing order which one could call an integrated theory.

Thus the problem of meaning of the term 'theory' seems to be a verbal one. I believe, however, that acceptance of the above terminological proposal has certain important consequences. Namely, it focuses our attention on looking for *theories sensu strictu*, that is, internally ordered systems of laws governing given aspects of educational processes, either those formulated on the basis of analysis of the functioning of various educational settings and institutions, or those formulated on the basis of other data (and in various social and behavioral sciences) but which are *indirectly* applicable to the study of educational processes. These theories should be formulated in such a way that they describe and explain the structure and functioning of phenomena which are subject to definite uniform regularities. Next, these so formulated and empirically tested theories should be applied to predictions and social practice in the area of educational processes.

These theories will have a varying scope and varying relations to each other. There may be theories that have no relation to others at all. They pertain to quite different phenomena which are not adjacent to any others either in time or in space. Sometimes they will be complementary to each other, pertaining to different aspects of the same sphere of phenomena. In some cases they may have a competitive character: they are logically inconsistent with each other, while at the same time there is a lack of sufficient empirical findings to settle the disagreement.

Let me give a few examples.

Learning theory or its sociological conclusion, the theory of socialization, can to a large extent explain why in certain systems of reward and punishment, or in certain systems of punishing and rewarding interactions between the pupil and the collectivity which educates him, there takes place the process of 'reinforcement', as the theory of learning says (or internalization, as the sociologists prefer to say), of the socially accepted patterns of behavior.[2] In order to explain the emergence and action of various patterns in different societies, in different social groups or systems, we must examine certain relevant chapters or parts of sociological theory, in particular those dealing with relations between the social structure and cultural patterns or ideological systems with the role of 'reference groups'[3] in this process, etc. In turn, in order to predict the behavior of our pupil when faced with real life situations we have to look to the theories which analyze relations between motivation and behavior, probably sometimes to the theory of motivational conflict, etc.

And here one additional remark. As I mentioned above, there is, and for a long time will be no homogeneous, all-inclusive theory of education. This is not to say, however, that there will be no growth of theoretical systematization and integration of the set of propositions pertaining to education. The development of our sciences indicates that we find more and more connections between the laws of various spheres of phenomena by deducing from them some additional general theoretical reasons or relating them in other ways. The theory of human behavior is no exception. We find more and more connections between different types of laws, explaining, for example, socialization in terms of learning theory, or the mechanisms of ethnic hostility in terms of psychoanalytic displacement of aggression or in projection mechanisms.[4]

By the same token we find connections between traditionally separated spheres of interests such as sociology and the domain of psychology. At the moment, however, we are very far from a logical unity of the sciences that would explain the shaping of man by society and reciprocally the shaping of society by man. Before such a unity is achieved there will have to be a lot of work done in seeking the connections between laws of various kinds. The theoretical unity of a scientific discipline can not be proclaimed through a vote.

But, there is no good reason for not using the term 'theoretical peda-gogics' just as other sciences — for example, 'theoretical physics' — consciously set our to discover laws which govern a certain sphere of phenomena and to order these laws into various theories, without prejudging the theoretical uniformity of the given science. The science of education is not an exception.

2. PROBLEMS OF SYSTEMATIZATION OF THEORETICAL KNOWLEDGE IN THE SCIENCE OF EDUCATION

A number of heterogeneous sociological and psychological factors, which we shall consider as independent variables, influence various aspects of the final shape and functioning of homo sapiens, which we shall call the dependent variables. Theoretical statements in the applied sciences may be systematized in many different ways. A science which wishes to keep its theoretical character while at the same time desiring application in social practice of the results of its investigations must consider (among others) the usefulness of the following two methods of systematization.

(a) Those methods of ordering theoretical statements which take as the point of departure a certain defined *independent variable* and state its consequences (in sufficiently large scope) for those dependent variables which are of interest to the theorist.

(b) Those methods of ordering statements which take as the point of departure a specific *dependent variable* and attempt to find a sufficiently large number of factors which influence this variable.

An example of the first kind of systematization is an 'inventory' of various propositions which state the role of various personality factors in educational processes. Another example of this category might be a collection of propositions about the educational consequences of the disintegration of the family, of alcoholism of the parents, of the juvenile's contacts with a gang or more generally the role of the peer group in shaping the personality and system of values. This method of systematization also entails investigation of the educational effects of participation in complex cultures which present to the person many alternative moral patterns to choose from. The method includes as well propositions that refer to the efficacy of specific techniques of teaching.

One can give many example but this is not the point. The principle is that here, in every case, we have a certain independent variable, defined as clearly as possible as a certain factor or a certain cause, and we want to state as completely as possible a set of its consequences in the sphere of educational phenomena.

Systematization of the second kind consists of attempts to find the complete set of causes, or different alternative causes, of a uniform consequence (effect) which is of interest, belonging to the area of results of educational processes. The investigation of the totality of factors which condition, for example, the learning of school material will point to causal factors in such spheres as the abilities of the child, his motivation, the methods of presenting the program, situations in which the learning takes place, as well as his wider contacts with the teacher and the class, including the out-of-school environment. Finally, the set will include some elements of the wider macro-social context.

I shall not refer to the diversity of conditions shaping specific values and norms, a fact which is not only obvious but beyond the scope of this discussion. In any case, the focus of our attention is, as we saw above, either a certain independent variable for which we define the totality of consequences, or a certain dependent variable for which we are trying to find the totality of causes.

3. CONSTRUCTION OF MODELS OF EDUCATIONAL SITUATIONS
FOR EXPLANATION AND PREDICTION

The schema of explanations and theoretical predictions necessitates, as is well known, knowledge in regard to two categories of premises.

(a) General propositions stating that among certain independent variables and certain defined dependent variables there exist relationships of a given type.

(b) The possibility of ascertaining whether the phenomena defined by the independent variables actually took place. If these variables are quantitative variables, what is their strength?

It should be noted here that only in very few cases can we find the propositions in 'ready-to-use' from in textbooks or monographs of sociology or psychology. In most cases the propositions we find there are too general for our practical use. They refer to *more abstract* variables defined in the language of the given theory. Before we apply them in our practical science we must first state which aspects of our concrete educational situation can be interpreted as subclasses (or exemplifications) of these more general variables. This applies both to the dependent and the independent variables of educational processes. In some cases such theoretical interpretation of our 'lower level' variables in terms of more general theoretical concepts may seem quite simple — in others it is not. Can we assume that students always perceive good marks as 'rewards' and bad ones as 'punishments'? How can we identify in our educational settings the action of such powerful mechanisms as 'approach-avoidance conflict'? How can we identify the antecedents of regularities discovered in laboratory experiments or the effects of different patterns on interpersonal relations in a natural setting of the family or of a classroom?

In a way the problem is even more serious with respect to the dependent variables of the educational process, i.e., the desired goals of the educational process. These goals are usually stated in terms of certain societal values and formulated in the common language. The main problem we face here is to find out which dependent variables of known psychological or sociological theories can be used for the theoretical 're-translation' of these values, so that we will use proper theories for the explanation or prediction of occurence of these values.

And here we should distinguish the kinds of such theories. One category would have the *dependent variables relatively open* to their concrete values. Learning theory will say that whatever pattern of behavior is properly 'reinforced' will be the one that is adopted. Theory of socialization will say approximately the same but with respect to a broader socializational impact. Other theories will be *much more specific* about the nature of the dependent variable, and will tend to describe certain specific predispositions as the result of the configuration of the antecedent variable.

It should be clear to us that in our attempts to 'translate' the educational process (with its antecedent and its dependent characteristics) we will never be completely successful. Only some aspects of the process can usually be identified in terms of more general theories, while much of it remains unexplainable in such abstract terms. Our first task therefore is to *delimit the applicability of more general* theories in our own field and to derive their limited consequences for our practical tasks, leaving the rest of our problems as unexplainable at the given moment of the development of these theories. On the other hand, it may well turn out that those aspects of the educational process which cannot be accounted for in terms of more general theories are nevertheless quite well handled by educational practitioners in their everyday practice. This means that these practitioners know more or less vaguely certain regularities, from which they can derive certain rules useful in the educationa process. It seems that these 'rules' should be formulated as clearly as possible in terms of the concrete language of the educational process, i.e., without pretending that they are of a very high level or generality or abstractness. Their being articulated in this way may constitute a challenge for the development of more general theories. At the same time such 'rules' may constitute an essential (even if theoretically less general) supplement to the general theories in the construction of models of concrete educational situations and their educational consequences. To disregard the knowledge and experience of social practitioners in favor of more abstract theories seems to be as unjustified in the theory of education as in any other field of applied social sciences.

Sometimes educational theories are applied, regardless of how general and abstract they are, to certain complex multi-factor systems in which there exist simultaneously many heterogeneous concrete phenomena that are subject to very different laws and which are connected with each other in many different ways. It seems to me that theoretical, abstract knowledge can be succesfully applied to explanation and prediction in such historically concrete educational situations via the construction of *complex multivariate models of educational processes, institutions and situations*. In these models would be placed all of the variables which we identify as existing and operating in a given situation, their importance having been assessed on the basis of our theoretical knowledge. But these variables would appear in the models *in their concrete values*, derived from empirical investigations. We know that they are somehow related to each other, that there are mutual connections between them, and that there are temporal sequences of their occurrence. The point of departure in the construction of a model of an educational process or an educational situation is, therefore, a system of initial conditions of a set of laws described as accurately as possible in their concreteness. Application of this set of laws will allow us to state the effects of the joint occurrence of a given large number of the initial conditions, that is, to state (to explain or to predict) the 'output' effect of the educational situations.

The relationship between such a theoretical model of an educational situation and the general theoretical knowledge about educational processes is the same as that between general theoretical knowledge about electricity and the theoretically analyzed model of the network which supplies an entire country with electricity. The word 'model' denotes:

(a) *the description of the initial conditions*, in a language which uniquely defines the values of certain variables including specification of the mutual couplings among these conditions;

(b) *ascriptions to the initial conditions of a set of consequences*, on the basis of some general laws of science, or at least historical generalizations important for this collectivity which connect the causes with their given consequences.

The multivariate, internally interrelated models of educational situations may serve two purposes.

(1) They can be used to explain the observed educational consequences of the operation of certain educational systems, understood either narrowly (as effects of the action of concrete institutions and educational processes) or more broadly (as when we are interested in the dependence of knowledge, values and the behavior of persons on the characteristics of the society in which they live).

Explanation of the observed educational effects must include the statement that in the given collectivity the events postulated by our model actually took place, and that the conditions stated actually imply (on the basis of known laws and generalizations which govern the educational process) the ascertained way of thinking, the system of values or the knowledge of the members of the collectivity. If the initial conditions are sufficiently numerous, then the general theoretical model will be so complex in its concreteness that it is probably historically unique, that is, it will be represented only once in the history of mankind. Even if the scope of the initial conditions is understood more narrowly our theoretical model will have, quite often, denotations which are clearly historically located, even if they are repeated a number of times. An example of analysis of a unique case is the Weberian analysis of the set of conditions that shaped the Protestant ethics, which was personified for Weber in the person of Benjamin Franklin.[5] An example of historically located analysis is the attempt at reconstruction of the educational model of the Prussian military schools. Finally, we can be interested in the mechanisms of educational functioning of a set of conditions having a very wide historical scope – for example, the patriarchal family or the division of roles as between professional teacher and student.

(2) These theoretical models of situations or educational processes can be used for predicting the effects of the operation of specified educational systems. The correctness of the prediction obviously depends on the correctness of our theoretical knowledge as well as on the accuracy of our knowledge of the initial conditions. But there is one other factor we must take into account when constructing a model for predictive purposes. It may happen that the effects of the joint operation of a certain set of antecedent phenomena are different than we expected on the basis of the theory, because the theory took into account the consequences of each of the causes separately whereas we unjustifiably assumed that the consequences have an 'additive' character. The effects of simultaneous occurrence of causative factors are not necssarily additive, but may 'interact' with each other according to any type of interactive relation.[6] Thus in all prognoses in complex systems of interrelated variables there is a certain element of risk due to the possible effect of interaction of some variables – a risk that is not eliminated even by a rather mature theoretical knowledge. The final test of the predictive value of the model is the observation of the real consequences of its operation.

The results of such systematization will be a model of a concrete situation which will reveal how the concrete values of different variables are related to each other with respect to an effect (or, a whole set of effects) they jointly produce. In this model we have to present (graphically or otherwise) which are direct causes of this effect and which influence it only indirectly by shaping or influencing more direct causes; which of them are acting additively and which are interacting with each other; and if there is interaction what its nature is. The model should also state which of the relations have been established only for concrete educational processes and which have been derived from more general theories. In the latter case, it should also be shown which aspects of the educational situation have been interpreted in terms of which more general theoretical concepts; if possible, the reasons for such theoretical interpretations should also be given.

Then we may say that we have a theoretical model valid both for explanations and for predictions of the results of educational processes.

A special case of the application of models for prediction is that of applying them for the purpose of conscious shaping of the personality of the pupil according to the goal of the educator, i.e., to a scientifically grounded educational practice. The following sections of this chapter are devoted to this problem.

4. CERTAIN CHARACTERISTICS OF PRACTICAL SCIENCES

Up to now we have considered educational theory as an analysis of certain educational processes which take place in reality. Now let us turn to the theory of education understood as a certain *applied discipline*, or, as one sometimes says, a *practical science*.

The practical sciences, as is well known, are those sciences which formulate the praxiological directives, i.e., 'instrumental norms' which state what means should be used in order to achieve a defined and desired goal (or avoid an undesired goal). They are the sciences which select their problems and adapt their research procedures and finally present the results of their work from the point of view of certain a priori assumed values. They characterize certain states, events and behaviors as valuable and desirable goals which should be realized. Thus, the selection of research problems in these sciences is always made from the standpoint of certain defined phenomena which interest the researcher on other than purely theoretical grounds. The research hypotheses are formulated in such a way that the phenomena on which our value system focuses our attention are usually the dependent variables.

A careful analysis of the desired goals is therefore the first task of any applied science.

The second important characteristic of practical sciences has already been described here. The goals to be achieved are often very complex systems of consequences, and are realized under conditions which are far from the homogeneity and simplicity of the laboratory, because they are life situations in their whole concrete complexity. In the theoretical sciences, as a rule, we strive to select as a dependent variable a phenomenon as uniform and clear in its abstractness as possible. In the practical sciences we must often use multivariate sets of propositions, and for practical predictions or in 'sociotechnical situations'[7] we may have to create complex multivariate models of educational situations and then on the basis of these derive the prognosis. Such models seem to be necessary also when we set out consciously to use theoretical knowledge in the practice of educational activity.

Here I must mention one further problem. The goals which a given practical science is supposed to help to achieve may be so important that we can not afford the luxury of waiting to improve our knowledge until it permits us to find means that lead unfailingly to the defined goals. Regardless of the degree of perfection of the agricultural or medical sciences people had to make efforts to fight for their bread and health. Regardless of how immature is the sociology of education or educational psychology the society cannot halt the educational process and throw out the unsatisfactory institutions which serve these actions; in these situations certain goals must be achieved because of their social necessity. Even everyday knowledge of the 'know how' type, theoretically limited and usually transmitted by the cultural heritage of simple practical skills available today, are for the practitioner more important than a mature knowledge which will be at our disposal in the future.

Therefore in our educational practices we must use knowledge of different levels of theoretical perfection; along with general theoretical knowledge, pedagogy, psychology and social psychology which use systems of propositions that are relatively well tested and relatively reliable, we must sometimes utilize generalizations that have a limited scope of applicability, i.e. certain historical generalizations. Finally, we must often use the individual experiences of educational practitioners. Since the limits of applicability and the efficacy of such procedures are not established, we fall back on the method of trial and error. But we still have to use them as long as we have nothing else.

Here the practitioner should be cautioned that the probability of educational failure is directly proportional to the level of imperfection of his theoretical knowledge, and especially that the risk of achieving effects other than the desired ones grows immeasurably when the educational processes take place under conditions which have not been previously encountered and thoroughly studied, or at least tested in the practice of education.

Let us move, then, to the methodological problems suggested by the practical applications of the science of education.

5. PROBLEMS CONNECTED WITH THE CHOICE OF GOALS OF EDUCATIONAL PROCESSES

The starting point for systematization of any practical science is clear recognition of the goals which the science is to serve. In the case of the theory of education these goals are usually defined as certain characteristics and dispositions of people, the means to which is a set of educational techniques that will contribute in a maximal way to the inculcation of the desired traits.

The educator is frequently seen as someone who should realize the values of his society in the choice of the goals of the educational process. He is often supposed to realize in his students those characteristics which in his society are considered good and desirable and to weed out those which are not socially accepted. If he is uncertain, he can take the advice of public opinion or the recognized authorities or the power holders for precise definition of these goals.

However, 'frequently' does not means 'always' or 'necessarily'. The history of education shows numerous cases of educators who joined the struggle against the existing system of values and for giving a new shape to human personality. These educators, sometimes representing only certain small subgroups of their society, promoted new revolutionary cultural values opposed to the old traditional ideals, or acted to further their own 'private definition' of educational goals different from the patterns established in society. It is worth remembering that regardless of whether the values realized in the educational process satisfy the values of the whole society, the final responsibility for the values determining the future development of the student is assumed by the person who realizes these values in the process of educational practice.

We can delineate the following categories of development which are usually especially important in the educational process:

(a) certain biological and physical characteristics of the organism;

(b) certain psychological characteristics related to personality;

(c) a certain knowledge about reality;

(d) certain norms, evaluations, wishes, desires or, more generally, certain systems of values of the student;

(e) certain predispositions to behave in a given way in certain situations;

(f) certain practical skills.

This is a typology rather than an exclusive classification. Various educational programs have varying scopes of interests. Teachers of foreign languages are interested only in one final product of the educational process, namely skill in using the foreign language. The athletic coach wants to shape physical ability and psychological dispositions so that students can achieve outstanding results in a given sport. The history teacher is primarily interested in the students' assimilating knowledge about a certain period of history of their own country and the world, although he is also, as a rule, interested in the degree to which they develop a certain set of attitudes and social values. These are examples of relatively narrow — one might even say, segmental — educational goals, which cover only selected spheres of the social personality of the student. But when we discuss the importance of politechnization in education, when we work on the development of innovative school programs, when we engage in ideological discussions about the desired personality structure in the 'atomic' or 'technotronic' era, then our interests are much more general and the educational goals we pose embrace, at least in general outline, alomst the totality of the personality of man. A clear recognition of the areas of personality that are involved when we concider the goals of a certain educational process is the first problem to be considered. Speaking somewhat technically, the *first stage of analysis is the choice and exact definition of the dependent variables of the educational process.*

Also important is the specification of the characteristics we want the student to acquire. In the technical language of logic this is the choice of desired *values of the chosen variables*. Intelligence is a variable that takes many values – from imbecile to genius; however, since this is a primary goal of the educational process, we are generally interested in giving it a high value. It is theoretically possible that there would be an educational program in which for certain groups of the population one would attempt to deliberately manipulate this variable so that it would take on a relatively small value. The pedagogy of Huxley's *Brave New World* was so conceived. An example somewhat better known from historical experience is the school policy of Nazi Germany in many countries of occupied Europe. Relations with other people is an example of a variable (nota bene, 'multi-dimensional'), with altruism, good will, and the search for warm contacts with people being examples of 'values' of this variable which the educational process might seek to enhance. Knowledge of contemporary physics is a variable which may take a wide range of values – from knowledge at the level of elementary school, through popular and superficial acquintance with nuclear physics and relativity theory (typical for the average reader of popular scientific journals), up to the degree of knowledge of the subject characteristic for a Ph.D. in theoretical physics. If without specification of the level of knowledge we state that the goal of the teaching process is the teaching of physics, then this variable has not been defined in its concrete value.

A precise formulation of the values of variables is not necessary and does not usually take place in socially functioning educational programs. Very often a range of either feasible or desirable values is given. A good educator likes to see in his students a deepening interest in *some* area of knowledge, but generally he doesn't express a preference that this interest be in physics, biology, or the history of literature, as any of these choices, along with many others, are deemed desirable. In cultivating appreciation of music, we may consider as equally desirable a taste for baroque music and that for contemporary music. In the sphere of customs we accept a variety of inclinations and types of interest. Even in the realm of social values we consider admissible both a devotion to helping other people in their daily lives and a dedication to the abstract ideals of knowledge, justice, beauty, etc., on the assumption that realization of these different postulates by different people may be socially desirable.

Educational goals sometimes afford an *infinite number of admissible possibilities*. This happens when goals are defined in *negative terms*, i.e., when we specify only that we are interested in the non-admissibility of certain characteristics without stating the characteristics which are of central *positive* importance. If we consider the norms of penal law as cultural values which state undesirable modes of behavior, then these norms, like all prohibitions, only exclude certain types of behavior, defining them as offenses or misdemeanors and attaching to them appropriate sanctions, and admitting all the remaining ones.

The foregoing is not to say that in educational programs we do not find precisely specified values. School programs, as a rule, define relatively precisely the scope of knowledge which should be assimilated by the student, the norms of professional ethics tend to a relatively precise description of the attitude of the doctor toward his patients, and the traffic laws learned during driving lessons define relatively precisely the behavior of the driver while driving. It is nevertheless worth remembering that along with the relatively specific values of the variables of the educational processes we often find broad alternatives of such goals. Sometimes these alternatives are internally ordered into a hierarchy of values of the given variable — from the most desired to the least.

One more thought on this. It may happen that in striving toward a defined value of a variable — e.g., to a high value — we tend in some instances to achieve a *maximal control* of the future behavior of the student, while in others the maximalization of a certain valued trait maximizes the *degree of uncertainty* with respect to his future behavior. When we try to teach a child the proper use of a knife and fork, or proper behavior in the company of adults, when we instruct him to learn exactly the available knowledge in a given group or to internalize fully the system of customs and moral norms operating in a given collectivity, then through the process of conforming his behavior, thinking and feeling we achieve a high degree of control over the behavioral results of the educational process. However, when the goal of the educational process is to maximize such characteristics as intelligence, criticism, creative abilities in science or the arts, sensitivity to social problems, the ability to react to new situations with appropriate moral or ideological proposals, then at the same time ex definitione we resign ourselves to *maximal uncertainty* as to the *specific* future behavior of the student. It is easy to predict the manner of painting of a conforming graduage of a fine arts academy; it is not possible to predict what will be created by a person whose innate dispositions and special educational conditions culminate in the personality of a Van Gogh or a Picasso. One can predict a conforming valedictorian's views about the structure of the universe; it is not possible to predict the views of an Einstein.

It is necessary for any society to create among its memebers a certain uniform way of thinking and behaving, without which any social order would not be possible. In every society there is a demand for creative talents in the areas of knowledge, art, morals and ideology. Generally speaking, there is a demand for people whose disposition to create new cultural values dominates their adaptational disposition. But the proportions, seen in the different categories of cultural values which are stated in the educational goals, vary from one society to another in the system of realized educational programs. The differences among these proportions seem to be good indicators as to whether the institutions and the educational programs serve the *adaptational stabilization* of the existing system or support its *dynamic creative development*.

6. ORDERING OF THE GOALS OF EDUCATIONAL PROGRAMS UNDERSTOOD AS SYSTEMS OF VALUES

We considered above the problem of choosing educational goals in terms of a set properties of the student. Regardless of whether our goals are stated categorically or alternatively, whether they have a *stereotypic-adaptational* or an *open-dispositional* character, we obtain a certain set of values, i.e. the set of desired characteristics of the student.

I believe that, generally speaking, such an *inventory of the desired characteristics* of students could be considered a *system of values*, that is, a coherent system of characteristics which are valued. Therefore we should attempt to order the inventory by means of theoretical tools which are proper for the analysis of a system of values.

First of all, we should try to determine which characteristics of the students are *autotelic* values, i.e. are valued for themselves, and which are *instrumental* values, i.e. are important because they are considered as means leading to realization of certain autotelic ends. The difference is essential inasmuch as the choice of autotelic values is not subject to the control of empirical knowledge. If according to his cultural ideal somebody consideres the development of this muscles as a goal in itself, then he can not be convinced by any statements about reality that his view is inappropriate. If, however, he considers it as a means leading to the conquest of women, then one may point out that this method does not operate equally well in all social circles, and as a single method not supported by some 'spiritual qualities' or attributes of social position is often clearly inefficient. Thus, *delimitation of the instrumental values within the whole field of the desired characteristics of the student points out the sphere of phenomena where choice is subject to the control of empirical knowledge*. This control means that choice of psychological or physical characteristics, knowledge and skills understood in a instrumental way, together with the system of accepted values, can be discussed in terms of empirical science and therefore questioned on the basis of some empirical premises.

As a result of such a division of educational goals into autotelic values and instrumental values, any educational program can be presented as a system in which:

(a) There are a small number of desired characteristics which the student (or the educator) values as autotelic goals.

(b) There are a defined number of external social requirements which have the character of autotelic goals for the educational process, i.e. tasks which the student must fulfill.

(c) All other characteristics of the student's personality, elements of the society's system of values, knowledge, etc., have an instrumental character, i.e. are subservient on the basis of empirical relations to the few personal and social autotelic values. These instrumental characteristics are also subject to scientific evaluation as to whether and to what degree they are 'truly functional' with respect to the goals and values they are meant to serve.

In practice the problem is much more complicated and such ordering of the system of educational goals can rarely be fully accomplished. I think that, for example, in distinction to technology where the difference between autotelic and instrumental values is relatively simple, it is almost universal in all cultural systems that *some autotelic goals are simultaneously means for something else*. As a result, deciding whether a desired characteristic has, in the system of values of the student, an autotelic value or is evaluated only instrumentally is very difficult.

A great majority of human characteristics are praised or considered undesirable *both for themselves and because of the consequences which they imply*. The mastery of an artisan, clearly an instrumental value, is often an object of disinterested evaluation or even esthetic fascination and may be an autotelic value to the artisan himself. With respect to the personality characteristics which seem to be of autotelic value we often add the remark, as if for their 'justification', that they imply certain positive or negative effects; we say, for example, that 'cruelty warps the character', or that 'virtue is its own reward'.

Thus the pedagog cannot define many of his goals as expressing definitively autotelic values, with all the others being purely instrumental and fully subservient to them. He must be aware that among the advocated educational goals some in larger and other in smaller proportions have *both the autotelic and the instrumental character*. This means that they are subject to evaluation with respect not only to the autotelic value attributed to them, but also to their consequences.

Another problem connected with the analysis of educational goals considered as a system of values is that of the *relative importance* of certain desired characteristics of the student. This matter is essential when, for some reason, it turns out that the recognized educational ideal cannot be fully achieved and we make the choice of giving up some of the desired characteristics in order to achieve others considered to be more important. This problem is well known to the authors of school programs. When faced with the limits of teaching time they must give up some segments of the program in order to realize others. This problem is enhanced because, as a rule, the projected profile of the student in all its optimal richness is for many reasons considered to be something that cannot be achieved, and choices need to be made without waiting for its realization: these are the acts of choice and resignation. It is extremely important that these choices and resignations are done in a conscious way in constant confrontation with our own system of values.

7. EVALUATION OF THE DISTANCE BETWEEN EDUCATIONAL GOALS AND STARTING POINT

When we know well what we would like to expect from our educational process, what the desired 'point of arrival' is with respect to certain specific characteristics of the student or his sufficiently full profile, including his knowledge, skills, values and personality traits, we must state or at least assume on the basis of justified diagnostic hypotheses what the present 'point of departure' is or, briefly, *what distance is between the existing and the desired state*.

This is essential for two reasons:

(1) First of all, the diagnosis tells us whether, and in which aspect of the 'profile' of the student how many, modifications must be made during the educational process.

(2) The diagnosis is necessary for the choice of appropriate educational means for transforming the defined present state into the desired state.

Every educational process is a process of transforming the individual, starting from a certain take-off point to a certain point of arrival where the latter, assuming educational success, is in as close agreement as possible with the planned educational program. Here it would seem obvious that the necessary condition of the programming and the realization of the educational process is the diagnosis of the characteristics of the student. This will tell us whether the 'educational intervention' is needed at all and then, in case of an affirmative answer, it permits us to choose the optimal means leading to the transformation of the existing state into the desired state. I do not think that such diagnosis (made for individuals, or at least for those assumed as model types) is often taken into account in the planning of educational programs. Even in the case of the transfer of knowledge the student cannot be treated as an 'unfilled vessel with a limited capacity and a small diameter at the opening'. This analogy is even more fallible when we are dealing with the shaping of the personality or the system of values of the student; here we know that the efficacy of educational procedures depends in a large degree on the correctness of our assumptions about the existing state. In certain cases it may turn out that for some specific goals the educational procedures are not needed at all, because the student is relatively close to the assumed goals. This is the case, for example, when the system of moral values cultivated in the family (and internalized succesfully by the student) is convergent with the values which the school wants to see in the student.

Besides, the means necessary to the realization of the process of education may vary considerably depending on the present personality dispositions or values and norms of the students. As I said before, this seems to be an elementary assumption of psychological or psychosocial knowledge, but I am afraid that it is taken into account only in the education of persons who are exceptional cases (for example, resocialization of 'deviant persons' or psychotherapy for persons clearly maladjusted, etc.). The extent of the social loss incurred as a result of not following this principle in building educational programs for 'normal' people is impossible to calculate. The point is not to test personality, knowledge, skills or value systems of every student before starting any educational process; but a least it is indispensable to formulate certain, perhaps simplified, model assumptions about initial conditions of the educational process, and to test the statistical validity of these assumptions for the groups from which the students are recruited and for categories corresponding to their individual characteristics (for example, sex and age). Otherwise we implicitly make the assumption of ideological plasticity of human nature – an assumption not believed in by any pedagog.

8. THEORETICAL EVALUATION OF ATTAINABILITY OF INTENDED EDUCATIONAL GOALS

When we know the internal structure of our educational goals, their hierarchical order and the functional-instrumental connections between them, and when we know (or at least assume with some probability) the initial conditions – the starting point of the educational process in individual cases or typical conditions for a certain category of students in a given society – then new problems appear, above all the problem of the *attainability* of our educational program.

This problem can be dissected into a sequence of specific questions. First of all we want to know whether it is at all possible to achieve the desired characteristics, whether it is possible that the dependent variables of the educational process can attain certain specific values. The second question: Under what conditions and by which means can the goals be achieved? The third question does not pertain to the specific traits of the individual but to a profile; we want to knoe whether it is theoretically possible to achieve a specific *configuration* of the desired characteristics of the student. Fourth: Which complex model of the educational situation and educational procedures can produce the proper complex profile of the educational consequences defined by our educational program? Last but not least, there is the question of whether practical realization of conditions sufficiently similar to the assumptions of the theoretical model can be achieved, or is beyond the range of practical possibilities of the educator.

Let us consider these questions in sequence.

The first is the question about the limits of plasticity of the human species, about the limits of man's adaptability to the educational influences of the social environment and the intended pedagogical procedures. This problem, at the present stage of social knowledge, may seem purely academic, as the problem of the plasticity and durability of iron must have seemed purely academic to the village blacksmith. As far as the capacity of the human brain is concerned we know that we are still far from any thresholds. Cultural anthropology and the history of culture describe a sufficiently wide range of values which successfully functioned as motives of human behavior, and at the same time we know with certainty that their transmission during the educational process is at least sometimes possible. Moreover, psychology describes a sufficiently rich collection of personality types so that we don't feel an inadequacy of available categories. Independently from what has been said above, quite often we face a situation in which the problem of the limits of human plasticity appears closely and sharply. Sometimes the problem appears in times when there is a need for exceptional people, in situations demanding non-average talents, abilities, personality traits or systems of values which have not occured before. In such moments it is worthwhile to remember a remark of Stanislaw Ossowski written during the war in Nazi-occupied Warsaw.

The psychological possibilities upon which one would like to base the vision of the new world come into existence even today: they are realized in exceptional individuals or in exceptional groups or in exceptional states of normal people. These relatively rare events point out that our tendencies are not far beyond reality. Exceptional people, and even more, exceptional moments in the lives of ordinary people, are an important argument, an important confirmation that the scope of human possibilities extends at least as far as the most exceptional *attitudes* and exceptional types of behavior, and this enables us to hold a maximalistic position without covering our eyes to reality.[8]

The history of the natural sciences points to other evaluations of the limits of possibility of certain attempts — limits often stretching beyond the limits of hitherto existing achievements. The whole history of the technological progress of mankind is the history of the creation of systems and situations based on theoretical assumptions, and of the consequences from hypotheses so general that they included also situations which had not existed in reality till then. Here again the possibility of creation of these systems depended on the creation of conditions in which they could appear.

And this brings up a new problem, namely, whether we know the general laws, historical generalizations, or even singular propositions of the type A is sometimes followed by B', where the variables which are of interest as the goals of our educational program appear in the consequent. These laws or, more loosely, these propositions must fulfill certain defined requirements; they must connect the concrete values of the antecedent with the concrete values of the consequent. Laws like 'social conditions shape the morality of members of the society', although true, are not very useful until we know *which* conditions shape *which* types of morality. Similarly of little use is the proposition stating that the personality of the student depends to a large extent on the interpersonal relations in the educational group, until we know which kinds of relations shape which types of personality.

If we do not know what situations shape in what way certain characteristics of the student, then the mere knowledge that these characteristics are attainable because at some place and at some time they were realized may be at most a *challenge to search for the appropriate regularities*. As soon as we know these regularities we can start to construct our model of the educational situations. The greater the number of alternative factors (functional equivalents) that can produce a certain desired characteristic of the student, the greater is our freedom both in theoretical and in practical attempts.

Every practical science, including the theory of education, faces the possibility of *conflicts among attempted educational goals* in the case of such a complex and in fact almost infinitely rich system of goals. These conflicts may be of various kinds.

First, our program may be *logically contradictory*, which means that we would like for our student to have a certain characteristic C_1, and at the same not have it, or that he should have some other characteristic C_2 which is a *logical contradiction* of the characteristic C_1 This matter would seem obvious and not worth mentioning here if it were not for the fact that in many published educational programs we postulate, for example, that the student should have the ability to make individual judgments about his own values and at the same time be obedient to group values; that he should love his neighbor as himself and should give the maximum of himself for society, but at the same time should know how to take for himsel as much as he can or be as competitive as possible.

Obviously, the logical contradiction of goals does not occur when the program assumes the alternative character of contradictory goals. If it is permissible for students either to like jazz or to dislike it, there is no contradiction in our program. If we welcome in the group both conformists who fully approve of the group's values and others who will be critical, there is no contradiction. Similarly, there is no contradiction when we recommend behaviors or traits which may differ under different conditions or in different situations. There is no logical contradiction between the prescription of intensive work on the job and full rest afterwards. Contradiction may occur in those elements of the program which:

(a) either are not relativized to the appropriate situations or defined categories of students but are formulated as universal recommendations, or

(b) being relativized for certain persons or situations, nevertheless jointly apply in mutually contradictory sets to the same persons in the same situations.

Specification of the limited validity of a desired attribute, rather than giving it universal sanction, may be one method of removing the contradiction. Such a procedure does not always correspond to our intuitions about our system of values. We do not want to give the norm 'thous shall not kill' the character of a norm with restricted scope, as in 'thou shall not kill except for the enemies of your country, dangerous criminals, people who threaten your life, etc.' We prefer to leave this norm in apparent contradiction to the norms that allow killing in some situations, rather than take away from it the educational power connected with its unconditional and universal character.

A second type of conflict among educational goals, especially well-known in the theory of school education, is that which is due to the *more or less limited capacity of the student or efficiency of the educational process*. This does not imply internal conflict in the program but it does force us to choose some goals and reject others. There is no logical reason why a person could not cultivate the performing of music and sports simultaneously, but great athletic champions are almost never the masters of piano; the limited capacity and the duration of the individual biography does not allow it. The same problem of choice causes difficulty for every person who designs school programs.

A third type of conflict is of a different kind. If we want our students to acquire two characteristics, B_1 and B_2, we might find it impossible because the realization of the first of these characteristics leads, *on the basis of certain causal regularities, to the liquidation or at least the severe weakening of the other desired characteristic. We might want our students to possess a wide knowledge of cultures and societies living in various parts of the world* and at various periods, while at the same time we would like them to accept without hesitation the values of their own group and to have no doubts that they are privileged to live in the most magnificent of all possible countries and cultures. But realization of the first of these goals, the broadening of comparative knowledge, almost inevitably leads to weakening the chances of realization of the second goal by weakening the belief that the society in which they live is the best of all possible. Similarly, in some cultures people are educated to believe that personal success, especially measured in terms of material gains and obtained in competition with others, is the highest measure of personal value. Then there is surprise that in these cultures there is no way to realize another desirable goal, namely unselfish, warm, spontaneous, non-exploitative relations between people! These are examples of situations where the contradictions within the educational program are based on the fact that it contains pairs of characteristics connected by 'negative feedback', such that realization of one of the characteristics implies the negation or at least a clear weakening of the second.

A fourth type of contradiction and conflict within the educational program has its source in the application of certain educational *means* leading to realization of some elements which simultaneously prevent the realization of other elements. Educational conditions in the Prussian army were conducive to the development of a specific type of discipline but did not influence positively the development of creativity or independence of thinking.

Except for the logical contradictions, all these other conflicts within the educational program may have only a *historical character*, being valid for a given stage of development of our knowledge, and for a given repertoire or scope of means at our disposal in the educational process. The broadening of our knowledge and the richer scope of means which will be at our disposal, and especially the knowledge of various alternative factors of the desired characteristics of the profile of the student, may enhance the efficiency of the educational process, eliminate the negative influence of some characteristics on other characteristics, and put at our disposal a means for the attainment of one goal which does not exclude — and may even help — the attainment of another.

9. THE MANIPULABLE VARIABLES OF THE EDUCATIONAL SITUATION

The problems of the previous section directed our attention toward theoretical matters. The problem of attainability of the intended educational program is a highly theoretical one because, in the first instance, we must decide whether we know the proper general laws or proper historical generalizations, the consequents of which contain the elements of the desired profile of the student, and whether from the antecedents of these laws or generalizations we are able to build an internally consistent model of the educational situation.

However, the problem of the degree to which we are able to attain our educational program is already a *practical* problem as well. It depends on the possibility of the practical realization of the system of elements that is consistent with the appropriate theoretical model of the educational process. This in turn depends on:

(a) the extent to which the *real situation* which is our starting point *is different* from the assumed theoretical model of the educational situation necessary for the realization of our goals; and

(b) the extent to which the *missing* elements that are necessary for the occurrence of the desired consequences, or the existing elements which are harmful for the intended goals, are amenable to our manipulation, i.e. can be *practically changed or shaped in the desired way*.

The term 'educational situation' can be understood more or less broadly and therefore there may be an infinite number of such models just as there are infinitely many types of concrete processes through which personality, knowledge, values and behavior are shaped.

There is, however, one category of situations that is worthy of particular attention, namely formal educational institutions. These institutions, typical for the majority of known civilizations, are specially called into existence for the purpose of teaching the members of the society so as to prepare them for their future social roles. As a rule, the educational institutions have a designated educational program, i.e. a system of more or less clearly formulated goals, and, as a rule, a system of clearly defined means, including special educational personnel. There is a defined segment of the life cycle of the student to which the educator has exclusive or at least dominating access, and there are certain defined procedures which the educator applies.

The theoretical analysis of an educational situation may lead to a number of different conclusions:

(a) The presently existing set of conditions is in principle identical with the postulated theoretical model and therefore the normal implementation of educational processes in the manner practiced up to now is quite sufficient; but this would mean that we observe in our educational system the realization of all desired educational goals, which seems unlikely in any contemporary society.

(b) The existing set of conditions demands only *certain modifications* which can in principle take place within the framework of existing educational institutions.

(c) It is necessary to create new educational institutions either because the present institutional arrangements are not amenable to transformation into a set of conditions which agree with our theoretical model, or because such transformation would involve a higher social cost than would the creation of a new institution.

(d) The attainment of the educational program is practically impossible (although it had seemed to be theoretically possible) because certain important variables characterizing the educational situation are beyond the reach of our manipulation.

The concept of 'manipulated variables' is very broad. It denotes all those characteristics of the objects, events, and systems of relations which can be changed in a direction consonant with the demands of the postualted theoretical model of the educational process. The scope of the variables obviously changes from situation to situation, from agent to agent. A rigid, centrally directed faulty educational program is not a manipulable variable from the point of view of the teacher in elementary education who is not permitted to modify this program. But the spatial distribution of buildings and the ecological factors in the teaching and housing of students are manipulable variables if we have sufficient capital or credits to design and build a new, experimental educational institution. Even the personality types of students and educators are manipulable if we are able (and permitted) to select the 'proper' students and teachers, whereas the old blackboard might be a 'nonmanipulable variable' (influencing the teaching process in the wrong direction) when there are no funds for buying a new one.

Determination of the scope of manipulable variables of the educational situation (in this case the educational institution) is an implicit statement of the degree to which the intended program can be attained.

In our model of concrete educational situations the different variables are related to each other in their concrete values, thus producing a common effect of this process. If we now state the practically (and morally, see below) possible range of manipulability of the variables of the model – and study the *consequences of different configurations of the values of all the manipulable variables for their joint effect* – we can estimate the degree to which we will be able to obtain the desired goal, or the different alternative outcomes of our manipulation of the antecedent conditions within the limits of their manipulability. In short, we will see what *really can be done* within the practical range of our possibilities.[9]

Finally, we face the problem of *deciding*, on moral or social grounds, to what extent we shall manipulate the variables. This again is a problem of our system of values. Every teacher knows that if he had only a few, very able students he could achieve much better results, but generally he would not condone the exclusion of less able students inasmuch as he believes that all students should have the right to education. In many contemporary armies the system of interpersonal relations has been changed and the iron discipline abandoned; a considerable factor in this decision was the belief that a soldier has the right to humane conditions even if the mechanical efficiency of the units were to be somewhat impaired. Not only the goals of the educational process but also the means which lead to them are subject to our evaluation, and the principle that the attainment of the intended educational goals justifies any and all means is as inadmissible here as in any other area of socio-techniques.

10. EDUCATIONAL INSTITUTIONS AND THE WIDER EDUCATIONAL ENVIRONMENT

As I stated above, an educational institution as a rule controls only certain, smaller or larger, segments of a person's life cycle. The *educational environment* is much broader and in reality embraces the totality of influences and actions upon the student during the ongoing educational process. The amounts of influence exerted under these categories are quite different. In the case of evening school for adults, a great part of their education comes from outside the school. In the case of a closed rehabilitation unit, isolation from the external world means that the educational environment is almost equivalent to the 'total' educational institution, although even there one has to cope incessantly with the influence of the outside world in the form of press, radio, or letters from the family.[10]

The practitioner in education who prepares the educational program also has to deal with the problem of defining the direction of influence of the broader environment. The problem is to find the way in which this environment will shape those personality traits of the student which are the object of interest of the educational program. The definition is, of course, contingent on the circumstances:

(a) From the standpoint of the goals to be achieved in the educational situation, these outside influences may be irrelevant.

(b) From the standpoint of the goals of the program, the influences may be *negative*; i.e. they are in opposition to the desired traits.

(c) These external influences may be assessed as having a *positive* impact, i.e. they work in the same social direction as the more narrowly understood educational setting.

(d) Finally and most commonly, the influence of the broader environment varies from the standpoint of the various intended elements of the educational effect, being irrelevant for some, negative for others, and positive for still others.

Evaluation of the direction and the strength of influence of the educational environment external to the educational institution is of essential practical importance. The educator must know which of his actions will be reinforced by the external influences, and which may be inhibited so as to demand especially close attention. But even in the latter case a certain manipulation of the variables is possible. The educator, keeping in mind the intended educational goals, can broaden or narrow the sphere of contacts with the external environment in such a way that the educational effect is optimally achieved.

The sphere of manipulation, real or postulated, with respect to the external environment can obviously be quite broad. History shows instances where in the name of the struggle for a new image of man there have been postulated or even achieved some basic transformations of cultural values or changes of entire social systems.

A final comment. In this chapter I have tried to look at various interrelationships between values and empirically testable knowledge in the theory of education. It would seem, however, that the applicability of the paradigms discussed above is much broader than in the educational area alone, and that with some modifications it could be applied to most areas of social reality where we would like to use our theoretical knowledge for the implementation of our moral or social values.

NOTES

* Paper presented at the conference on methodological problems of theory of education held in Warsaw in February 1966. Published in *Biuletyn Pedagogiczny* (1966) and translated from that edition.

[1] See Chapter VIII.

[2] See, e.g., G.T. Buswell, 'Educational Theory and the Psychology of Learning', *Journal of Educational Psychology* 47, 1956; See also J. Aronfreed, *Conduct and Conscience, The Socialization of Internalized Control over Behavior*, New York, 1968.

[3] See R.K. Merton and A. Rossi, 'Contributions to the Theory of Reference Group Behavior', in R.K. Merton (ed.), *Social Theory and Social Structure*, New York, 1968.

[4] For the analysis of this kind of relation see Chapter XI.

[5] See M. Weber, *Protestant Ethics and the Spirit of Capitalism*, translated by T. Parsons, New York, 1958.

[6] See Chapter V.

[7] See A. Podgórecki, 'Sociotechnique,' *The Polish Sociological Bulletin* 2 (8), 1963.

[8] S. Ossowski, 'Z natrojów manichejskich' (Manichean Moods), in: S. Ossowski, *Dziela* (*Works*) vol. III, p. 196, Warsaw, 1967.

[9] For the analysis of more concrete problems related to the manipulability of educational variables see: N. Gross and J.A. Fishman, 'The Management of Educational Establishment,' in: P.F. Lazarsfeld, W.H. Sewell, and H.L. Wilenski (eds.), *The Uses of Sociology*, New York, 1967.

[10] For the more detailed analysis of this problem see: E. Litwak and H.J. Meyer, 'The School and the Family; Linking Organizations and External Primary Groups', in: P.F. Lazarsfeld, W.H. Sewell, and H.L. Wilenski (eds.), *The Uses of Sociology*.

BIBLIOGRAPHY

Aronsfreed, J., *Conduct and Conscience, The Socialization of Internalized Control over Behavior*, New York, 1968.

Buswell, G.T., 'Educational Theory and the Psychology of Learning', *Journal of Educational Psychology* 47, 1956.

Gross, N., and J.A. Fishman, 'The Management of Educational Establishment', in: P.F. Lazarsfeld, W.H. Sewell, and H.L. Wilenski (eds.), *The Uses of Sociology*, New York, 1967.

Litwak E., and H.J. Meyer, 'The School and the Family; Linking Organizations and External Primary Groups', in: P.F. Lazarsfeld, W.H. Sewell, and H.L. Wilenski (eds.), *The Uses of Sociology*, New York, 1967.

Merton R.K., and A. Rossi, 'Contributions to the Theory of Reference Group Behavior', in: R.K. Merton (ed.), *Social Theory and Social Structure*, New York, 1968.

Nowak, S., 'Empirical Knowledge and Social Values in the Cumulative Development of Sociology', in this volume, p. 285.

Nowak, S., 'The Logic of Reductive Systematizations of Social and Behavioral Theories', in this volume, p. 376.

Nowak, S., 'Causal Interpretation of Statistical Relationships in Social Research', in this volume, p. 165.

Ossowski, S., 'Z nastrojów manichejskich' (Manichean Moods), in: S. Ossowski, *Dziela* (*Works*) vol. III, p. 196, Warsaw, 1967.

Podgórecki, A., 'Sociotechnique', *The Polish Sociological Bulletin* 2 (8), 1963.

Weber, M., *Protestant Ethics and the Spirit of Capitalism*, translated by T. Parsons, New York, 1958.

INDEX OF NAMES

SYNTHESE LIBRARY

Monographs on Epistemology, Logic, Methodology,
Philosophy of Science, Sociology of Science and of Knowledge, and on the
Mathematical Methods of Social and Behavioral Sciences

Managing Editor:
Jaakko Hintikka (Academy of Finland and Stanford University)

Editors:
Robert S. Cohen (Boston University)
Donald Davidson (The Rockefeller University and Princeton University)
Gabriël Nuchelmans (University of Leyden)
Wesley C. Salmon (University of Arizona)

Colloquium for the Philosophy of Science 1964–1966, in Memory of Norwood Russell Hanson. Boston Studies in the Philosophy of Science (ed. by Robert S. Cohen and Marx W. Wartofsky), Volume III. 1967, XLIX + 489 pp.

15. C.D. Broad, *Induction, Probability, and Causation. Selected Papers.* 1968, XI + 296 pp.
16. Günther Patzig, *Aristotle's Theory of the Syllogism. A Logical-Philosophical Study of Book A of the Prior Analytics.* 1968, XVII + 215 pp.
17. Nicholas Rescher, *Topics in Philosophical Logic.* 1968, XIV + 347 pp.
18. Robert S. Cohen and Marx W. Wartofsky (eds.), *Proceedings of the Boston Colloquium for the Philosophy of Science 1966–1968,* Boston Studies in the Philosophy of Science (ed. by Robert S. Cohen and Marx W. Wartofsky), Volume IV. 1969, VIII + 537 pp.
19. Robert S. Cohen and Marx W. Wartofsky (eds.), *Proceedings of the Boston Colloquium for the Philosophy of Science 1966–1968,* Boston Studies in the Philosophy of Science (ed. by Robert S. Cohen and Marx W. Wartofsky), Volume V. 1969, VIII + 482 pp.
20. J.W. Davis, D.J. Hockney, and W.K. Wilson (eds.), *Philosophical Logic.* 1969, VIII + 277 pp.
21. D. Davidson and J. Hintikka (eds.), *Words and Objections: Essays on the Work of W. V. Quine.* 1969, VIII + 366 pp.
22. Patrick Suppes, *Studies in the Methodology and Foundations of Science. Selected Papers from 1911 to 1969,* XII + 473 pp.
23. Jaakko Hintikka, *Models for Modalities. Selected Essays.* 1969, IX + 220 pp.
24. Nicholas Rescher *et al.* (eds.), *Essays in Honor of Carl G. Hempel. A Tribute on the Occasion of his Sixty-Fifth Birthday.* 1969, VII + 272 pp.
25. P. V. Tavanec (ed.), *Problems of the Logic of Scientific Knowledge.* 1969, XII + 429 pp.
26. Marshall Swain (ed.), *Induction, Acceptance, and Rational Belief.* 1970, VII + 232 pp.
27. Robert S. Cohen and Raymond J. Seeger (eds.), *Ernst Mach; Physicist and Philosopher*, Boston Studies in the Philosophy of Science (ed. by Robert S. Cohen and Marx W. Wartofsky), Volume VI. 1970, VIII + 295 pp.
28. Jaakko Hintikka and Patrick Suppes, *Information and Inference.* 1970, X + 336 pp.
29. Karel Lambert, *Philosophical Problems in Logic. Some Recent Developments.* 1970, VII + 176 pp.
30. Rolf A. Eberle, *Nominalistic Systems.* 1970, IX + 217 pp.
31. Paul Weingartner and Gerhard Zecha (eds.), *Induction, Physics, and Ethics, Proceedings and Discussions of the 1968 Salzburg Colloquium in the Philosophy of Science.* 1970, X + 382 pp.
32. Evert. W. Beth, *Aspects of Modern Logic.* 1970, XI + 176 pp.
33. Risto Hilpinen (ed.), *Deontic Logic: Introductory and Systematic Readings.* 1971, VII + 182 pp.
34. Jean-Louis Krivine, *Introduction to Axiomatic Set Theory.* 1971, VII + 98 pp.
35. Joseph D. Sneed, *The Logical Structure of Mathematical Physics.* 1971, XV + 311 pp.
36. Carl R. Kordig, *The Justification of Scientific Change.* 1971, XIV + 119 pp.

37. Milič Čapek, *Bergson and Modern Physics*, Boston Studies in the Philosophy of Science (ed. by Robert S. Cohen and Marx W. Wartofsky), Volume VII. 1971, XV + 414 pp.

38. Norwood Russell Hanson, *What I do not Believe, and other Essays*, (ed. by Stephen Toulmin and Harry Woolf). 1971, XII + 390 pp.

39. Roger C. Buck and Robert S. Cohen (eds.), *PSA 1970. In Memory of Rudolf Carnap*, Boston Studies in the Philosophy of Science (ed. by Robert S. Cohen and Marx W. Wartofsky), Volume VIII. 1971, LXVI + 615 pp. Also available as a paperback.

40. Donald Davidson and Gilbert Harman (eds.), *Semantics of Natural Language*. 1972, X + 769 pp. Also available as a paperback.

41. Yehoshua Bar-Hillel (ed.), *Pragmatics of Natural Languages*. 1971, VII + 231 pp.

42. Sören Stenlund, *Combinators, λ-Terms and Proof Theory*. 1972, 184 pp.

43. Martin Strauss, *Modern Physics and Its Philosophy. Selected Papers in the Logic History, and Philosophy of Science*. 1972, X + 297 pp.

44. Mario Bunge, *Method, Model and Matter*. 1973, VII + 196 pp.

45. Mario Bunge, *Philosophy of Physics*. 1973, IX + 248 pp.

46. A. A. Zinov'ev, *Foundations of the Logical Theory of Scientific Knowledge (Complex Logic)*, Boston Studies in the Philosophy of Science (ed. by Robert S. Cohen and Marx W. Wartofsky), Volume IX. Revised and enlarged English edition with an appendix, by G. A. Smirnov, E. A. Sidorenka, A. M. Fedina, and L. A. Bobrova. 1973, XXII + 301 pp. Also available as a paperback.

47. Ladislav Tondl, *Scientific Procedures*, Boston Studies in the Philosophy of Science (ed. by Robert S. Cohen and Marx W. Wartofsky), Volume X. 1973, XII + 268 pp. Also available as a paperback.

48. Norwood Russell Hanson, *Constellations and Conjectures*, (ed. by Willard C. Humphreys, Jr.), 1973, X + 282 pp.

49. K. J. J. Hintikka, J. M. E. Moravcsik, and P. Suppes (eds.), *Approaches to Natural Language. Proceedings of the 1970 Stanford Workshop on Grammar and Semantics*. 1973, VIII + 526 pp. Also available as a paperback.

50. Mario Bunge (ed.), *Exact Philosophy – Problems, Tools, and Goals*. 1973, X + 214 pp.

51. Radu J. Bogdan and Ilkka Niiniluoto (eds.), *Logic, Language, and Probability*. A selection of papers contributed to Sections IV, VI, and XI of the Fourth International Congress for Logic, Methodology, and Philosophy of Science, Bucharest, September 1971. 1973, X + 323 pp.

52. Glenn Pearce and Patrick Maynard (eds.), *Conceptual Chance*. 1973, XII + 282 pp.

53. Ilkka Niiniluoto and Raimo Tuomela, *Theoretical Concepts and Hypothetico-Inductive Inference*. 1973, VII + 264 pp.

54. Roland Fraïssé, *Course of Mathematical Logic* – Volume 1: *Relation and Logical Formula*. 1973, XVI + 186 pp. Also available as a paperback.

55. Adolf Grünbaum, *Philosophical Problems of Space and Time*. Second, enlarged edition, Boston Studies in the Philosophy of Science (ed. by Robert S. Cohen and Marx W. Wartofsky), Volume XII. 1973, XXIII + 884 pp. Also available as a paperback.

56. Patrick Suppes (ed.), *Space, Time, and Geometry*. 1973, XI + 424 pp.

57. Hans Kelsen, *Essays in Legal and Moral Philosophy*, selected and introduced by

Ots Weinberger. 1973, XXVIII + 300 pp.

58. R. J. Seeger and Robert S. Cohen (eds.), *Philosophical Foundations of Science. Proceedings of an AAAS Program, 1969.* Boston Studies in the Philosophy of Science (ed. by Robert S. Cohen and Marx W. Wartofsky), Volume XI. 1974, X + 545 pp. Also available as paperback.

59. Robert S. Cohen and Marx W. Wartofsky (eds.), *Logical and Epistemological Studies in bontemporary Physics,* Boston Studies in the Philosophy of Science (ed. by Robert S. Cohen and Marx W. Wartofsky), Volume XIII. 1973, VIII + 462 pp. Also available as paperback.

60. Robert S. Cohen and Marx W. Wartofsky (eds.), *Methodological and Historical Essays in the Natural and Social Sciences. Proceedings of the Boston Colloquium for the Philosophy of Science, 1969–1972,* Boston Studies in the Philosophy of Science (ed. by Robert S. Cohen and Marx W. Wartofsky), Volume XIV. 1974, VIII + 405 pp. Also available as paperback.

61. Robert S. Cohen, J. J. Stachel and Marx W. Wartofsky (eds.), *For Dirk Struik. Scientific, Historical and Political Essays in Honor of Dirk. J. Struik,* Boston Studies in the Philosophy of Science (ed. by Robert S. Cohen and Marx W. Wartofsky), Volume XV. 1974, XXVII + 652 pp. Also available as paperback.

62. Kazimierz Ajdukiewicz, *Pragmatic Logic,* transl. from the Polish by Olgierd Wojtasiewicz. 1974, XV + 460 pp.

63. Sören Stenlund (ed.), *Logical Theory and Semantic Analysis. Essays Dedicated to Stig Kanger on His Fiftieth Birthday.* 1974, V + 217 pp.

64. Kenneth F. Schaffner and Robert S. Cohen (eds.), *Proceedings of the 1972 Biennial Meeting, Philosophy of Science Association,* Boston Studies in the Philosophy of Science (ed. by Robert S. Cohen and Marx W. Wartofsky), Volume XX. 1974, IX + 444 pp. Also available as paperback.

65. Henry E. Kyburg, Jr., *The Logical Foundations of Statistical Inference.* 1974, IX + 421 pp.

66. Marjorie Grene, *The Understanding of Nature: Essays in the Philosophy of Biology,* Boston Studies in the Philosophy of Science (ed. by Robert S. Cohen and Marx W. Wartofsky), Volume XXIII. 1974, XII + 360 pp. Also available as paperback.

67. Jan M. Broekman, *Structuralism: Moscow, Prague, Paris.* 1974, IX + 117 pp.

68. Norman Geschwind, *Selected Papers on Language and the Brain,* Boston Studies in the Philosophy of Science (ed. by Robert S. Cohen and Marx W. Wartofsky), Volume XVI. 1974, XII + 549 pp. Also available as paperback.

69. Roland Fraïssé, *Course of Mathematical Logic – Volume II: Model Theory.* 1974, XIX + 192 pp.

70. Andrzej Grzegorczyk, *An Outline of Mathematical Logic. Fundamental Results and Notions Explained with All Details.* 1974, X + 596 pp.

71. Franz von Kutschera, *Philosophy of Language.* 1975, VII + 305 pp.

72. Juha Manninen and Raimo Tuomela (eds.), *Essays on Explanation and Understanding. Studies in the Foundations of Humanities and Social Sciences.* 1976, VII + 440 pp.

73. Jaakko Hintikka (ed.), *Rudolf Carnap, Logical Empiricist. Materials and Perspectives.* 1975, LXVIII + 400 pp.

74. Milič Čapek (ed.), *The Concepts of Space and Time. Their Structure and Their*

Development. Boston Studies in the Philosophy of Science (ed. by Robert S. Cohen and Marx W. Wartofsky), Volume XXII. 1976, LVI + 570 pp. Also available as paperback.

75. Jaakko Hintikka and Unto Remes, *The Method of Analysis. Its Geometrical Origin and Its General Significance.* Boston Studies in the Philosophy of Science (ed. by Robert S. Cohen and Marx W. Wartofsky), Volume XXV. 1974, XVIII + 144 pp. Also available as paperback.

76. John Emery Murdoch and Edith Dudley Sylla, *The Cultural Context of Medieval Learning. Proceedings of the First International Colloquium on Philosophy, Science, and Theology in the Middle Ages – September 1973.* Boston Studies in the Philosophy of Science (ed. by Robert S. Cohen and Marx W. Wartofsky), Volume XXVI. 1975, X + 566 pp. Also available as paperback.

77. Stefan Amsterdamski, *Between Experience and Metaphysics. Philosophical Problems of the Evolution of Science.* Boston Studies in the Philosophy of Science (ed. by Robert S. Cohen and Marx W. Wartofsky), Volume XXXV. 1975, XVIII + 193 pp. Also available as paperback.

78. Patrick Suppes (ed.), *Logic and Probability in Quantum Mechanics.* 1976, XV + 541 pp.

80. Joseph Agassi, *Science in Flux.* Boston Studies in the Philosophy of Science (ed. by Robert S. Cohen and Marx W. Wartofsky), Volume XXVIII. 1975, XXVI + 553 pp. Also available as paperback.

81. Sandra G. Harding (ed.), *Can Theories Be Refuted? Essays on the Duhem-Quine Thesis.* 1976, XXI + 318 pp. Also available in paperback.

84. Marjorie Grene and Everett Mendelsohn (eds.), *Topics in the Philosophy of Biology.* Boston Studies in the Philosophy of Science (ed. by Robert S. Cohen and Marx W. Wartofsky), Volume XXVII. 1976, XIII + 454 pp. Also available as paperback.

85. E. Fischbein, *The Intuitive Sources of Probabilistic Thinking in Children.* 1975, XIII + 204 pp.

86. Ernest W. Adams, *The Logic of Conditionals. An Application of Probability to Deductive Logic.* 1975, XIII + 156 pp.

89. A. Kasher (ed.), *Language in Focus: Foundations, Methods and Systems. Essays dedicated to Yehoshua Bar-Hillel.* Boston Studies in the Philosophy of Science (ed. by Robert S. Cohen and Marx W. Wartofsky), Volume XLIII. 1976, XXVIII + 679 pp. Also available as paperback.

90. Jaakko Hintikka, *The Intentions of Intentionality and Other New Models for Modalities.* 1975, XVIII + 262 pp. Also available as paperback.

93. Radu J. Bogdan, *Local Induction.* 1976, XIV + 340 pp.

95. Peter Mittelstaedt, *Philosophical Problems of Modern Physics.* Boston Studies in the Philosophy of Science (ed. by Robert S. Cohen and Marx W. Wartofsky), Volume XVIII. 1976, X + 211 pp. Also available as paperback.

96. Gerald Holton and William Blanpied (eds.), *Science and Its Public: The Changing Relationship.* Boston Studies in the Philosophy of Science (ed. by Robert S. Cohen and Marx W. Wartofsky), Volume XXXIII. 1976, XXV + 289 pp. Also available as paperback.

SYNTHESE HISTORICAL LIBRARY

Texts and Studies
in the History of Logic and Philosophy

Editors:

N. Kretzmann (Cornell University)
G. Nuchelmans (University of Leyden)
L. M. de Rijk (University of Leyden)

1. M. T. Beonio-Brocchieri Fumagalli, *The Logic of Abelard.* Translated from the Italian. 1969, IX + 101 pp.
2. Gottfried Wilhelm Leibnitz, *Philosophical Papers and Letters.* A selection translated and edited, with an introduction, by Leroy E. Loemker. 1969, XII + 736 pp.
3. Ernst Mally, *Logische Schriften,* ed. by Karl Wolf and Paul Weingartner. 1971, X + 340 pp.
4. Lewis White Beck (ed.), *Proceedings of the Third International Kant Congress.* 1972, XI + 718 pp.
5. Bernard Bolzano, *Theory of Science,* ed. by Jan Berg. 1973, XV + 398 pp.
6. J. M. E. Moravcsik (ed.), *Patterns in Plato's Thought. Papers arising out of the 1971 West Coast Greek Philosophy Conference.* 1973, VIII + 212 pp.
7. Nabil Shehaby, *The Propositional Logic of Avicenna: A Translation from al-Shifa: alQiyas,* with Introduction, Commentary and Glossary. 1973, XIII + 296 pp.
8. Desmond Paul Henry, *Commentary on De Grammatico: The Historical-Logical Dimensions of a Dialogue of St. Anselm's.* 1974, IX + 345 pp.
9. John Corcoran, *Ancient Logic and Its Modern Interpretations.* 1974, X + 208 pp.
10. E. M. Barth, *The Logic of the Articles in Traditional Philosophy.* 1974, XXVII + 533 pp.
11. Jaakko Hintikka, *Knowledge and the Known. Historical Perspectives in Epistemology.* 1974, XII + 243 pp.
12. E. J. Ashworth, *Language and Logic in the Post-Medieval Period.* 1974, XIII + 304 pp.
13. Aristotle, *The Nicomachean Ethics.* Translated with Commentaries and Glossary by Hypocrates G. Apostle. 1975, XXI + 372 pp.
14. R. M. Dancy, *Sense and Contradiction: A Study in Aristotle.* 1975, XII + 184 pp.
15. Wilbur Richard Knorr, *The Evolution of the Euclidean Elements. A Study of the Theory of Incommensurable Magnitudes and Its Significance for Early Greek Geometry.* 1975, IX + 374 pp.
16. Augustine, *De Dialectica.* Translated with the Introduction and Notes by B. Darrell Jackson. 1975, XI + 151 pp.